Introductory Algebra for College Students

Introductory Algebra for College Students

EUGENE NICHOLS
Florida State University

25746

HOLT, RINEHART AND WINSTON, INC.
New York   Chicago   San Francisco   Atlanta   Dallas
Montreal   Toronto   London
Sydney

Copyright © 1971 by Holt, Rinehart and Winston, Inc.
All rights reserved
Library of Congress Catalog Card Number: 77-136557
SBN: 03-084689-7
Printed in the United States of America
4 3 2 1 038 9 8 7 6 5 4 3 2 1

PREFACE

This is an introductory algebra text. It contains all of the topics necessary for studying college algebra; they are developed in a way as to maintain proper balance between theory and application. The book is intended for those college students who have not had an algebra course in high school or who have had a strictly traditional high-school algebra course.

Some of the more prominent features of this textbook are the following:

1. The basic concepts of sets are introduced in the first chapter and used throughout the book. This approach provides an elegant way of expressing many mathematical relations and helps clarify and unify algebraic concepts. The first chapter should be used as a reference or review by those students who have already studied sets.
2. Throughout, there is emphasis on reasoning and proof; but the idea of proof and practice in writing proofs are introduced very gradually. The student is first taught to recognize valid proofs and to supply reasons for the statements in proofs.

3. A consistent effort has been made to provide exposition which is simple and clear so that the student is encouraged to read the textbook on his own.
4. An unusually large number of problems is supplied so that the instructor has a flexibility in deciding on the lengths of assignments.

A glossary of new terms is provided at the end of each chapter. The student should use it to review the vocabulary of the chapter. Chapter review problems are also provided at the end of each chapter. They can be used as review or as a test.

Eugene Nichols

Tallahassee, Florida
November 1970

CONTENTS

PREFACE v

1 Sets 1

1.1 Membership, Finite Sets, Infinite Sets, and Equivalent Sets 1
1.2 Numbers and Numerals 6
1.3 Subsets and Equality 9
1.4 Disjoint Sets and Venn Diagrams 11
1.5 Operations with Sets 14
1.6 Counting the Subsets of a Finite Set, Subsets of Infinite Sets 17
1.7 Complement of a Set 20
Glossary 24
Chapter Review Problems 25

2 The System of Real Numbers: Addition and Subtraction 29

2.1 Order of Operations 29
2.2 The Number Line 31

2.3   Integers   32
2.4   Absolute Value   36
2.5   Addition of Integers   38
2.6   Rational Numbers   41
2.7   Repeating Decimal Numerals   43
2.8   Irrational Numbers   48
2.9   Real Numbers and the Number Line   50
2.10 Subtraction of Real Numbers   56
      Glossary   61
      Chapter Review Problems   62

3 The System of Real Numbers: Multiplication and Division   65

3.1   Multiplication of Real Numbers   65
3.2   Commutative and Associative Properties   70
3.3   Closure   73
3.4   The Distributive Property   76
3.5   Special Properties of 1 and 0   78
3.6   Division of Real Numbers   82
3.7   Simpler Numerals for Real Numbers   86
3.8   The Set of Real Numbers as a System   87
3.9   From Word Phrases to Mathematical Phrases   91
      Glossary   93
      Chapter Review Problems   94

4 Exponents   98

4.1   The Product of Powers   98
4.2   The Power of a Power   102
4.3   The Power of a Product   102
4.4   The Power of a Quotient   104
4.5   The Quotient of Powers   105
4.6   Zero and Negative Integers as Exponents   108
4.7   Other Patterns in Exponents   110
4.8   Scientific Notation   113
4.9   Word Phrases to Mathematical Phrases   116
      Glossary   117
      Chapter Review Problems   117

Contents ix

5 Open Expressions 120

5.1 Expressions 120
5.2 Proving Expressions Equivalent 126
5.3 Distributivity and Rearrangement 127
5.4 Subtraction, Multiplication by -1 134
Glossary 139
Chapter Review Problems 139

6 Solution Sets: Equations and Inequalities 142

6.1 What is an Equation? 142
6.2 From Word Sentences to Equations 143
6.3 Applications of Equations 146
6.4 Equations Involving Absolute Value 152
6.5 The Addition Properties for Equations 155
6.6 The Multiplication Properties for Equations 158
6.7 Using All Equation Properties 163
6.8 Writing Equivalent Expressions 164
6.9 Simple Inequalities 167
6.10 Binomials 170
6.11 Simplification of Expressions 171
6.12 Theorems on Subtraction of Binomials 176
Glossary 180
Chapter Review Problems 181

7 Open Rational Expressions 185

7.1 Patterns in Multiplication 185
7.2 Multiplicative Inverses Theorem 191
7.3 Multiplicative Identity Theorem 192
7.4 Rational Expressions—Division 196
7.5 Theorems about Addition and Subtraction 203
7.6 Complex Rational Expressions 209
7.7 Equations with Rational Expressions 213
Glossary 215
Chapter Review Problems 215

8 Applications of Equations and Inequalities   218

8.1 Equation Solving   218
8.2 Uses of Mathematics in Science   222
8.3 Formulas as Equations   227
8.4 Using Equations to Solve Problems   229
8.5 Inequalities   234
8.6 Addition and Subtraction Theorems for Inequalities   235
8.7 Multiplication Theorems for Inequalities   237
8.8 Absolute Value in Inequalities   241
8.9 Other Patterns in Inequalities   244
  Glossary   248
  Chapter Review Problems   248

9 Relations and Functions   252

9.1 Solution Sets of Open Sentences   252
9.2 Ordered Pairs   253
9.3 The Coordinate Plane and Ordered Numbered Pairs   254
9.4 Open Sentences with Two Variables   259
9.5 Relations and Functions   262
9.6 Graphing Equations of the Form $Ax + By = C$ 267
9.7 Slope   269
9.8 Intercepts   274
9.9 Slope-intercept Form   276
9.10 Horizontal and Vertical Lines   277
9.11 Linear Functions   281
9.12 Inverses   283
9.13 Naming Functions   292
9.14 Quadratic Functions   293
  Glossary   297
  Chapter Review Problems   297

10 Systems of Equations and Inequalities   301

10.1 Equations in Two Variables   301
10.2 Solving Systems of Equations by Graphing   304
10.3 Independent, Dependent, Inconsistent Systems   308

Contents

10.4 Graphing Inequalities 314
10.5 Graphing Unions and Intersections of Sets 319
10.6 Linear Programming 322
10.7 Solving Number-theory Problems 329
10.8 Problems Involving Distance and Rate 332
Glossary 334
Chapter Review Problems 335

11 Algebraic Methods of Solving Systems 338

11.1 Weakness of Graphic Method 338
11.2 Comparison Method 339
11.3 Substitution Method 342
11.4 Addition Method 345
11.5 Problems Leading to Systems of Equations 348
11.6 Solving Problems about Angle Measure 349
11.7 Solving Problems about Perimeters 353
11.8 Solving Mixture Problems 355
Glossary 358
Chapter Review Problems 358

12 Roots and Radical Equations 362

12.1 Square Root 362
12.2 Products of Roots 366
12.3 Simplifying Expressions; Rationalizing Denominators 370
12.4 Exponents of the Form $\frac{1}{n}$ 376
12.5 Rational-number Exponents 377
12.6 Square Root and Absolute Value 381
12.7 Radical Equations 381
Glossary 384
Chapter Review Problems 385

13 Polynomials; Factoring 387

13.1 Multiplication of Two Binomials 387
13.2 Multiplication of Binomials of the Form $x + a$ 388
13.3 Special Products 391

- 13.4 Factoring Trinomials of the Form $x^2 + px + q$ 394
- 13.5 Factoring Trinomials of the Form $ax^2 + bx + c$ 397
- 13.6 Complete Factorization   402
- 13.7 Factoring Techniques   404
- 13.8 Use of Factoring to Simplify Rational Expressions   410
- 13.9 Rationalizing Binomial Denominators   412
- 13.10 Polynomials   416
- 13.11 Dividing Polynomials   417
  - Glossary   421
  - Chapter Review Problems   421

14 Quadratic Equations   424

- 14.1 Solving Quadratic Equations   424
- 14.2 Patterns in Squaring Binomials   427
- 14.3 Perfect Squares and Quadratic Equations   428
- 14.4 The Quadratic Formula   433
- 14.5 The Discriminant   435
- 14.6 Using Equation Properties   438
- 14.7 Quadratic Equations in Solving Problems   442
- 14.8 Solving Problems about Area   444
  - Glossary   450
  - Chapter Review Problems   450

TABLE OF ROOTS AND POWERS   453
INDEX   455

*Introductory Algebra
for
College
Students*

1

Sets

1.1 Membership, Finite Sets, Infinite Sets, and Equivalent Sets

Two mathematicians, George Boole (1815–1864) and Georg Cantor (1845–1918), are credited with the development of many ideas of sets. Cantor is considered to be the founder of set theory. In honor of Boole, who was the first to introduce some ideas of sets, the algebra of sets is frequently called *Boolean Algebra*.

A set is simply a collection of things. Some examples of sets are:

A: the set consisting of the numbers 1, 5, 10, 20
B: the set consisting of all students in this course
C: the set of all citizens of the United States

Now consider three other sets.

D: the set of all natural numbers; that is, 1, 2, 3, 4, . . .
E: the set of all even natural numbers; that is, 2, 4, 6, 8, . . .
F: the set of all odd natural numbers; that is, 1, 3, 5, 7, . . .

In sets D, E, and F, the three dots indicate that each sequence of numbers continues according to the pattern suggested by the listed numbers. It is easy to discover each pattern and continue listing subsequent numbers, if one should desire to do so.

Each object which belongs to a set is said to be a *member* or an *element* of the set. The following notation is used to say that 5 is an element of set A:

$$5 \in A,$$

which is read:

five is a member of A,

or

five is an element of A,

or

five belongs to A.

To say that 3 is not a member of A, the following notation is used:

$$3 \notin A.$$

This is read:

three does not belong to A.

There is one very essential difference between sets A, B, C, and D, E, F. Each of the sets A, B, and C is a *finite* set and each of the sets D, E, and F is an *infinite* set.

To simplify writing, braces are used to indicate a set. For example,

$$A = \{1, 5, 10, 20\}$$
$$D = \{1, 2, 3, \ldots\}.$$

In referring to infinite sets, it is not possible to list all of their elements. It is important to list enough elements to establish the pattern according to which the subsequent elements are arranged.

There is one important relation which exists between pairs of some sets. It can be illustrated by using a pair of finite sets.

$$X = \{1, 2, 3\} \quad Y = \{100, 101, 102\}.$$

A *one-to-one correspondence* can be established between sets X and Y by matching each element of X with exactly one element of Y. This can be done in several ways. Three of these ways are the following:

$$
\begin{array}{ccccc}
1 \leftrightarrow 100 & & 1 \leftrightarrow 100 & & 2 \leftrightarrow 100 \\
3 \leftrightarrow 101 & \text{or} & 2 \leftrightarrow 101 & \text{or} & 1 \leftrightarrow 101 \\
2 \leftrightarrow 102 & & 3 \leftrightarrow 102 & & 3 \leftrightarrow 102.
\end{array}
$$

It is possible to reason out the total number of different ways in which the sets X and Y can be matched. 1 in X can be matched in three different

1.1 Membership, Finite Sets, Infinite Sets, and Equivalent Sets

ways with an element in Y: with 100, with 101, or with 102. After that, 2 can be matched in two different ways and 3 in one way. Thus, the total number of matchings is $3 \times 2 \times 1$, or 6.

DEFINITION 1.1 Two sets between which there exists a one-to-one correspondence are said to be *equivalent* sets.

It is easy to conclude that two finite equivalent sets have the same number of elements. For example, each of the sets X and Y has three elements. The notation $n(X) = 3$ (read: n of X is three) will be used to say that the number of elements in set X is 3.

Consider next the matter of a one-to-one correspondence between two infinite sets. Let N be the set of natural numbers and E the set of even natural numbers.

$$N = \{1, 2, 3, \ldots\}$$
$$E = \{2, 4, 6, \ldots\}.$$

Since these sets are not finite, it is not possible to display the complete matching. To show that there is a one-to-one correspondence between two infinite sets, it is necessary to begin matching according to a pattern which will continue. For the sets N and E above, this pattern is indicated by the following:

$$N: \quad \{1, 2, 3, 4, 5, 6, 7, 8, \ldots\}$$
$$\updownarrow \updownarrow \updownarrow \updownarrow \updownarrow \updownarrow \updownarrow \updownarrow$$
$$E: \quad \{2, 4, 6, 8, 10, 12, 14, 16, \ldots\}.$$

Each arrow above shows the two elements which are matched with each other. The pattern suggested by this matching is the following:

To each k in N there corresponds $2k$ in E.

To each t in E there corresponds $\frac{t}{2}$ in N.

Note that t is an even number, thus $\frac{t}{2}$ is a natural number.

Let us consider another pair of infinite sets.

$$X: \quad \{1, 2, 3, 4, 5, 6, 7, 8, \ldots\}$$
$$\updownarrow \updownarrow \updownarrow \updownarrow \updownarrow \updownarrow \updownarrow \updownarrow$$
$$Y: \quad \{100, 101, 102, 103, 104, 105, 106, 107, \ldots\}.$$

The set X is the set of all natural numbers and the set Y is the set of natural numbers which are greater than 99. Each set is infinite. In general, n in X is matched with $n + 99$ in Y and k in Y is matched with $k - 99$ in X.

For some exercises below it will be necessary to know the concept of divisibility. It is defined as follows:

DEFINITION 1.2 The natural number x is *divisible* by the natural number y if and only if $x \div y$ is a natural number; y is said to be a *divisor* of x.

Divisibility of natural numbers by 3 and by 9 is particularly interesting. It can be shown that a natural number is divisible by 3 if and only if the sum of its digit values is divisible by 3. For example, the sum of the digit values of 73,062 is $7 + 3 + 0 + 6 + 2 = 18$ and since 18 is divisible by 3, so is 73,062. Similarly, a natural number is divisible by 9 if and only if the sum of its digit values is divisible by 9. For example, the sum of the digit values of 1854 is divisible by 9, therefore 1854 is divisible by 9. Since the sum of the digit values of 265 is divisible by neither 3 nor 9, 265 is divisible by neither 3 nor 9.

It is easy to observe that if a number is divisible by 9, then it is also divisible by 3.

EXERCISE 1.1

1. Are the following pairs of sets matching sets?
   - **a.** S: {2, 3, 4, 5, 6}
     T: {Mike, Bob, Hal, Leroy, Ed, Bill}
   - **b.** C: {Scranton, Florida, California}
     D: {St. Paul, Chicago, Denver, Vermont}
   - **c.** G: {1, 3, 5, 7, 9, . . .}
     H: {2, 4, 6, 8, 10}

2. Give examples of three pairs of *finite* sets such that the two sets in each pair are matching sets.

3. Give examples of three pairs of finite sets such that *no* two sets in a pair are matching sets.

4. Give examples of three pairs of *infinite* sets such that the two sets in each pair are matching sets.

5. For each pair of sets, a one-to-one correspondence is shown. Describe the one-to-one correspondence between the elements of A and those of B.

Example A: {1, 2, 3, 4, . . .}
       ↕ ↕ ↕ ↕
     B: {1, 4, 9, 16, . . .}.

1.1 Membership, Finite Sets, Infinite Sets, and Equivalent Sets 5

Description: Each element of A is matched with its square ($1^2 = 1$; $2^2 = 4$; $3^2 = 9$; and so on).

a. A: $\{\frac{1}{2}, \frac{1}{3}, \frac{1}{4}, \frac{1}{5}, \ldots\}$
   $\updownarrow \updownarrow \updownarrow \updownarrow$
   B: $\{2, 3, 4, 5, \ldots\}$

b. A: $\{\frac{1}{2}, \frac{1}{3}, \frac{1}{4}, \frac{1}{5}, \ldots\}$
   $\updownarrow \updownarrow \updownarrow \updownarrow$
   B: $\{3, 4, 5, 6, \ldots\}$

c. A: $\{\frac{1}{4}, \frac{1}{6}, \frac{1}{8}, \frac{1}{10}, \ldots\}$
   $\updownarrow \updownarrow \updownarrow \updownarrow$
   B: $\{3, 5, 7, 9, \ldots\}$

d. A: $\{1, 2, 3, 4, \ldots\}$
   $\updownarrow \updownarrow \updownarrow \updownarrow$
   B: $\{1, 8, 27, 64, \ldots\}$

e. A: $\{\frac{1}{2}, \frac{1}{3}, \frac{1}{4}, \frac{1}{5}, \ldots\}$
   $\updownarrow \updownarrow \updownarrow \updownarrow$
   B: $\{\frac{1}{4}, \frac{1}{9}, \frac{1}{16}, \frac{1}{25}, \ldots\}$

f. A: $\{1, 2, 3, 4, \ldots\}$
   $\updownarrow \updownarrow \updownarrow \updownarrow$
   B: $\{2, 5, 8, 11, \ldots\}$

g. A: $\{\frac{1}{2}, \frac{1}{4}, \frac{1}{8}, \frac{1}{16}, \ldots\}$
   $\updownarrow \updownarrow \updownarrow \updownarrow$
   B: $\{\frac{3}{4}, \frac{3}{8}, \frac{3}{16}, \frac{3}{32}, \ldots\}$

h. A: $\{1, 2, 3, 4, 5, \ldots\}$
   $\updownarrow \updownarrow \updownarrow \updownarrow \updownarrow$
   B: $\{2, 3, 5, 7, 11, \ldots\}$

i. A: $\{1, 2, 3, 4, 5, \ldots\}$
   $\updownarrow \updownarrow \updownarrow \updownarrow \updownarrow$
   B: $\{2, 8, 18, 32, 50, \ldots\}$

j. A: $\{1, 2, 3, 4, 5, \ldots\}$
   $\updownarrow \updownarrow \updownarrow \updownarrow \updownarrow$
   B: $\{3, 9, 19, 33, 51, \ldots\}$

k. A: $\{1, 2, 3, 4, 5, \ldots\}$
   $\updownarrow \updownarrow \updownarrow \updownarrow \updownarrow$
   B: $\{3, 10, 29, 66, 127, \ldots\}$

l. A: $\{\frac{1}{2}, \frac{3}{4}, \frac{5}{6}, \frac{7}{8}, \ldots\}$
   $\updownarrow \updownarrow \updownarrow \updownarrow$
   B: $\{3, 7, 11, 15, \ldots\}$

6. List each of the following sets.
   a. the set of natural numbers which are less than 6
   b. the set of months whose names start with the letter D
   c. the set of natural numbers which are less than 9 and greater than 6
   d. the set of odd natural numbers which are less than 20
   e. the set of the planets of the solar system
   f. the set of the seasons of the year
   g. the set of odd natural numbers which are less than 20 and are divisible by 5
   h. the set of natural numbers between 50 and 100 which are divisible by 6

7. What do the three sets described below have in common? [*Hint:* First list their elements.]

   the natural numbers less than 7 which are also greater than 5

   the odd natural numbers less than 20 which are also divisible by 7

   the states of the United States entirely surrounded by water

8. Let T be the set of natural numbers divisible by 3. Label each statement with T for true and F for false.

a. $39 \in T$ **b.** $17 \notin T$
c. $117 \notin T$ **d.** $237 \in T$
e. $112 \in T$ **f.** $10{,}001 \in T$
g. $522 \notin T$ **h.** $92{,}322 \in T$
i. $92.321 \in T$

9. Let S be the set of natural numbers divisible by 9. Label each statement as either true or false.

a. $81 \notin S$ **b.** $89 \in S$
c. $207 \in S$ **d.** $88{,}092 \in S$
e. $34{,}560 \notin S$ **f.** $88{,}776 \in S$

10. Give an example of a number which is divisible by 3, but is not divisible by 9.

*11. Prove the divisibility rule by 3.

*12. Prove the divisibility rule by 9.

1.2 Numbers and Numerals

Consider the question, "how many numbers *five* are there?" The answer is: there is only *one* number five. Yet the number five has many names. When we want to refer to the number five, we may use any one of many names, such as

$$5 \qquad \tfrac{10}{2} \qquad 3+2 \qquad V \qquad \text{five} \qquad 2\tfrac{1}{2} \times 2.$$

A name of a number is called a *numeral*. In algebra, numbers are the subject of study most of the time. Occasionally, however, reference may be made to numerals. For example, the sentence

V is a Roman numeral for the number 5

is about a name of the number 5.

Sometimes it is a little awkward to be careful about distinguishing between numbers and numerals. For example, when saying "Give an example of a two-digit number divisible by 7," we are somewhat careless. It is because, when saying "a two-digit number," we really mean a two-digit numeral, since we tell the number of digits by looking at a symbol. For example, "43" has two digits, "4" and "3." To be quite careful, we would have to rephrase the statement above to read something like this:

Give an example of a two-digit numeral which names a number divisible by 7.

At times in this book, we shall allow ourselves the luxury of avoiding the awkwardness due to the distinction between number and numeral.

1.2 Numbers and Numerals

We shall use the less careful language, and assume that the distinction between the two concepts is understood.

Most of the time, base-ten names for numbers are used. On occasion, however, other bases are more convenient. For example, base two is used in computations carried out by electronic computers. Base eight is also used in some situations.

In base ten the place values are powers of ten as is illustrated by the following:

$$3492 = 3 \times 1000 + 4 \times 100 + 9 \times 10 + 2 \times 1$$
$$= 3 \times 10^3 + 4 \times 10^2 + 9 \times 10^1 + 2 \times 10^0.$$

The above is called an *expanded form* of 3492. To see that it is logical to conclude that $10^0 = 1$, observe the following pattern:

$$10^4 = 10,000$$
$$10^3 = 1000 \quad \text{divide by } 10$$
$$10^2 = 100 \quad \text{divide by } 10$$
$$10^1 = 10 \quad \text{divide by } 10$$

As the power of 10 decreases by 1, the number is divided by 10. The next entry in this sequence is 10^0, because if we continue the pattern, the following is obtained:

$$10^0 = 10 \div 10 = 1.$$

The following shows the general pattern for base-ten numerals:

... 10,000	1000	100	10	1
□	□	□	□	□
... ten^4	ten^3	ten^2	ten^1	ten^0.

In base-eight numerals the same pattern is used, except eight occurs in place of ten. The notation 623_{eight} is used to indicate that the numeral is in base eight.

$$623_{\text{eight}} = 6 \times 8^2 + 2 \times 8^1 + 3 \times 8^0$$
$$= 6 \times 64 + 2 \times 8 + 3 \times 1$$
$$= 384 + 16 + 3$$
$$= 403_{\text{ten}}.$$

That $8^0 = 1$ is true can be shown in the same way that $10^0 = 1$ was shown to be true.

In the previous section divisibility of natural numbers was discussed. This concept can be used to partition the set of natural numbers into three distinct subsets. Consider the following definitions.

DEFINITION 1.3 A natural number which has exactly two divisors is called a *prime number*.

DEFINITION 1.4 A natural number which has more than two divisors is called a *composite number*.

For example, 6 is a composite number because it has more than two divisors; 7 is a prime number because it has exactly two divisors, 1 and 7. There is one natural number left which does not fit either of the two definitions. It is the number 1. It has exactly one divisor. Thus, N, the set of natural numbers can be partitioned into the following three subsets:

$$A = \{1\}$$
$$P = \{2, 3, 5, 7, \ldots\} \quad \text{(prime numbers)}$$
$$C = \{4, 6, 8, 9, \ldots\} \quad \text{(composite numbers)}.$$

EXERCISE 1.2

1. Each set of five names below contains four names for the same number. Identify the one name which does not fit with the other four.

   a. $\frac{18}{2}$ IX $2\frac{1}{2} \times 4$ $4\frac{1}{2} \times 2$ $99 \div 11$

   Recall: 3^2 (read: three squared or three to the second power) means 3×3.

   2^3 (read: two cubed or two to the third power) means $2 \times 2 \times 2$.

   2^4 (read: two to the fourth power) means $2 \times 2 \times 2 \times 2$.

   b. 2^4 4^2 $\frac{32}{2}$ $\frac{48}{2}$ $29 - 13$

   Recall: $\sqrt{16}$ (read: the square root of sixteen) is 4 because $4 \times 4 = 16$.

   $\sqrt[3]{27}$ (read: the cube root of twenty-seven) is 3 because $3 \times 3 \times 3 = 27$.

   c. $\sqrt{64}$ $\sqrt{24}$ $2 \times \sqrt{16}$ $4 \div \frac{1}{2}$ VIII

   d. $\frac{1}{2}$ $\sqrt{\frac{1}{4}}$ $\frac{1}{4} + \frac{1}{4}$ $\sqrt{2}$ .5

   e. .25 $\frac{1}{4}$ $\sqrt{\frac{1}{2}}$ $\sqrt{\frac{1}{16}}$ $\frac{1}{8} + \frac{1}{8}$

   Recall: 24_{five} means 2 fives + 4 ones, which is fourteen.

   f. 21_{seven} 30_{five} 15_{ten} 32_{six} 13_{twelve}

2. Show the expansion of 2753_{eight} using powers of eight.

3. There is only one prime number which is also an even number. What number is it?

4. Write a brief argument showing that any even number greater than 2 is not a prime number.
5. List the first ten members of the set of prime numbers.
6. Let P be the set of prime natural numbers less than 11. Label the following statements with T for true and F for false.

   a. $6 \in P$ b. $9 \notin P$
   c. $3 \in P$ d. $7 \in P$
   e. $5 \notin P$ f. $13 \in P$
   g. $1 \notin P$ h. $0 \in P$
   i. $11 \in P$

7. Build a pattern beginning with 8^4 and, moving to lower powers, show that it is logical to conclude that $8^0 = 1$.
8. Prove that for all nonzero numbers x and y, $x^0 = y^0$.

1.3 Subsets and Equality

Frequently one set is chosen and its elements are used to form other sets.

DEFINITION 1.5 The larger set from which members to form new sets are chosen is called the *universe* or *universal set*.

Considering the set of natural numbers, N, to be the universe, one can choose the set of even numbers, E. Since every element of E is also in N, we say that E is a *subset* of N.

DEFINITION 1.6 Set X is said to be a *subset* of set Y if and only if every element of X is also an element of Y.

The notation
$$X \subseteq Y$$
is used to say that set X is a subset of set Y. The definition of a subset leads to the conclusion that every set is a subset of itself. That is, it is true for every set X that
$$X \subseteq X.$$
The symbol $\subseteq$ consists of two symbols, $\subset$ and $=$. The symbol $\subset$ is used to say that a set is a *proper* subset of another set.

DEFINITION 1.7 X is said to be a *proper subset* of Y, written $X \subset Y$, if and only if $X \subseteq Y$ and $X \neq Y$.

If a set M *is not a subset* of set N, that is, there is at least one element in M which is not in N, we write $M \not\subseteq N$, which is read: M is not a subset of N.

DEFINITION 1.8 If set A and set B are two names for the same set, we say that sets A and B are *equal* and write $A = B$.

For example, if $A = \{1, 2, 3\}$ and $B = \{1 + 2, 1 + 1, 1\}$, then $A = B$.

EXERCISE 1.3

1. a. Give examples of two sets A and B such that A is a subset of B and B is not a subset of A.
   b. Give examples of two sets C and D such that C is not a subset of D and D is not a subset of C.
   c. Give examples of sets E and F such that E is a subset of F and F is a subset of E. (Do you end up with two different sets E and F?)
2. On the basis of problem 1c, complete the following statement:
   For all sets A and B, if $A \subseteq B$ and $B \subseteq A$, then ___?___ .
3. Show that if $A \subset B$, then $B \not\subset A$.
4. Given the following sets

   $$E = \{1, 3, 4, 6\} \quad G = \{1, 6\} \quad H = \{1, 3, 4, 6, 8, 9\},$$

   label each statement with T for true and F for false.

   a. $E \subseteq G$ b. $H \subseteq E$
   c. $G \subseteq H$ d. $G \subset H$
   e. $H \subseteq G$ f. $H \subseteq H$
   g. $H \subset H$ h. $E \not\subset H$
   i. $E \not\subseteq G$ j. $H \not\subseteq H$
   k. $G \not\subseteq H$ l. $G \not\subset G$

5. Let the universal set U be the set of those states of the United States which are located west of the Mississippi River.
   a. If P is the set of the states (chosen from U) touching the Pacific Ocean, label each of the following as either true or false.
      i. $P \subseteq U$ ii. $P \subset U$ iii. $P = U$
   b. If C is the set of the states (chosen from U) touching the border of Canada, label each of the following as either true or false.
      i. $C \subseteq U$ ii. $C \subset U$ iii. $C = U$ iv. $C \subset P$
   c. If T is the set of the states (chosen from U) directly adjoining

1.4 Disjoint Sets and Venn Diagrams

the Mississippi River, label each of the following as either true or false.

  i. $T \subset U$ ii. $T \subset C$ iii. $T \subset P$

6. Regroup the following list of sets into groups of equal sets. (If a base of a numeral is not given, assume that it is ten.)

  a. $\{1, 2, 3, 4, \ldots\}$ b. $\{2, 4, 6, 8, \ldots\}$
  c. $\{1, 3, 5, 7, \ldots\}$ d. the set of all natural numbers
  e. $\{\frac{1}{2}, \frac{1}{4}\}$ f. $\{10_{\text{two}}, 10_{\text{three}}, 10_{\text{four}}\}$
  g. $\{.25, .5\}$ h. $\{4, 3, 2\}$
  i. $\{10_{\text{two}}, 11_{\text{two}}, 100_{\text{two}}\}$ j. $\{\frac{1}{4} + \frac{1}{4}, \frac{1}{8} + \frac{1}{8}\}$
  k. the set of all even natural numbers
  l. the set of all odd natural numbers
  m. the set of all natural numbers which are divisible by 2
  n. the set of all natural numbers which are not divisible by 2

1.4 Disjoint Sets and Venn Diagrams

Given two sets, they are related in one of two ways. Either they have no elements in common or they have one or more elements in common.

DEFINITION 1.9 Two sets are *disjoint* if and only if they have no elements in common.

Consider the following pair of sets:

  A = the set of all natural numbers less than 12.
  B = the set of all natural numbers between 50 and 100.

Sets A and B have no elements in common. Figure 1.1 is a Venn diagram showing this. The diagram shows that A and B are disjoint.

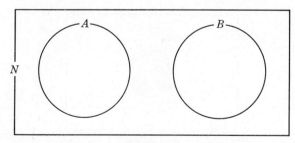

Figure 1.1

Now consider two other sets.

C = the set of all natural numbers between 25 and 37 not including 25 and 37.
D = the set of all natural numbers less than 32.

Sets C and D have some elements in common. For example, $30 \in C$ and $30 \in D$.

The relationship between sets C and D may be portrayed as in Figure 1.2.

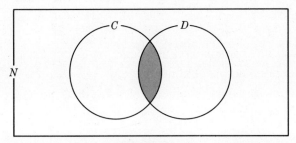

Figure 1.2

In Figure 1.2, we designate the points within circle C to be the elements of set C. The points within circle D represent the elements of set D. The elements of each C and D are contained in set N, the set of all natural numbers, which is represented by the interior of rectangle N. The shaded part represents the elements which belong to *both* C and D.

Now consider the following two sets.

E = the set of all natural numbers less than 20.
F = the set of all natural numbers between 10 and 17, not including 10 and 17.

Observe that $F \subseteq E$ since every element of F also belongs to E. Figure 1.3 portrays this relationship between sets F and E. Set F is shaded to indicate that all of the elements of F belong to both sets, E and F.

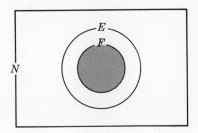

Figure 1.3

1.4 Disjoint Sets and Venn Diagrams

EXERCISE 1.4

1. In this problem, the universal set U is the set of all whole numbers, $W = \{0, 1, 2, 3, \ldots\}$. A and B are subsets of U. In each instance, tell whether or not

   i. $A \subseteq B$ ii. $A \subset B$ iii. $A = B$ iv. A and B are disjoint sets.

   a. $A = \{1, 2, 3\}$; $B = \{1, 2, 3, 4, 5\}$
   b. $A = \{0, 1\}$; $B = \{0\}$
   c. $A = \{1, 4, 6\}$; $B = \{1, 6, 7\}$
   d. A is the set of all natural numbers divisible by 10.
      B is the set of all natural numbers.
   e. A is the set of all natural numbers less than 100.
      B is the set of all natural numbers greater than 55.
   f. $A = \{1, 2\}$; $B = \{1, 2\}$
   g. $A = \{0\}$; $B = \{0\}$

2. In this problem, let U be the universal set given in each case. For each problem state whether or not

   i. $A \subseteq B$ ii. $A \subset B$ iii. $A = B$ iv. A and B are disjoint sets.

Example $U = \{4, 6, 8, 10, 12\}$
   $A = \{4\}$; $B = \{4, 6\}$.
   i. $A \subseteq B$ ii. $A \subset B$ iii. $A \neq B$ iv. not disjoint

   a. $U = \{0, 1, 2, 3, 4, 5\}$
      $A = \{1, 3, 4, 5\}$; $B = \{1, 4, 5\}$
   b. $U = \{\frac{1}{2}, \frac{1}{3}, \frac{1}{4}\}$
      $A = \{.5\}$; $B = \{.5, .25\}$
   c. $U = \{0, 1\}$
      $A = \{0\}$; $B = \{1\}$
   d. $U = \{$Arizona, California, New Mexico, Oregon$\}$
      $A = \{$Arizona, California$\}$; $B = \{$Oregon, California, Arizona$\}$
   e. $U = \{$New York, Los Angeles, Chicago, New Orleans$\}$
      $A = \{$Los Angeles, Chicago$\}$; $B = \{$Chicago$\}$
   f. $U = \{1, 2, 3, 4, 5, 6, 7, 8, 9\}$
      $A = \{1, 3, 7, 9\}$; $B = \{3, 7, 8, 9\}$
   g. $U = $ the set of all natural numbers which are less than 20
      $A = $ the set of natural numbers which are less than 9
      $B = $ the set of natural numbers which are greater than 5 and less than 8

h. $U = $ the set of all natural numbers
   $A = $ the set of all even natural numbers
   $B = $ the set of all odd natural numbers
3. **a.** Suppose you know that $5 \in X$ and $5 \in Y$, where X and Y are some sets. Are you justified in concluding that $X \subseteq Y$? $Y \subseteq X$? X and Y are not disjoint?
   b. Suppose you know that sets M and P have three common elements. Give one conclusion you are justified in drawing from this fact.
4. Let U, the universal set, be the set of all natural numbers. Using the interior of a rectangle to represent U, draw a Venn diagram showing the correct relationships between the following sets.

$$U = \{1, 2, 3, 4, \ldots\}$$
$$A = \{4\} \quad B = \{6, 5, 4\} \quad C = \{5\} \quad D = \{10\}$$

1.5 Operations with Sets

The four basic operations with numbers are addition, subtraction, multiplication, and division. Each of these is called a *binary* operation because each requires a pair of numbers.

Consider the following two sets:

$$A = \{1, 5, 7, 16\}; \quad B = \{7, 16, 21\}.$$

We shall obtain a third set, C, by combining the elements of the set A with the elements of the set B. Thus, $C = \{1, 5, 7, 16, 21\}$.

The set C is called the *union* of sets A and B. The following notation will be used for the operation of union:

$$A \cup B = C,$$

or

$$\{1, 5, 7, 16\} \cup \{7, 16, 21\} = \{1, 5, 7, 16, 21\}.$$

Here is another example,

$$X = \{5, 10, 15\}; \quad Y = \{7, 9, 11, 13\}$$
$$\{5, 10, 15\} \cup \{7, 9, 11, 13\} = \{5, 7, 9, 10, 11, 13, 15\}.$$

DEFINITION 1.10 The *union* of any pair of sets X and Y is a set consisting of all the elements which belong to X or Y.

The word "or" is being used here in a sense which may be different from the sense to which you are accustomed. The phrase "is a member of X or Y" may imply any one of the following:

1.5 Operations with Sets

a is a member of X,
a is a member of Y,
a is a member of both X and Y.

Note one interesting difference between the pairs of sets A and B, and X and Y, given above. Sets A and B have elements which are common to both sets, whereas the sets X and Y have no common elements. That is, sets X and Y are disjoint sets, whereas sets A and B are not disjoint sets.

Note that the set $X = \{5, 10, 15\}$ has three elements. The set $Y = \{7, 9, 11, 13\}$ has four elements. The union of the two sets, $X \cup Y = \{5, 7, 9, 10, 11, 13, 15\}$, has seven elements. And $3 + 4 = 7$.

Now observe the set $A = \{1, 5, 7, 16\}$. It has four elements. The set $B = \{7, 16, 21\}$ has three elements. The union, $A \cup B = \{1, 5, 7, 16, 21\}$, has five elements. But $4 + 3 \neq 5$.

With a little deliberation it is possible to conclude that, for all finite sets A and B, if A and B are disjoint, then

$$n(A) + n(B) = n(A \cup B).$$

On the other hand, if A and B are not disjoint, then

$$n(A) + n(B) > n(A \cup B).$$

Let us now obtain a third set from the sets A and B above in a different way. We will take this third set to be the set of only those elements which are in *both* A and B.

$A = \{1, 5, 7, 16\}$; $B = \{7, 16, 21\}$. The only elements which are in both A and B are 7 and 16. The set D, then, consisting of elements belonging to both A and B, is $D = \{7, 16\}$.

The set $D = \{7, 16\}$ is the intersection of sets $A = \{1, 5, 7, 16\}$ and $B = \{7, 16, 21\}$. A intersection B is equal to D is written as

$$A \cap B = D.$$

The symbol for the operation of intersection is $\cap$.

DEFINITION 1.11 The *intersection* of a pair of sets is the set consisting of all those elements which are common to both sets.

Sets X and Y given above do not have any elements in common. Thus, there are no elements in their intersection. It would be advantageous to have a set which is equal to the intersection of two disjoint sets. We define such a set as follows.

DEFINITION 1.12 A set which has no elements is called the *empty* set. The symbol for the empty set is ϕ.

Thus, for the sets X and Y above, we have

$$X \cap Y = \phi.$$

Observe that it takes a pair of sets to perform the operations of union and intersection. Therefore, *union and intersection are binary operations.*

Given any pair of sets, there is a set which is their union. Also, given any pair of sets, there is a set which is their intersection.

EXERCISE 1.5

1. Name the set which is the intersection of each pair of sets below.
   a. A: all natural numbers
      B: all odd natural numbers
   b. A: all natural numbers
      B: all even natural numbers
   c. A: all odd natural numbers
      B: all even natural numbers
   d. A: all natural numbers
      B: all prime natural numbers

2. Name the set which is the union of each pair of sets in Problem 1.
3. Give an example of a pair of finite disjoint sets.
4. Give an example of a pair of infinite disjoint sets.
5. Give an example of a pair of finite sets which are not disjoint.
6. Give an example of a pair of infinite sets which are not disjoint.
7. Find the intersection of each pair of sets in Problems 3–6.
8. Find the union of each pair of sets in Problems 3–6.
9. Explain why the intersection of the empty set with any set is the empty set. This may be stated as follows:

   For every set A, $A \cap \phi = \phi$; therefore, $\phi \cap \phi = \phi$.

10. Explain why the union of the empty set with any set is that set. This may be stated as follows:

    For every set A, $A \cup \phi = A$; therefore, $\phi \cup \phi = \phi$.

11. Let $A = \{1, 3, 5\}$, $B = \{3, 5, 7, 9\}$, and $C = \{4, 6, 8, 10\}$. Find
    a. $A \cup B$
    b. $A \cap B$
    c. $A \cup C$
    d. $A \cap C$
    e. $B \cup C$
    f. $B \cap C$
    g. $(A \cup B) \cup C$

[*Hint:* Parentheses mean to first find the set $A \cup B$, then find the union of the resulting set with C.]

h. $A \cup (B \cup C)$ **i.** $(A \cap B) \cap C$
j. $A \cap (B \cap C)$ **k.** $A \cup (B \cap C)$
l. $(A \cup B) \cap (A \cup C)$ **m.** $A \cap (B \cup C)$

12. Complete each statement. For each set A,
    a. $A \cup A = ?$ **b.** $A \cap A = ?$
    c. $(A \cup A) \cup A = ?$ **d.** $(A \cap A) \cap A = ?$
    e. $(A \cup A) \cap A = ?$ **f.** $(A \cap A) \cup A = ?$
    g. $A \cup \phi = ?$ **h.** $A \cap \phi = ?$

13. Each of the following two phrases describes the empty set:

    the set of all polar bears whose natural habitat is Africa
    the set of all girls who are members of Baltimore Colts.

    Give three other phrases each describing the empty set.

1.6 Counting the Subsets of a Finite Set, Subsets of Infinite Sets

When studying various sets, it is good practice to specify the universe from which the elements of sets are selected. For example, we may agree that for the purposes of our study for this particular moment, we shall select the elements of our sets from the set of natural numbers: $\{1, 2, 3, 4, \ldots\}$. That is, the set of natural numbers is, at this time, the universal set or the universe. Thus

$$U = \{1, 2, 3, 4, \ldots\}.$$

Of course, there is no limit to the number of subsets we can obtain from the set of natural numbers.

For example, the set $A = \{1\}$ is a subset of U. So is $B = \{3, 7, 10\}$. Both A and B are subsets of U. Thus, $A \subseteq U$ and $B \subseteq U$.

To show that some set, say A, is not a subset of another set, B, it is necessary to produce at least one element of A which is not in B. Since the empty set, ϕ, has no elements, it is not possible to do this for ϕ in relation to any other set. Therefore, we shall agree that the empty set is a subset of each set.

For each set X, $\phi \subseteq X$.

Now consider the universe L, where $L = \{1, 2\}$; that is, the only members in the universe are the numbers 1 and 2. Let us form all possible subsets of L. In doing this, we must remember that the empty set is a subset of every set. Furthermore, we must remember that every set is a subset of itself. Now, we are ready to list all subsets of $L = \{1, 2\}$.

ϕ $\{1\}$ $\{1, 2\}$
$\{2\}$

Thus, the set $L = \{1, 2\}$ has *two* elements and *four* subsets.

Now, take the universe $K = \{1, 2, 3\}$. Form all possible subsets of K.

ϕ $\{1\}$ $\{1, 2\}$ $\{1, 2, 3\}$
$\{2\}$ $\{1, 3\}$
$\{3\}$ $\{2, 3\}$

Thus, the set K has *three* elements and *eight* subsets.

Now, take the universe $T = \{1, 2, 3, 4\}$. Form all possible subsets of T.

ϕ $\{1\}$ $\{1, 2\}$ $\{1, 2, 3\}$ $\{1, 2, 3, 4\}$
$\{2\}$ $\{1, 3\}$ $\{1, 2, 4\}$
$\{3\}$ $\{1, 4\}$ $\{1, 3, 4\}$
$\{4\}$ $\{2, 3\}$ $\{2, 3, 4\}$
$\{2, 4\}$
$\{3, 4\}$

Thus, the set T has *four* elements and *sixteen* subsets.

Let us tabulate our results.

Number of elements in a set	Number of all possible subsets
2	$4 = 2 \times 2$
3	$8 = 2 \times 2 \times 2$
4	$16 = 2 \times 2 \times 2 \times 2$

The table above reveals a pattern which points to the relationship between the number of elements in the universal set and the number of all possible subsets. According to this pattern, there should be $2 \times 2 \times 2 \times 2 \times 2$, or 32, subsets that can be formed from a universal set with *five* elements. And there should be 2^6, or 64, subsets that can be formed from a set with six elements. To fit into the pattern, a set with one element would have to have two subsets. Let us check.

Let $A = \{5\}$. The subsets of A are ϕ and $\{5\}$. Thus $A = \{5\}$ has two subsets.

Since there is no end to the number of elements in an infinite set, there is no end to the number of subsets one can obtain. This is easily seen if we consider only those subsets which have one element. For example, take the set of all natural numbers $\{1, 2, 3, 4, 5, 6, \ldots\}$. Some subsets are

1.6 Counting the Subsets of a Finite Set, Subsets of Infinite Sets

$\{1\}, \{2\}, \{3\}, \{4\}, \{5\}, \{6\}$, and so on.

Each of these subsets has only one element, and there is no end to the number of these subsets.

There is at least one interesting thing about subsets of infinite sets. Given any finite set, say $A = \{5, 7, 10\}$, and deleting at least one element of it, say 5, we obtain a subset $B = \{7, 10\}$ which cannot be matched with the original set A. This is because the set A has more elements than the set B. Let us now see whether the same is true for infinite sets. In the set of all natural numbers $N = \{1, 2, 3, 4, 5, \ldots\}$, let us delete the number 1. Set $K = \{2, 3, 4, 5, 6, \ldots\}$ is obtained such that $K \subset N$, because every member of K is also a member of N and $K \neq N$.

It is possible to establish a one-to-one correspondence between the set N and its subset K.

$$N: \{1, 2, 3, 4, 5, 6, \ldots\}$$
$$\updownarrow \updownarrow \updownarrow \updownarrow \updownarrow \updownarrow$$
$$K: \{2, 3, 4, 5, 6, 7, \ldots\}.$$

That is, to each number x which is a member of the set N, we assign $x + 1$ in K. And to each y in K we assign $y - 1$ in N.

Thus, there is a way of establishing a one-to-one correspondence between the set of all natural numbers and its proper subset, all natural numbers except 1. This example proves that it is possible for a proper subset of an infinite set and the infinite set to be equivalent.

EXERCISE 1.6

Given: $A = \{\frac{1}{2}, \frac{1}{3}, \frac{1}{4}, \frac{1}{5}\}$ and $B = \{\frac{1}{4}, \frac{1}{5}, \frac{1}{6}, \frac{1}{7}, \frac{1}{8}\}$ (to be used in Problems 1–6).

1. Form all subsets of A. How many subsets are there?
2. How many subsets of set B are there?
3. Determine $A \cup B$.
4. How many subsets may be formed from $A \cup B$?
5. Determine $A \cap B$.
6. a. Form all subsets of $A \cap B$. How many are there?
   b. How many of these are proper subsets of $A \cap B$?
   c. How many of these are proper nonempty subsets of $A \cap B$?
7. Let the universal set be the set of all natural numbers, $U = \{1, 2, 3, 4, 5, 6, \ldots\}$.
   a. Show one way of establishing a one-to-one correspondence between U and the set of all natural numbers greater than 100.

b. Is it possible to establish a one-to-one correspondence between U and the set of all even numbers? If your answer is yes, show one way of doing it. If your answer is no, explain.

c. Is it possible to establish a one-to-one correspondence between U and the set of all natural numbers less than 1001? If your answer is yes, show one way of doing it. If your answer is no, explain.

8. Let $U_1 = \{1, 2, 3, 4, 5, \ldots\}$
$U_2 = \{1, 3, 5, 7, 9, \ldots\}$
$U_3 = \{100, 101, 102, 103, 104, 105, \ldots\}$.

Find

a. $U_1 \cup U_2$
b. $U_1 \cup U_3$
c. $U_2 \cup U_3$
d. $U_1 \cap U_2$
e. $U_1 \cap U_3$
f. $U_2 \cap U_3$
g. $(U_1 \cup U_2) \cup U_3$
h. $(U_1 \cap U_2) \cap U_3$
i. $(U_1 \cup U_2) \cap U_3$

1.7 Complement of a Set

Take the universe to be the set $U = \{1, 2, 3, 4, 5, 6, 7, 8, 9, 10\}$; that is, the natural numbers from 1 through 10. Now take a subset of U, say $A = \{5, 6, 7\}$. Consider the set of elements remaining in U after the set A has been removed. This set is $B = \{1, 2, 3, 4, 8, 9, 10\}$. Note two things about A and B. First, they are disjoint; that is,

$$A \cap B = \phi.$$

Second, the union of A and B is equal to the universe; that is,

$$A \cup B = U = \{1, 2, 3, 4, 5, 6, 7, 8, 9, 10\}.$$

The set B is called the *complement* of the set A in the set U. The symbol $\overline{A}$ is used for the complement of A.

The following two generalizations concerning the complement of a set are true for each set A:

$$A \cap \overline{A} = \phi \qquad A \cup \overline{A} = U.$$

The following Venn diagram, Figure 1.4, shows these relationships. If the interior of the rectangle represents the universe and the interior of the circle the set A, then $\overline{A}$, the complement of A, is represented by the shaded portion. The Venn diagram makes it easy to see that a set and its complement are always disjoint sets and that the union of a set and its complement is the universe.

1.7 Complement of a Set

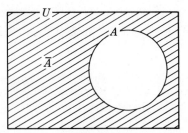

Figure 1.4

EXERCISE 1.7

1. Let $U = \{1, 2, 3, 4, 5, 6, 7, 8\}$. Determine $\overline{A}$ for the given set A.

Example $A = \{1, 2, 6\}$.
$\overline{A}$ is the set of elements in U other than the elements of A. Therefore, $\overline{A} = \{3, 4, 5, 7, 8\}$.

   a. $A = \{2, 4, 6, 8\}$ **b.** $A = \{1, 3, 5, 7\}$ **c.** $A = \{1, 8\}$

2. Let $U = \{1, 2, 3, 4, 5, 6, 7\}$, $A = \{1, 2\}$, and $B = \{2, 4, 5\}$. Determine the following sets.

   a. $\overline{A}$ **b.** $\overline{B}$
   c. $A \cap B$ **d.** $\overline{A \cap B}$
   e. $\overline{A} \cap \overline{B}$ **f.** $\overline{A \cup B}$
   g. $\overline{A} \cup \overline{B}$ **h.** $\overline{A} \cap A$
   i. $\overline{B} \cap B$ **j.** $\overline{A} \cap B$
   k. $A \cap \overline{B}$ **l.** $\overline{A} \cup B$

3. Let $U = \{0, 1, 2\}$, and $A = \{0\}$. Label each statement with T for true and F for false.

   a. $\overline{A} = U$ **b.** $A \cup U = \{1, 2\}$
   c. $A \cap U = \phi$ **d.** $\overline{A} \cap U = \{1, 2\}$
   e. $\overline{A} = \{1, 2\}$ **f.** $\overline{U} = \{0\}$

4. $U = \{1, 3, 5, 7, 9, 11\}$, $A = \{1, 5\}$, $B = \{7, 9, 11\}$, $C = \{3, 7, 9, 11\}$, and $D = \{1, 11\}$. Determine the following sets.

   a. $\overline{B}$ **b.** $\overline{C}$
   c. $\overline{D}$ **d.** $\overline{A}$
   e. $A \cup U$ **f.** $A \cap U$
   g. $\overline{A \cap U}$ **h.** $C \cap \overline{D}$

i. $U \cap \overline{C}$ j. $\overline{D \cap \phi}$
k. $\overline{B} \cap \phi$ l. $\overline{U \cap A}$
m. $\overline{D \cup B}$ n. $\overline{D} \cap \overline{B}$
o. $\overline{U}$

5. Let U = the set of all natural numbers. Describe in words the complement of each of the following sets.

Example $X = \{1, 2\}$.
$\overline{X}$ is the set of all natural numbers greater than 2.

a. $A = \{1, 2, 3\}$
b. $B = \{100, 101, 102, 103, \ldots\}$
c. $C = \{17, 18, 19, 20, \ldots, 58, 59, 60\}$
d. $D = \{2, 4, 6, 8, 10, \ldots\}$
e. $E = \{1, 3, 5, 7, 9, 11, \ldots\}$
f. $F = \{2, 3, 5, 7, 11, 13, 17, 19, 23, \ldots\}$
g. $G = \{7, 14, 21, 28, 35, 42\}$
h. $H = \{10, 20, 30, 40, \ldots, 210\}$
i. $I = \{1000, 900, 800, \ldots, 100\}$
j. $J = \{1, 4, 7, 10, 13, 16, \ldots\}$

Example Given: A universal set U containing six elements and two subsets A and B. A contains three elements, and B contains four elements. $B \cap A$ contains two elements. How many elements are there in each of the following sets?
a. $\overline{A}$ b. $\overline{B}$
c. $\overline{A} \cup B$ d. $\overline{B} \cap \overline{A}$
e. $\overline{A \cup B}$

How to Solve. For the purpose of making a pictorial representation (see Figure 1.5) the six elements of U are named 1, 2, 3, 4, 5, and 6; or $U = \{1, 2, 3, 4, 5, 6\}$.

$B \cap A$ contains two elements. We draw A and B so that they have two elements in common. $B \cap A = \{2, 3\}$, so $B \cap A$ contains two elements.

$\overline{A}$ consists of all elements in U which are not in A. Therefore, $\overline{A} = \{4, 5, 6\}$, or $\overline{A}$ contains three elements.

$\overline{B}$ is the set of all elements in U that are not in B. Therefore, $\overline{B} = \{1, 6\}$, or $\overline{B}$ contains two elements.

1.7 Complement of a Set

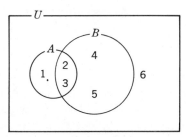

Figure 1.5

A contains three elements and B contains four elements, but $\overline{A \cup B}$ does not necessarily contain seven elements. Remember that $A = \{4, 5, 6\}$, $B = \{2, 3, 4, 5\}$. Since 2 and 3 are common to both A and B, $A \cup B = \{1, 2, 3, 4, 5\}$, or $A \cup B$ contains five elements.

$\overline{B} = \{1, 6\}, \overline{A} = \{4, 5, 6\}$; so $\overline{B} \cap \overline{A} = \{6\}$, or $\overline{B} \cap \overline{A}$ contains one element.

$A = \{1, 2, 3\}, B = \{2, 3, 4, 5\}, A \cup B = \{1, 2, 3, 4, 5\}$. The elements in $\overline{A \cup B}$ are all of the elements in U that are not in $A \cup B$. Therefore, $\overline{A \cup B} = \{6\}$, or $\overline{A \cup B}$ contains one element.

6. Given: A universal set U containing thirteen elements and two subsets C and D; C contains four elements and D contains seven elements; $C \cap D$ contains two elements.

Indicate how many elements there are in the following sets. (Venn diagrams will prove helpful.)

a. $C \cup D$
b. $\overline{C}$
c. $\overline{D}$
d. $\overline{C} \cup D$
e. $D \cup C$
f. $C \cap \overline{D}$
g. $\overline{C \cap D}$
h. $\overline{C \cup D}$
i. $C \cap \overline{D}$

7. Given: A universal set U containing nine elements and three subsets G, F, and K; G contains four elements, F contains six elements, K contains two elements, $G \cup F$ contains seven elements, $K \cap F$ contains one element, and $G \cap K$ contains two elements.

How many elements are there in each of the following sets?

a. $G \cap F$
b. $\overline{G \cap F}$
c. $\overline{F} \cap G$
d. $\overline{F} \cup \overline{G}$
e. $\overline{F \cap G}$
f. $K \cup F$
g. $G \cup K$
h. $\overline{K}$
i. $\overline{K} \cap \overline{G}$
j. $\overline{K \cup G}$

GLOSSARY

Binary operation: An operation which is performed on two elements.

Boolean algebra: Algebra of sets.

Complement of a set: Set A is a complement of set B in a given universal set if and only if A consists of all elements of the universal set which are not in B. The symbol for the complement of B is $\overline{B}$.

Composite number: A natural number which has more than two divisors.

Disjoint sets: Sets which have no members in common.

Divisible: The natural number x is divisible by the natural number y if and only if $x \div y$ is a natural number.

Divisor: If $x \div y$ is a natural number for the natural numbers x and y, then y is said to be a divisor of x.

Element of a set: An object belonging to a set.

Empty set: A set which has no members. The symbol for the empty set is ϕ.

Equivalent sets: Two sets between which there is a one-to-one correspondence.

Euler circles: A pictorial way of representing relationships between sets.

Finite set: A set which has a finite number of members.

Infinite set: A set which has an infinite number of members.

Intersection of two sets: The set which consists of only those members which belong to both sets.

Member of a set: The same as element of a set.

Numeral: A symbol which names a number.

One-to-one correspondence (between two sets): Matching in which there is exactly one element in one set assigned to each element in the other set and vice versa.

Prime number: A natural number which has exactly two divisors.

Proper subset: Set A is a proper subset of B ($A \subset B$) if and only if every member of A also belongs to B and $A \neq B$.

Subset: Set A is a subset of B ($A \subseteq B$) if and only if every member of A also belongs to B.

Union of two sets: The set which consists of the members which belong to one or the other set.

Universal set: The set of all objects which are chosen in a particular context.

Venn diagrams: The same as Euler circles.

CHAPTER REVIEW PROBLEMS

1. Let P be the set of prime numbers less than 100. Label each statement with T for true and F for false.
   a. The number 2 is the only even number such that $2 \in P$.
   b. $P \subset T$, where T is the set of all prime numbers less than 200.
   c. $101 \in P$
   d. $169 \in P$
   e. $\{2, 11, 21\} \subset P$
   f. $\{37, 41, 43\} \not\subset P$
   g. $81 \notin P$
   h. $13 \notin P$

2. Label each statement with T for true and F for false.
   a. The set of even natural numbers and the set of prime numbers are disjoint sets.
   b. The set of even natural numbers and the set of odd natural numbers are matching sets.
   c. If $U =$ the set of natural numbers and $A = U$, then $\overline{A} = \phi$.
   d. Every set with four elements has sixteen subsets.
   e. $\{1, 3, \frac{1}{2}\} \cap \{\frac{1}{4}, \frac{1}{2}\} = \{1\}$.
   f. $X \cap X = \phi$ for each set X.
   g. $Y \cap \phi = Y$ for each set Y.
   h. $M \cup \phi = M$ for each set M.
   i. If $A = \{0, \frac{1}{2}, \frac{1}{4}\}$ and $B = \{0, \frac{1}{3}, \frac{1}{9}\}$, then $A \cap B = \phi$.
   j. $\{\frac{1}{2}, \frac{1}{3}, \frac{1}{4}\} \subset \{\frac{1}{2}, \frac{1}{4}, \frac{1}{8}, \frac{1}{16}\}$.

3. a. How many subsets of $\{10, 15, 20, 25\}$ have exactly one element? Two elements? Three elements? Four elements? Five elements? No elements?
   b. List the elements of the set described by "all natural numbers less than 17 and greater than 11."
   c. Which of the following phrases describe the empty set?
      i. the set of all natural numbers greater than 15 and less than 16
      ii. the set of all fractions greater than $\frac{1}{2}$ and less than $\frac{1}{3}$
      iii. the set of all prime numbers less than 3

iv. the set of all odd natural numbers greater than 21 and less than 23

v. the set of all voters in the United States less than 19 years of age

4. a. $A = \{1, 7, 13, 41, 67\}$; $B = \{41, \frac{1}{2} + \frac{1}{2}, 6 + 7, 7, 70 - 3\}$. Are A and B equivalent sets?

   b. Give an example of one pair of infinite sets which are equivalent.

5. For each pair of sets tell whether the set A is a subset of B.

   a. $A = \{0, 1, 2\}$; $B = \{0, 1, 2, 3, 4, 5\}$
   b. $A = \{5, 6, 7\}$; $B = \{6, 7\}$
   c. $A = \{2, 4, 6, \ldots\}$; $B = \{1, 2, 3\}$
   d. $A = \{1, 3, 5, \ldots\}$; $B = \{1, 2, 3, \ldots\}$
   e. $A = \{$all prime numbers$\}$; $B = \{1, 2, 3, \ldots\}$

6. For each pair of sets tell whether the set X is a proper subset of Y.

   a. $X = \{2, 4, 6, \ldots\}$; $Y = \{1, 2, 3, \ldots\}$
   b. $X = \{1, 3, 5, \ldots\}$; $Y = \{1, 2, 3, \ldots\}$
   c. $X = \{$all prime numbers$\}$; $Y = \{1, 2, 3, \ldots\}$
   d. $X = \{1, 2, 3, \ldots\}$; $Y = \{1, 2, 3, \ldots\}$
   e. $X = \{10, 11, 12\}$; $Y = \{1, 2, 3, \ldots\}$
   f. $X = \{1, 2, 3, \ldots, 98, 99, 100\}$;
      $Y = \{1, 2, 3, \ldots, 198, 199, 200\}$
   g. $X = \{1, 2, 3\}$; $Y = \{1, 2, 3, 4\}$
   h. $X = \phi$; $Y = \{0\}$

7. Find both the intersection and the union of each pair of sets.

   a. $A = \{5, 6, 7, 8, 9\}$; $B = \{7, 8, 9, 10\}$
   b. $C = \{1, 3, 5, \ldots\}$; $D = \{2, 4, 6, \ldots\}$

8. If the universal set is the set of natural numbers, give the complement of each of the following sets.

   a. $A = \{1, 2, 3, 4, 5\}$ b. $B = \{1, 3, 5, \ldots\}$ c. $C = \{5, 10, 15, \ldots\}$

9. List the elements of the following sets.

   a. natural numbers less than 11 and greater than 7
   b. even natural numbers less than 13
   c. prime natural numbers less than 12

10. Suppose $A = \{0, 1, 2\}$. List *all* subsets of A.

11. List the elements of the set of the following:

Chapter Review Problems

    a. all composite numbers less than 15
    b. all prime numbers less than 15

12. How many natural numbers are neither prime nor composite?
13. How many even natural numbers are also prime numbers?
14. Which of the following numerals name the number five?
    a. $\sqrt{25}$ **b.** 10_{five} **c.** 11_{four} **d.** 21_{three} **e.** $\sqrt[3]{15}$
15. Given: the universal set U and subsets C, F, and G.
$U = \{1, 2, 3, 4, 5, 6, 7, 8\}$
$C = \{2, 6, 7, 8\}$; $F = \{1, 2, 4, 6\}$; $G = \{1, 2, 3, 5, 8\}$
List the elements of each of the following sets.
    a. $C \cap F$ **b.** $F \cup G$
    c. $\overline{C}$ **d.** $(C \cup F) \cap G$
    e. $\overline{G \cap F}$ **f.** $U \cap F$
    g. $U \cup C$ **h.** $C \cup \overline{C}$
    i. $\overline{F} \cap \overline{G}$
16. Draw and shade Venn diagrams to verify the truth of each of the following statements for any two sets A and B that are subsets of some universe U.
    a. $U \cap A = A$ **b.** $A \cup \overline{A} = U$
    c. $A \cap \overline{A} = \phi$ **d.** $\overline{A \cup B} = \overline{A} \cap \overline{B}$
    e. $\overline{A \cap B} = \overline{A} \cup \overline{B}$ **f.** $\overline{\overline{A}} = A$
17. If the sets A, B, and C are as pictured in Figure 1.6, name the elements of each of the following.

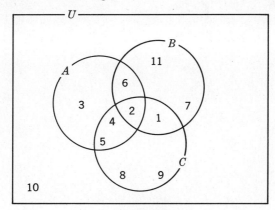

Figure 1.6

    a. $C \cap B$ **b.** $A \cap B$

c. $B \cap C$
d. $A \cup B$
e. $(A \cup B) \cup C$
f. $A \cup (B \cup C)$
g. $\overline{A}$
h. $\overline{B}$
i. $\overline{C}$
j. $\overline{(A \cup B) \cup C}$
k. $(A \cap B) \cap C$
l. $\overline{(A \cap B) \cap C}$
m. $A \cap (B \cap C)$
n. $\overline{(A \cup B) \cap C}$

18. Which of the following describes correctly what is pictured in the Venn diagram in Figure 1.7?

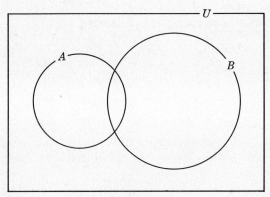

Figure 1.7

a. $A \subset B$
b. $A \cap B = \phi$
c. $(A \cup B) \subset A$
d. $(A \cap B) \subset A$
e. $A \cup B = A$

2

The System of Real Numbers: Addition and Subtraction

2.1 Order of Operations

In this chapter our concern is with the so-called *real numbers* and with operations upon them. We shall shortly clarify what is meant by real numbers. First it is necessary to establish some agreement as to the order in which operations are to be performed in cases where more than one operation is involved. Consider the following story.

An employee was out traveling for his employer. His task was to investigate the merit of buying certain stock on a stock market. He found the price rather high; therefore, he decided to wire his employer before deciding whether or not to buy the stock. The return wire read as follows:

NO PRICE TOO HIGH.

The employee went ahead and bought the stock not knowing, of course, that his employer intended to say

NO, PRICE TOO HIGH.

The punctuation symbol (,) was of crucial importance in that case.

In the English language, as in other languages, we use punctuation marks such as the comma and period in order to make clear statements.

Punctuation marks are very important, as the preceding example illustrates.

In mathematics, there are similar situations in which it is necessary to use mathematical punctuation symbols such as ( ) parentheses, [ ] brackets, and { } braces. These symbols are often necessary to make clear what is to be done in mathematics problems. Consider, for example, the following problems: give the simplest name for $2 + 3 \times 7$.

There are two ways that one might do this simplification:

$$2 + 3 \times 7 = 5 \times 7 = 35$$

or

$$2 + 3 \times 7 = 2 + 21 = 23.$$

Obviously, both ways cannot be intended, because 35 is not equal to 23. The expression $2 + 3 \times 7$ *must* have only one meaning. It is customary to use parentheses, which are mathematical punctuation marks, to make the meaning of such phrases clear.

We shall agree that one should perform the operation inside the parentheses first. Thus,

$$(2 + 3) \times 7 = 5 \times 7 = 35$$
$$2 + (3 \times 7) = 2 + 21 = 23.$$

For convenience and brevity, parentheses are frequently omitted in some expressions. In this case, it is necessary for us to reach some kind of an agreement on the unique meaning of $2 + 3 \times 7$. Agreement: Whenever addition, subtraction, multiplication, and division are involved in a problem and there are no parentheses to guide us, we shall perform the multiplication and division first, then the addition and subtraction.

According to this agreement,

$$2 + 3 \times 7 = 2 + 21 = 23.$$

Similarly,

$$2 + 6 \div 3 = 2 + 2 = 4$$
$$\tfrac{1}{2} \times 7 - 16 \div 8 = 3\tfrac{1}{2} - 2 = 1\tfrac{1}{2}$$
$$3 \times 5 - 2 \times 3 + 7 \times 11 = 15 - 6 + 77 = 9 + 77 = 86.$$

Whenever divisions and multiplications are indicated, these two operations are performed in the order in which they occur, from left to right.

Example $3 \times 4 \div 2 \div 3 \times 2 = 12 \div 2 \div 3 \times 2$
$ = 6 \div 3 \times 2$
$ = 2 \times 2$
$ = 4.$

2.2 The Number Line

When additions and subtractions appear, we perform these operations as they occur, from left to right.

Example
$$10 - 8 + 4 + 6 - 3 + 5 = 2 + 4 + 6 - 3 + 5$$
$$= 6 + 6 - 3 + 5$$
$$= 12 - 3 + 5$$
$$= 9 + 5$$
$$= 14.$$

We will use various ways of expressing the product of a pair of numbers. Instead of writing 3×4, we may also indicate this product by $3 \cdot 4$, $(3)4$, $3(4)$, or $(3)(4)$.

2.2 The Number Line

The number line will serve as a convenient model on which to base the development of rules for certain operations upon numbers. It is a line on which numbers have been assigned to points so that we can refer to the points by means of numbers. To do this, we choose one point and assign to it the number 0. Then we mark off equal distances to the left and right of the point corresponding to the number 0 and assign numbers to them, as in the picture below. It is not necessary to have a line in a horizontal position; but we shall agree to place it in this position so that we can conveniently refer to points as being to the *left* or to the *right* of a given point.

In order to establish a one-to-one correspondence between the points on the number line and the numbers, we make a distinction between the points to the left of the point corresponding to 0 and those to its right. We put letters of identification above the points on the line, and agree to write the number names below the line preceded by the *raised symbol* "⁺" for the points to the right of zero and by the *raised symbol* "⁻" for the points to the left of zero, as in Figure 2.1.

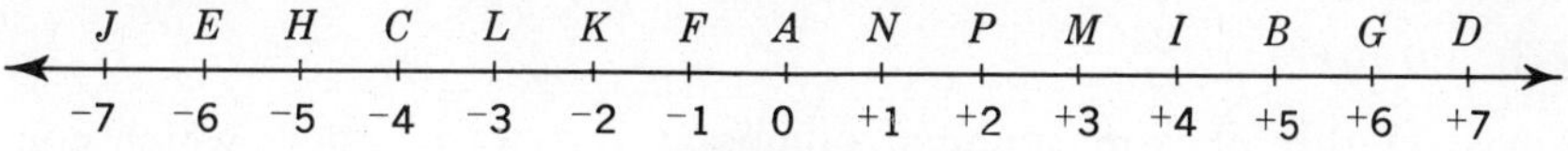

```
  J    E    H    C    L    K    F    A    N    P    M    I    B    G    D
◄─┼────┼────┼────┼────┼────┼────┼────┼────┼────┼────┼────┼────┼────┼────┼─►
 ⁻7   ⁻6   ⁻5   ⁻4   ⁻3   ⁻2   ⁻1    0   ⁺1   ⁺2   ⁺3   ⁺4   ⁺5   ⁺6   ⁺7
```
Figure 2.1

There is exactly one point corresponding to a given number. For example, point N corresponds to the number $^+1$ (*positive 1*) and point H corresponds to the number $^-5$ (*negative 5*). Conversely, there is exactly one number corresponding to a given point. For example, the number $^-2$ (negative 2) corresponds to the point K and the number $^+7$ (positive 7) corresponds to the point D.

We call numbers of this kind *directed numbers*. By means of directed numbers, we can answer not only the question "how many?" but also the question "in what direction?"

For example, instead of saying "5 degrees above zero" we may say "⁺5 degrees" (positive 5 degrees); or, instead of saying "17 degrees below zero" we may say "⁻17 degrees" (negative 17 degrees).

We are now in a position to discuss more fully a one-to-one correspondence between the points on the number line and the directed numbers.

We must understand that the choice of the direction to the right to be the positive direction is quite arbitrary. We could choose the direction to the left as the positive direction; we could even choose to picture the line in a vertical position, or slanted in some other direction. But once a direction is chosen as positive, the opposite direction must be the negative direction, as illustrated in Figure 2.2.

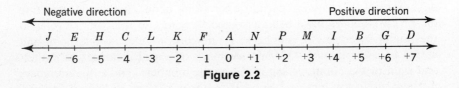

Figure 2.2

Note that the symbols "+" and "−" are now used to mean either of two different things; "+" may mean either to add or that a positive number is used and "−" may mean either to subtract or that a negative number is used. For example, ⁺3 − ⁻4 means "subtract negative four from positive three."

To make this distinction quite clear, notice that we are elevating the symbols "+" and "−" whenever they are used to indicate a positive or a negative number, respectively. To indicate addition or subtraction, the symbols will not be elevated.

2.3 Integers

Consider the following familiar numbers: a set of numbers which consists of positive whole numbers ⁺1, ⁺2, ⁺3, ⁺4, ⁺5, ⁺6, and so on; negative whole numbers ⁻1, ⁻2, ⁻3, ⁻4, ⁻5, ⁻6, and so on; and the number 0. We say that all of these numbers make up the set of numbers called the *integers*. Thus, the set of all integers, I, is the union of the set of positive integers, I_P, the set of negative integers, I_N, and the set whose only member is 0.

$$I = I_P \cup I_N \cup \{0\}.$$

2.3 Integers

Given any two integers, we shall need to know which of the two is the greater. The examples below use the number line and show how to compare two integers.

```
  J    E   (H)   C    L    K    F    A   (N)   P    M    I    B    G    D
──┼────┼────┼────┼────┼────┼────┼────┼────┼────┼────┼────┼────┼────┼────┼──▶
 ⁻7   ⁻6   ⁻5   ⁻4   ⁻3   ⁻2   ⁻1    0   ⁺1   ⁺2   ⁺3   ⁺4   ⁺5   ⁺6   ⁺7
```

Example Suppose the two integers to be compared are ⁻5 and ⁺1. We locate the two points corresponding to the two numbers: point H corresponds to ⁻5, point N corresponds to ⁺1. Point N is to the *right of* the point H, and therefore ⁺1 is greater than ⁻5. Similarly, ⁻3 is greater than ⁻5 because point L (point corresponding to ⁻3) is to the right of point H (point corresponding to ⁻5).

We shall use the symbol ">" to mean *is greater than,* and the symbol "<" to mean *is less than.* Thus, ⁺1 > ⁻5 is read "positive one is greater than negative five." We can also write ⁻5 < ⁺1, which is read "negative five is less than positive one."

The number line comes in very handy when we want to compare two numbers. Of course, it would be impractical to have a picture of a line long enough to locate on it points that are very far apart, but once one gets the principle well in mind, he should have no trouble telling which one of two given numbers is the greater.

From the two symbols of inequality, "<" and ">," we can obtain two more symbols: "≮," which means *is not less than,* and "≯," which means *is not greater than.* The symbol "≠" means *is not equal to.* For example, ⁺3 ≠ ⁺4, ⁺6 + ⁺3 ≠ ⁺6 + ⁺7 are true statements, while ⁺6 ≠ ⁺6 is a false statement.

We can obtain two more symbols by combining "<" with "=" and ">" with "=". The two symbols thus obtained are "≤" (read: is less than *or* equal to), and "≥" (read: is greater than *or* equal to). For example, ⁺5 ≤ ⁺7 (read: positive five is less than or equal to positive seven) and ⁺3 ≥ ⁺3 (read: positive three is greater than or equal to positive three) are true statements.

Note that each statement above consists of two parts. For example, the first statement has the parts

positive five is less than positive seven

and

positive five is equal to positive seven.

The two parts are connected by the word "or." We shall agree to consider the whole statement true if either one of the parts is true. Also, the whole statement is true if both parts are true. If both parts are false, the entire statement is false. Thus, $^+4 \geq {}^+7$ (read: positive four is greater than or equal to positive seven) is false, because "positive four is greater than positive seven" is false and "positive four is equal to positive seven" is false.

EXERCISE 2.1

1. For each of the following, find the simplest name.

Example 1 $7 \times (\frac{1}{2} - \frac{1}{3}) = 7 \times (\frac{3}{6} - \frac{2}{6})$
$= 7 \times \frac{1}{6}$
$= \frac{7}{6}.$

Example 2 $\frac{1}{2} \times [\frac{1}{3} \times (.7 - \frac{1}{2})] = \frac{1}{2} \times [\frac{1}{3} \times (.7 - .5)]$
$= \frac{1}{2} \times [\frac{1}{3} \times .2]$
$= \frac{1}{2} \times [\frac{1}{3} \times \frac{1}{5}]$
$= \frac{1}{2} \times [\frac{1}{15}]$
$= \frac{1}{30}.$

Example 3 $\{[7 + (6 \times 3)] \times 2\} \times 5 = \{[7 + 18] \times 2\} \times 5$
$= \{25 \times 2\} \times 5$
$= 50 \times 5$
$= 250.$

a. $5 \times (\frac{2}{5} - \frac{1}{5})$
b. $2 \times (\frac{5}{7} - \frac{2}{7})$
c. $\frac{1}{2} \times (\frac{2}{3} - \frac{1}{3})$
d. $.5 \times (.3 - .1)$
e. $50 \times (.2 - .1)$
f. $(17 - 13) \times (20 + 5)$
g. $(\frac{3}{5} - .2) \times (.6 - \frac{1}{2})$
h. $3 \times [4 \times (7 - 5)]$
i. $5 \times [3 \times (2 + 7)]$
j. $5 \times (7 + 2 - 8) + 3 \times \frac{1}{2}$
k. $1737 + 14.75 + 13.6$
l. $14.75 + 1737 + 13.6$
m. $1737 + 13.6 + 14.75$
n. $.675 \times 10$
o. $10 \times .675$
p. $(15.7 + 198.5) + 3.6$
q. $15.7 + (198.5 + 3.6)$
r. $(6.2 \times 3) \times 2$
s. $6.2 \times (3 \times 2)$
t. $(3 \times 6.2) \times 2$
u. $5 \times (6 + 7)$
v. $5 \times 6 + 5 \times 7$
w. $(11 + 12) \times 10$
x. $11 \times 10 + 12 \times 10$
y. $3 \times [(4 \times 6) + (7 \times 9)]$
z. $\{[4(3) + 14] \div 13\} + 4 \times 7$

2.3 Integers

2. Think of five situations in which the idea of a direction and its opposite can be used. Then choose one of the directions to be positive and the opposite direction to be negative. Describe a specific case in words, then restate it using directed numbers.

Example Situation: business transactions in which gain or loss may take place
Directions: gain—positive, loss—negative
A case: a gain of 5 dollars, a loss of $12\frac{1}{2}$ dollars
Directed numbers: $^+5$ dollars, $^-12\frac{1}{2}$ dollars

3. List the letters **a** through **p** on your paper; then write T for each true statement and F for each false statement.

a. $^-5 \neq {}^+5$ \qquad b. $0 < {}^-17$
c. $^-8 > {}^+11$ \qquad d. $^-7 \not> {}^+8$
e. $^-1 \not< {}^+1$ \qquad f. $^-10 \not< {}^-15$
g. $^+5 \leq {}^+7$ \qquad h. $^+6 \geq {}^+8$
i. $^-3 \leq {}^-3$ \qquad j. $^-2 \geq {}^-2$
k. $^-100 \leq {}^+100$ \qquad l. $0 \not> 0$
m. $0 < {}^-1150$ \qquad n. $0 > {}^+25$
o. $^+10 \not< {}^+10$ \qquad p. $^-1 \not> 0$

4. Use the set $U = \{^-2, {}^-1, 0, {}^+1, {}^+2\}$ as the universal set from which to obtain the replacements for x. Give all replacements for x which will produce true statements in each of the following:

a. $x > {}^+1$ \qquad b. $x < 0$
c. $x \leq {}^-1$ \qquad d. $x \neq 0$
e. $x \leq {}^-2$ \qquad f. $x < {}^-2$
g. $x \geq {}^+2$ \qquad h. $x > {}^+2$
i. $x \neq {}^-2$ \qquad j. $x < {}^+2$
k. $x \leq {}^+2$ \qquad l. $x \geq {}^-2$

5. Using the set $U = \{^-2, {}^-1, 0, {}^+1, {}^+2\}$ for each statement, tell whether it is true or false. Explanation: "$\{x|x > {}^+1\}$" means "the set of all x belonging to U such that x is greater than $^+1$."

a. $\{x|x > {}^+1\} = \{x|x \geq {}^+2\}$ \qquad b. $\{x|x < 0\} = \{x|x \leq 0\}$
c. $\{x|x > {}^-2\} = \{x|x > {}^-1\}$ \qquad d. $\{x|x \leq {}^+2\} = \{x|x \geq {}^-2\}$
e. $\{x|x > {}^-2\} = \{x|x < {}^+2\}$ \qquad f. $\{x|x \neq 0\} \subseteq U$
g. $\{x|x > {}^-2\} \subseteq \{x|x \leq {}^+2\}$ \qquad h. $\{x|x \neq 0\} \not\subseteq \{x|x < {}^+2\}$

6. For each true statement write T and for each false statement write F.

Example 1 $0 = {}^-1$ *or* $^+3 < {}^+4$.
True, because one of the two statements, $^+3 < {}^+4$, is true.

Example 2 $0 = {}^-1$ *or* $^+3 > {}^+4$.
False, because each of the two statements: $0 = {}^-1$, $^+3 > {}^+4$, is false.

Example 3 $^-1 < 0$ *or* $^+1 > 0$.
True, because each of the two statements: $^-1 < 0$, $^+1 > 0$, is true.

Example 4 $0 = {}^-1$ *and* $^+3 < {}^+4$.
False, because one of the two statements, $0 = {}^-1$, is false.

Example 5 $0 = {}^-1$ *and* $^+3 > {}^+4$.
False, because each of the two statements: $0 = {}^-1$, $^+3 > {}^+4$, is false.

Example 6 $^-1 < 0$ *and* $^+1 > 0$.
True, because each of the two statements: $^-1 < 0$, $^+1 > 0$, is true.

a. $^+3 = {}^+4$ or $^+7 < {}^+9$
b. $^-5 < {}^+3$ and $^+9 \neq {}^-7$
c. $^-13 \leq {}^+4$ or $^+5 \leq {}^+5$
d. $^+6 > {}^+2$ and $^-6 < {}^+3$
e. $^+6 \neq {}^+5$ or $^-3 = {}^+3$
f. $^-11 < {}^+9$ and $^+1 < {}^+2$
g. $^-9 \leq {}^-8$ or $^-12 > {}^-3$
h. $0 < {}^+3$ and $^+4 \leq {}^+4$
i. $^+7 \not< {}^+6$ or $^-15 > {}^-19$
j. $^+3 < {}^+4$ and $^+1 \leq {}^+2$
k. $^-9 \not> {}^+3$ or $^-7 \neq {}^-6$
l. $^+4 > {}^+12$ and $^-6 < {}^-7$

7. Which of the following are true for all specified values of the variables?
   a. If x is negative and y is positive, then $x > y$.
   b. If x is negative and y is positive, then $x \neq y$.
   c. If x is negative and y is positive, then $x \not< y$.
   d. If x is positive, then $x > 0$.
   e. If x is negative, then $x > 0$.

2.4 Absolute Value

Look at the pairs of numbers listed below.

$$^+5, {}^-5$$
$$^-3, {}^+3$$
$$^+100, {}^-100$$
$$^-17, {}^+17$$
$$0, 0$$

2.4 Absolute Value

In each pair there is a number and its *additive inverse:*

$^-5$ is the additive inverse of $^+5$.
$^+3$ is the additive inverse of $^-3$.
$^-100$ is the additive inverse of $^+100$.
$^+17$ is the additive inverse of $^-17$.
0 is the additive inverse of 0.

It follows that the additive inverse of a negative number is a positive number and the additive inverse of a positive number is a negative number. 0 is the only number which is its own additive inverse.

It is frequently useful to associate with each directed number its *absolute value.* For example, the absolute value of $^-5$ is $^+5$, the absolute value of $^-7$ is $^+7$, the absolute value of $^+3$ is $^+3$, the absolute value of $^+5$ is $^+5$, and the absolute value of 0 is 0.

Thus, the absolute value of a positive number is the number itself, whereas the absolute value of a negative number is its additive inverse, which is a positive number. To abbreviate *the absolute value of x,* we write $|x|$; for example:

$|^-5| = {}^+5$ (read: the absolute value of negative five is positive five).
$|^+5| = {}^+5$ (read: the absolute value of positive five is positive five).
$|^+3| = {}^+3$ (read: the absolute value of positive three is positive three).
$|0| = 0$ (read: the absolute value of zero is zero).

In general,

for each $x \geq 0$, $|x| = x$.
for each $x < 0$, $|x| =$ the additive inverse of x.

To simplify writing, we shall use the symbol "—" for "the additive inverse of." Thus,

the additive inverse of x

will be written

$-x$.

EXERCISE 2.2

1. For each true statement write T and for each false statement write F.
    a. $^-2 \neq {}^+2$
    b. $|^-2| \neq |^+2|$
    c. $^-7 > {}^+8$
    d. $|^-7| > {}^+8$
    e. $|^+17| < 0$
    f. $|^-21| < |^+20|$
    g. $|0| > {}^-2$
    h. $^-3 \not> {}^-5$
    i. $^-1 < |^-17|$
    j. $^-3 > {}^-256$

k. $|^-10| \neq |^+10|$ l. $|^-156| = |^+156|$
m. $0 < |^-1137|$ n. $|^+11| > |^+2|$
o. $|^-211| \not< |^+10|$ p. $|^-100| \not> |^+100|$
q. $|^-337| < |^-2|$ r. $0 \neq |0|$
s. $|^-1| < |^-100|$ t. $|^-2| \not> |^+2|$

2. Tell which of the following are true for all replacements of the variables under specified conditions. If not all instances of replacements are true, give an example of a false statement. Use the set of integers as the universal set.

  a. For each $x > y$, $|x| > |y|$
  b. For each $x \neq y$, $|x| \neq |y|$
  c. For each $x < y$, $|x| < |y|$
  d. For each $x = y$, $|x| = |y|$
  e. For each $x \leq y$, $|x| \leq |y|$
  f. For each $x \geq y$, $|x| \geq |y|$
  g. For each $x \leq {}^-2$, $|x| \geq {}^+2$
  h. For each $x \geq {}^+2$, $|x| \geq {}^+2$
  i. For each $x \leq 0$, $|x| \leq 0$
  j. For each $x \geq 0$, $|x| \geq 0$

2.5 Addition of Integers

We cannot use numbers to good advantage unless we at least know how to add, subtract, multiply, and divide them.

We shall first learn to add directed numbers. For this, we shall use the number line.

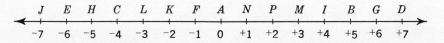

We shall imagine making moves along the number line. A move to the right will be described by a positive number and a move to the left by a negative number. Study the following examples.

        A move from A to M is described by $^+3$.
        A move from N to B is described by $^+4$.
        A move from C to F is described by $^+3$.
        A move from E to B is described by $^+11$.
        A move from A to C is described by $^-4$.
        A move from L to E is described by $^-3$.
        A move from B to I is described by $^-1$.
        A move from M to H is described by $^-8$.

2.5 Addition of Integers

We now use the idea of moves in adding directed numbers. Suppose we have the following problem:

$$^+2 + {}^+3 = ?$$

Study the picture below until you understand what has been done.

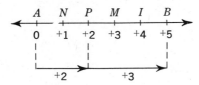

Therefore, $^+2 + {}^+3 = {}^+5$.

Study each of the problems and pictures below to see how the answers are obtained.

Example 1 $\quad ^-1 + {}^-4 = ?$

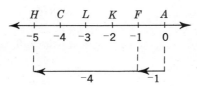

Therefore, $^-1 + {}^-4 = {}^-5$.

Example 2 $\quad ^+3 + {}^-1 = ?$

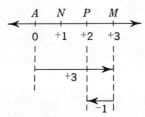

Therefore, $^+3 + {}^-1 = {}^+2$.

Example 3 $\quad ^+2 + {}^-5 = ?$

Therefore, $^+2 + {}^-5 = {}^-3$.

EXERCISE 2.3

1. Find the sum in each of the following, using the number line as in the examples above. Make a picture of the moves for each problem.
   a. ⁺1 + ⁺5
   b. ⁺3 + ⁺3
   c. 0 + ⁺5
   d. 0 + ⁺7
   e. 0 + ⁻5
   f. 0 + ⁻3
   g. ⁻1 + ⁻2
   h. ⁻2 + ⁻4
   i. ⁻1 + ⁺7
   j. ⁻7 + ⁺2
   k. ⁺3 + ⁻4
   l. ⁻4 + ⁺3
   m. ⁻4 + ⁻7
   n. ⁻7 + ⁻4
   o. ⁻2 + ⁺6
   p. ⁺6 + ⁻2
   q. 0 + ⁻7
   r. ⁻7 + 0
   s. ⁻8 + ⁻1
   t. ⁻1 + ⁻8

2. Study the answers in Problem 1 in an attempt to formulate a rule for adding integers. Now give the answers to the following without the use of the number line.
   a. ⁻19 + ⁺13
   b. ⁺91 + ⁻47
   c. ⁻16 + ⁻19
   d. ⁻16 + ⁺42
   e. ⁺37 + ⁻13
   f. ⁺42 + ⁺38
   g. ⁻41 + ⁻32
   h. ⁺176 + ⁺133
   i. ⁺573 + ⁺335
   j. ⁺13 + ⁻173
   k. ⁺105 + ⁻205
   l. ⁻366 + ⁻42
   m. ⁻378 + ⁻112
   n. ⁻85 + ⁺12
   o. ⁻105 + ⁺15

3. Replace the letters with numerals to obtain true statements.

Example ⁺3 + x = ⁻4.
   On the number line we would go seven units to the left from ⁺3 to reach ⁻4. Therefore, if x is replaced by ⁻7, we obtain a true statement: ⁺3 + ⁻7 = ⁻4.

   a. n + ⁺3 = ⁺7
   b. ⁺5 + m = ⁻3
   c. ⁻6 + ⁺14 = p
   d. y + ⁻9 = ⁻13
   e. ⁺7 + x = ⁺2
   f. ⁻13 + t = ⁺22
   g. s + ⁺12 = ⁻3
   h. ⁺13 + ⁻7 = v
   i. ⁻19 + k = ⁻11
   j. z + ⁺12 = ⁺3

2.6 Rational Numbers

k. $w + {}^-5 = {}^-10$
l. $c + {}^-9 = {}^-9$
m. ${}^+11 + r = 0$
n. ${}^-4 + s = {}^-4$
o. $0 + x = 0$

4. For each true statement write T and for each false statement write F.
   a. $|{}^+2| + |{}^-3| < {}^-3 + {}^+2$
   b. ${}^+4 + {}^-8 \neq {}^-4$
   c. ${}^+6 + {}^-3 > {}^-2$
   d. ${}^+6 + |{}^-5| > {}^+11$
   e. ${}^-3 + {}^+2 < |{}^-5|$
   f. $|{}^-5| + {}^-5 \neq 0$
   g. $(|{}^+6| + {}^-5) + {}^-1 \not< {}^+2$
   h. $|{}^-3| + |{}^-2| \not< {}^-3 + {}^-2$

5. Tell which of the following are true for all replacements of the variables under specified conditions. If not all instances of replacements are true, give an example of a false statement. Use the set of integers as the universal set.
   a. If $x > 0$ and $y > 0$, then $x + y > 0$.
   b. If $x < 0$ and $y < 0$, then $x + y < 0$.
   c. For each x, $x + 0 = x$.
   d. If $x > 0$ and $y < 0$, then $x + y < 0$.
   e. If $x > 0$ and $y < 0$, then $x + y > 0$.
   f. If $x > 0$, $y < 0$, and $|x| > |y|$, then $x + y < 0$.
   g. If $x > 0$, $y < 0$, and $|x| > |y|$, then $x + y > 0$.
   h. If $x > 0$, $y < 0$, and $|x| < |y|$, then $x + y < 0$.
   i. If $x > 0$, $y < 0$, and $|x| < |y|$, then $x + y > 0$.
   j. If $x > 0$, $y < 0$, and $|x| = |y|$, then $x + y = 0$.

2.6 Rational Numbers

Let us return to a picture of the number line.

```
  J   E   H   C   L   K   F   A   N   P   M   I   B   G   D
◄─┼───┼───┼───┼───┼───┼───┼───┼───┼───┼───┼───┼───┼───┼───┼─►
 -7  -6  -5  -4  -3  -2  -1   0  +1  +2  +3  +4  +5  +6  +7
```

Remember that there is a one-to-one correspondence between some points on the number line and the integers. So far we have paid attention only to integers and their associated points. But there are many more points on the number line. Let us enlarge the portion of the number line between the points A and N, as in Figure 2.3.

Figure 2.3

Now let us mark some points corresponding to numbers between 0 and $^{+}1$, as in Figure 2.4.

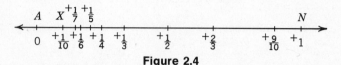

Figure 2.4

We could go on marking more points because there is always at least one more point between any two that are marked. For example, to find a point between A and X, we could take 0 and $^{+}\frac{1}{10}$ and find a number between the two. In this case, $^{+}\frac{1}{20}$ will do because $0 < {^{+}\frac{1}{20}} < {^{+}\frac{1}{10}}$. We agree that $0 < {^{+}\frac{1}{20}} < {^{+}\frac{1}{10}}$ is an abbreviation for $0 < {^{+}\frac{1}{20}}$ and $^{+}\frac{1}{20} < {^{+}\frac{1}{10}}$. It is possible to devise a method which will never fail to produce a number which is between two given numbers.

Numbers such as those described above are called *rational numbers*. Each rational number has a name which can be expressed in the form

$$\frac{^{+}a}{b} \quad \text{or} \quad \frac{^{-}a}{b} \quad \text{or} \quad \frac{0}{b},$$

where a and b can be replaced by names for natural numbers. Let us make two replacements to obtain some nonzero rational numbers:

$$3 \text{ for } a, 7 \text{ for } b: {^{+}\tfrac{3}{7}}, {^{-}\tfrac{3}{7}},$$
$$4 \text{ for } a, 1 \text{ for } b: {^{+}\tfrac{4}{1}}, {^{-}\tfrac{4}{1}}.$$

A simpler name for $^{+}\frac{4}{1}$ is $^{+}4$; a simpler name for $^{-}\frac{4}{1}$ is $^{-}4$.

Any natural-number replacement for b in $\frac{0}{b}$ results in 0. For example,

$$\tfrac{0}{2} = 0 \qquad \tfrac{0}{25} = 0 \qquad \tfrac{0}{596} = 0.$$

Notice that the set of rational numbers includes the integers. Thus, *the set of integers is a subset of the set of rational numbers.*

On the number line in Figure 2.5 are marked a few points which correspond to some of the rational numbers.

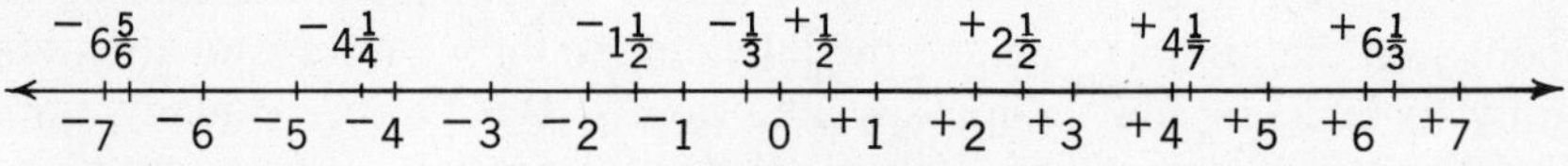

Figure 2.5

Recall from arithmetic that each rational number has a decimal name. For example

$$^{+}\tfrac{1}{2} = {^{+}.5} \qquad {^{+}2\tfrac{1}{2}} = {^{+}2.5} \qquad {^{-}4\tfrac{1}{4}} = {^{-}4.25}.$$

2.7 Repeating Decimal Numerals

Some rational numbers have more complicated decimal names than those above.

To find a decimal name for $\frac{1}{3}$, we divide 1 by 3.

$$\begin{array}{r} .333\ldots \\ 3\overline{\smash{)}1.000} \\ \underline{9} \\ 10 \\ \underline{9} \\ 10 \\ \underline{9} \\ 1 \end{array}$$

We see that $\frac{1}{3} = .333\ldots$. We shall write $.\overline{3}$, the bar meaning that 3 is repeated on and on without end. Thus, $\frac{1}{3} = .\overline{3}$.

Dividing 31 by 99, we find that $\frac{31}{99} = .\overline{31}$. The bar over 31 means that 31 repeats again and again; that is, $.\overline{31} = .313131\ldots$.

The examples above demonstrate that a rational number may be represented in the form of a decimal which either ends (*terminating decimal*), or is a *repeating decimal;* that is, it has a block of one or more digits which repeat continuously in the same pattern.

Rational numbers are added according to the same patterns that integers are added. Study the examples below.

Example 1 $\frac{^+1}{3} + \frac{^-1}{6} = \frac{^+2}{6} + \frac{^-1}{6} = \frac{^+1}{6}.$

Example 2 $\frac{^+1}{8} + \frac{^-3}{4} = \frac{^+1}{8} + \frac{^-6}{8} = \frac{^-5}{8}.$

Example 3 $\frac{^-1}{4} + \frac{^-7}{12} = \frac{^-3}{12} + \frac{^-7}{12} = \frac{^-10}{12} = \frac{^-5}{6}.$

Example 4 $^+3.12 + {^-1.08} = {^+2.04}.$

Example 5 $^-12.7 + {^+9.8} = {^-2.9}.$

2.7 Repeating Decimal Numerals

It is easy to conclude that every terminating decimal names a rational number. Given a terminating decimal, it is easy to find a name of the form $\frac{a}{b}$.

$$.175 = \frac{175}{1000} \qquad .16507 = \frac{16{,}507}{100{,}000}.$$

◀ In general $.n_1n_2 \ldots n_k = \dfrac{n_1n_2 \ldots n_k}{\underbrace{10 \times 10 \times \ldots \times 10}_{\text{10 used } k \text{ times}}}$

where each of $n_1, n_2, \ldots, n_k$ is replaceable by the digits from 0 through 9.

Does a repeating decimal name a rational number? Let us develop a way of finding names of the form $\dfrac{a}{b}$ for numbers given by repeating decimals.

Follow each step.
Given $.\overline{7}$, we write $n = .\overline{7}$.
Since $.\overline{7} = .777\ldots$, 10 times $.\overline{7}$ is equal to $\overline{7.7}$.
Thus,

$$10n = 7.777\ldots$$
$$n = .777\ldots$$

Now, if $a = b$ and $c = d$, then $a - c = b - d$.
Knowing that $10n - n = 10 \times n - 1 \times n = (10 - 1)n = 9n$, we have

$$10n = 7.777\ldots$$
$$\underline{n = .777\ldots}$$
$$10n - n = 7$$
$$9n = 7$$
$$n = \tfrac{7}{9}.$$

Therefore, $.\overline{7} = \tfrac{7}{9}$. Divide 7 by 9 to verify this.

We now carry out the work in finding a name of the form $\dfrac{a}{b}$ for $.\overline{135}$. Study each step.

$$1000n = 135.135135\ldots$$
$$\underline{n = .135135\ldots}$$
$$1000n - n = 135$$
$$999n = 135$$
$$n = \tfrac{135}{999}.$$

Therefore, $.\overline{135} = \tfrac{135}{999}$. Divide 135 by 999 to verify this.

The procedure above can be generalized to show that there is a name of the form $\dfrac{a}{b}$ for each repeating decimal. Study each step below.

Each repeating decimal is of the form

$$.\overline{n_1n_2 \ldots n_k}.$$

2.7 Repeating Decimal Numerals

We are ignoring the whole-number parts, since we know that every whole number w has a name of the form $\frac{w}{1}$.

Since we will be multiplying by multiples of 10, we observe the following abbreviation.

$$10 \times 10 = 10^2 \text{ (read: ten to the second power).}$$
$$10 \times 10 \times 10 = 10^3 \text{ (read: ten to the third power).}$$
$$10 \times 10 \times 10 \times 10 = 10^4 \text{ (read: ten to the fourth power).}$$
$$10 \times 10 \times 10 \times \ldots \times 10 = 10^k \text{ (read: ten to the } k\text{th power).}$$

Now returning to the problem,

$$10^k \times x = n_1 n_2 \ldots n_k . n_1 n_2 \ldots n_k \ldots$$
$$x = . n_1 n_2 \ldots n_k \ldots$$
$$\overline{10^k \times x - x = n_1 n_2 \ldots n_k}$$
$$(10^k - 1)x = n_1 n_2 \ldots n_k$$
$$x = \frac{n_1 n_2 \ldots n_k}{10^k - 1}.$$

Now observe that $n_1 n_2 \ldots n_k$ is a k-digit numeral naming a natural number and $10^k - 1$ is also a natural number. Thus, x is given in the form $\frac{a}{b}$, where a and b are natural numbers.

EXERCISE 2.4

1. How many points which correspond to rational numbers on the number line are there between the points corresponding to 0 and $^+1$?

2. What replacements would you make in $\frac{a}{b}$ to obtain all positive integers?

3. What replacements would you make in $\frac{^-a}{b}$ to obtain all negative integers?

4. Give five different replacements for a and for b in $\frac{a}{b}$, each resulting in 0.

5. For each of the following rational numbers give a decimal and state whether it is a terminating or a repeating decimal.

Example 1 $\frac{3}{8}$.

$$\begin{array}{r} .375 \\ 8\overline{)3.000} \\ \underline{2\ 4} \\ 60 \\ \underline{56} \\ 40 \\ \underline{40} \\ 0 \end{array}$$

The decimal name for $\frac{3}{8}$ is a terminating decimal, namely .375.

Example 2 $\frac{1}{7}$.

$$\begin{array}{r} .142857 \\ 7\overline{)1.000000} \\ \underline{7} \\ 30 \\ \underline{28} \\ 20 \\ \underline{14} \\ 60 \\ \underline{56} \\ 40 \\ \underline{35} \\ 50 \\ \underline{49} \\ 1 \end{array}$$ ← From here on we have repetition.

The decimal name for $\frac{1}{7}$ is a repeating decimal, namely .142857.

a. $\frac{5}{8}$ **b.** $\frac{4}{9}$
c. $\frac{2}{3}$ **d.** $\frac{3}{8}$
e. $\frac{1}{6}$ **f.** $\frac{7}{9}$
g. $\frac{6}{11}$ **h.** $\frac{2}{11}$
i. $\frac{5}{6}$ **j.** $\frac{4}{15}$
k. $\frac{2}{15}$ **l.** $\frac{6}{17}$
m. $\frac{8}{9}$ **n.** $\frac{3}{11}$
o. $\frac{5}{9}$ **p.** $\frac{5}{18}$

6. Find the sums.

  a. $\frac{^+1}{2} + \frac{^+1}{5}$ **b.** $\frac{^+2}{3} + \frac{^-1}{9}$
  c. $\frac{^+7}{8} + \frac{^-1}{2}$ **d.** $\frac{^-4}{5} + \frac{^-7}{15}$
  e. $\frac{^+4}{7} + \frac{^-1}{14}$ **f.** $\frac{^-1}{3} + \frac{^-5}{4}$

2.7 Repeating Decimal Numerals

g. $\frac{^+2}{9} + \frac{^-7}{3}$
h. $\frac{^-29}{7} + \frac{^+29}{7}$
i. $^+1.9 + {}^+12.6$
j. $^+3.9 + {}^-2.6$
k. $^+10.3 + {}^-20.6$
l. $^-7.12 + {}^-8.13$
m. $^-3.9 + {}^+9.1$
n. $^+12.1 + {}^-14.0$
o. $^+20.6 + {}^-28.9$
p. $^-1.2 + {}^-13.9$
q. $^-7.9 + {}^+2.8$
r. $^-100.1 + {}^+99.3$
s. $^-17.6 + {}^+117.6$
t. $^-261.5 + {}^+261.5$
u. $^+199.69 + {}^-199.69$

7. Give an example of a repeating decimal in which a block of
   a. two digits repeats   b. three digits repeats   c. five digits repeats

8. For each number, determine a name of the form $\frac{a}{b}$, where a and b are natural numbers.

Example 1 $2.35 = 2 + \frac{35}{100}$
$= \frac{200}{100} + \frac{35}{100}$
$= \frac{235}{100}$
$= \frac{47}{20}.$

Example 2 $.6\overline{3}.$
Let $\qquad\qquad n = .633\ldots$
Then $\qquad\quad 10n = 6.333\ldots$
$\quad 10n - n = 6.3\overline{3} - .6\overline{3}$
$\qquad\quad 9n = 5.7$
$\qquad\quad\; n = \frac{5.7}{9} = \frac{57}{90} = \frac{19}{30}.$
Thus, $\qquad .6\overline{3} = \frac{19}{30}.$

a. .7
b. 2.97
c. $.8\overline{3}$
d. $.1\overline{6}$
e. $2.\overline{3}$
f. 3.17
g. 5.9
h. $.\overline{1}$
i. $4.\overline{6}$
j. 1.3
k. $2\frac{1}{6}$
l. .875
m. $.\overline{4}$
n. $3\frac{1}{9}$
o. 7.06
p. $.\overline{7}$
q. .365
r. $1\frac{2}{9}$
s. $2\frac{3}{4}$
t. .125
u. $3.\overline{2}$
v. 1.36
w. $2.\overline{6}$
x. $.\overline{307692}$
y. $2.\overline{18}$
z. $3.\overline{12}$
a'. $4.\overline{27}$

2.8 Irrational Numbers

We observed that between any two rational numbers there is another rational number. This is true no matter how close the two rational numbers are. For example, consider the rational numbers $\frac{2}{37}$ and $\frac{2}{38}$. To find a number which is between the two, we can simply find the average (arithmetic mean) of the two numbers.

$$\frac{\frac{2}{37} + \frac{2}{38}}{2} = \frac{\frac{76}{1406} + \frac{74}{1406}}{2} = \frac{\frac{150}{1406}}{2} = \frac{75}{1406}.$$

To see that $\frac{75}{1406}$ is indeed between $\frac{2}{37}$ and $\frac{2}{38}$, all we need to do is observe that $\frac{2}{37}$ is the same as $\frac{76}{1406}$ and $\frac{2}{38}$ is the same as $\frac{74}{1406}$. Now it is easy to see that

$$\frac{74}{1406} < \frac{75}{1406} < \frac{76}{1406}.$$

Since between any two rational numbers there is another rational number, the set of rational numbers is called *dense*. In fact, it might appear that every point on the number line has a rational number associated with it, but this is not so.

Let us consider a number which when multiplied by itself gives 2. Multiplying a number by itself is called "squaring the number." For example, $3^2 = 9$ is read "three squared is equal to nine" or "the square of three is nine." We will see later that in the set of directed numbers there are two numbers whose square is positive nine; they are $^+3$ and $^-3$.

The positive number that makes the statement $x^2 = 2$ true is $^+\sqrt{2}$. We will keep in mind that we are dealing with a positive number, and omit the raised $+$ sign to simplify writing. Thus, we have

$$(\sqrt{2})^2 = \sqrt{2} \times \sqrt{2} = 2.$$

In the following discussion we shall make use of the fact that for all positive numbers a, b, and c

$$\text{if } a < b < c, \text{ then } a^2 < b^2 < c^2$$

and

$$\text{if } a^2 < b^2 < c^2, \text{ then } a < b < c.$$

Note that

$1.4 < \sqrt{2} < 1.5$, since $(1.4)^2 = 1.96$ and $(1.5)^2 = 2.25$ and
$$1.96 < 2 < 2.25.$$

2.8 Irrational Numbers

$1.41 < \sqrt{2} < 1.42$, since $(1.41)^2 = 1.9881$ and $(1.42)^2 = 2.0164$ and $1.9881 < 2 < 2.0164$.
$1.414 < \sqrt{2} < 1.415$, since $(1.414)^2 = 1.999396$ and $(1.415)^2 = 2.002225$ and $1.999396 < 2 < 2.002225$.

Continuing further, we can find rational numbers that are closer to $\sqrt{2}$; but we can never find a rational number that is equal to $\sqrt{2}$.

A method for determining rational numbers which are very close to $\sqrt{2}$ can be devised. We start with 1 as a trial divisor of 2.

```
      2
   1 |2
      2
      ─
      0
```

$1 < \sqrt{2}$ and $2 > \sqrt{2}$; thus, $1 < \sqrt{2} < 2$; so we choose a trial divisor between 1 and 2, say the average of 1 and 2, which is 1.5.

```
        1.3
  1.5 |2.00
       1 5
       ───
        50
        45
        ──
         5
```

$1.3 < \sqrt{2} < 1.5$; so we choose a trial divisor between 1.3 and 1.5, the average of 1.3 and 1.5, which is 1.4.

```
        1.42
  1.4 |2.000
       1 4
       ───
        60
        56
        ──
         40
         28
         ──
         12
```

$1.4 < \sqrt{2} < 1.42$; so we choose a trial divisor between 1.4 and 1.42, the average of 1.4 and 1.42, which is 1.41.

```
         1.418
 1.41 |2.00000
       1 41
       ────
        590
        564
        ───
         260
         141
         ────
         1190
         1128
         ────
           62
```

$1.41 < \sqrt{2} < 1.418$; so we choose the next trial divisor, the average of 1.41 and 1.418, which is 1.414.

```
            1.4144
    1.414 ⟌2.0000000
           1 414
           5860
           5656
           2040
           1414
           6260
           5656
           6040
           5656
            384
```

$1.414 < \sqrt{2} < 1.4144$; so we know that $\sqrt{2}$ is 1.414 correct to the nearest .001.

This process may be continued as far as we please to obtain *rational approximations* which are as close to $\sqrt{2}$ as we want to make them. But no rational number is equal to $\sqrt{2}$.

Continuing a few more times, we find that to seven decimal places $\sqrt{2}$ is 1.4142135. To check to see how close the number 1.4142135 is to $\sqrt{2}$, we multiply it by itself:

$$1.4142135 \times 1.4142135 = 1.99999982358225.$$

The result differs from 2 by

$$.00000017641775,$$

which is a very small number! The square of each successive estimate is closer to 2.

Since no rational number is equal to $\sqrt{2}$, $\sqrt{2}$ has no name of the form $\frac{a}{b}$, where a and b are natural numbers. A number which is not a rational number is called an *irrational number*.

2.9 Real Numbers and the Number Line

We have discovered that, in addition to rational numbers, there are also irrational numbers. For example, $\sqrt{2}$ is an irrational number. We shall now show that there is a point on the number line which corresponds to $\sqrt{2}$. Thus, we shall show that there are points on the number line in addition to those corresponding to the rational numbers.

To do this, we need to recall a property of right triangles. Consider, for example, the right triangle whose sides have lengths of 3, 4, and 5 in. Note that $3^2 + 4^2 = 5^2$. Figure 2.6 is an illustration of the Pythagorean Theorem, which holds for each right triangle.

2.9 Real Numbers and the Number Line

Figure 2.6

PYTHAGOREAN THEOREM If a, b, and c are the lengths of the two legs and the hypotenuse, respectively, of a right triangle, then $a^2 + b^2 = c^2$.

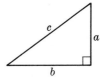

The side opposite the right angle in any right triangle is called the *hypotenuse* and the other two sides are called the *legs*. What is the length of the hypotenuse in a triangle in which each leg is 1 in. long?

$x^2 = 1^2 + 1^2 = 1 + 1 = 2$
$x = \sqrt{2}$
Thus, the hypotenuse is $\sqrt{2}$ in. long.

We are now ready to demonstrate the existence of a point on the number line which corresponds to the irrational number $\sqrt{2}$.

Consider the part of the number line between the points A and N, the points corresponding to the numbers 0 and 1, respectively.

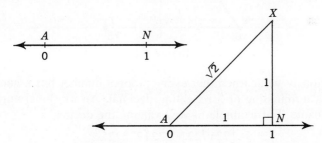

Now, the length of the line segment $\overline{AN}$ is 1. We draw the line segment $\overline{NX}$ perpendicular to $\overline{AN}$, and also of one unit length.

From the computations above we know that the length of the hypotenuse $\overline{AX}$ is $\sqrt{2}$. Now, we swing an arc with a radius of $\sqrt{2}$ from X, using point A as a center, down on the number line, as in Figure 2.7. Since $\overline{AX}$ and $\overline{AY}$ are two radii of the same circle, they are of the same

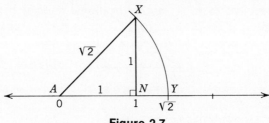

Figure 2.7

length. Thus, the length of $\overline{AY}$ is $\sqrt{2}$ and the number $\sqrt{2}$ corresponds to the point Y.

Since we know that $\sqrt{2}$ is an irrational number, we have demonstrated *the existence on the number line of a point which corresponds to an irrational number*. Of course, there are many more points on the number line which correspond to other irrational numbers.

We can use the same method as above to find the point which corresponds to $\sqrt{3}$ on the number line. Study Figure 2.8 and do a similar construction using a ruler and a compass.

Point T corresponds to the irrational number $\sqrt{3}$.

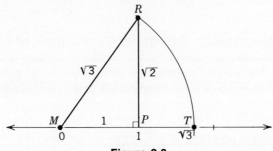

Figure 2.8

Previously we learned that every rational number has a name in the form of a terminating or a repeating decimal. An irrational number has no such name. For example, if we examine the decimal

$$.172172217222172222\ldots$$

we discover that it is a nonrepeating, nonterminating decimal; that is, no block of digits is repeating continuously. Thus

$$.172172217222172222\ldots$$

is an irrational number.

Let us take another irrational number:

$$.261261126111261111\ldots.$$

2.9 Real Numbers and the Number Line

It is easy to see that there is another irrational number between these two. For example, .182182218222182222 . . . is such a number, because
.172172217222172222 . . . < .182182218222182222 . . .
< .261261126111261111 . . . .

In fact, it is possible to find an irrational number between any two irrational numbers.

◂ If we take the set of rational numbers and the set of irrational numbers and form the union of these two sets, the result is the set of *real* numbers. Thus, the set of all real numbers consists of all rational and all irrational numbers. Since no rational number is irrational, and no irrational number is rational, the two sets are disjoint sets.

Note that we use the phrase "directed numbers" and the phrase "real numbers" to mean the same thing.
If
$$Q = \text{the set of rational numbers,}$$
$$S = \text{the set of irrational numbers,}$$
$$R = \text{the set of real numbers,}$$
then
$$Q \cup S = R,$$
$$Q \cap S = \phi.$$

EXERCISE 2.5

1. Find the rational number midway between the following pairs of numbers.
   a. $\frac{1}{2}$ and $\frac{6}{4}$
   b. $\frac{5}{8}$ and $\frac{4}{16}$
   c. $\frac{28}{37}$ and $\frac{13}{74}$
   d. 4.17 and 4.96
   e. $\frac{11}{2}$ and $\frac{5}{6}$
   f. 2.19 and 1.11
   g. 3.12 and 3.76

2. The table on page 453 supplies squares and approximations of square roots, correct to the nearest .001, of numbers from 1 through 100. Record the positive square root of each of the following numbers from the table. Each of these numbers has a whole number for its square root.

54   The System of Real Numbers: Addition and Subtraction

Example   9216; $\sqrt{9216} = 96$.

    a. 2401   **b.** 8836
    c. 4489   **d.** 2809
    e. 1225   **f.** 6084
    g. 576   **h.** 4761

3. Compute an approximation of the positive square root. In each case, carry out the computations to four decimal places and round off to three decimal places; if the fourth place is less than 5, just drop the fourth decimal place. Compare your result with that given in the table on page 453.

Example   Approximate $\sqrt{3}$ correct to the nearest .001. Choose 2 as the first trial divisor of 3.

$$\begin{array}{r} 1.5 \\ 2 \overline{)3.0} \\ \underline{2} \\ 10 \\ \underline{10} \\ 0 \end{array}$$

Therefore, $1.5 < \sqrt{3} < 2$.

Choose a rational number between 1.5 and 2, say 1.8, as the next trial divisor.

$$\begin{array}{r} 1.66 \\ 1.8 \overline{)3.000} \\ \underline{1\ 8} \\ 1\ 20 \\ \underline{1\ 08} \\ 120 \\ \underline{108} \\ 12 \end{array}$$

Therefore, $1.66 < \sqrt{3} < 1.8$.

Choose a number between 1.66 and 1.8, say 1.7, as the next trial divisor.

$$\begin{array}{r} 1.76 \\ 1.7 \overline{)3.00} \\ \underline{1\ 7} \\ 1\ 30 \\ \underline{1\ 19} \\ 110 \\ \underline{102} \\ 8 \end{array}$$

Therefore, $1.7 < \sqrt{3} < 1.76$.

2.9 Real Numbers and the Number Line

Choose a number between 1.7 and 1.76, say 1.73, as the next trial divisor.

$$
\begin{array}{r}
1.734 \\
1.73 \overline{\smash{)}3.00000} \\
\underline{1\,73} \\
1\,270 \\
\underline{1\,211} \\
590 \\
\underline{519} \\
710 \\
\underline{692} \\
18
\end{array}
$$

Therefore, $1.73 < \sqrt{3} < 1.734$.

Choose a number between 1.73 and 1.734, say 1.732, as the next trial divisor.

$$
\begin{array}{r}
1.7321 \\
1.732 \overline{\smash{)}3.0000000} \\
\underline{1\,732} \\
1\,2680 \\
\underline{1\,2124} \\
5560 \\
\underline{5196} \\
3640 \\
\underline{3464} \\
1760 \\
\underline{1732} \\
28
\end{array}
$$

Therefore, $1.732 < \sqrt{3} < 1.7321$.

Hence, the $\sqrt{3}$ is within .0001 of 1.732 or 1.732 is $\sqrt{3}$, correct to three decimal places.

$$1.732 \times 1.732 = 2.999824.$$

a. $\sqrt{5}$ **b.** $\sqrt{19}$ **c.** $\sqrt{27}$

4. Using the method of construction illustrated in this section, locate the points on the number line corresponding to the following numbers.

a. $\sqrt{5}$ **b.** $\sqrt{6}$ **c.** $\sqrt{7}$

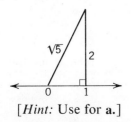

[*Hint:* Use for **a.**]

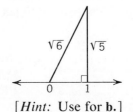

[*Hint:* Use for **b.**]

5. For each true statement write T and for each false statement write F.

Q = the set of rational numbers,
S = the set of irrational numbers,
R = the set of real numbers.

a. $Q \subseteq R$ b. $Q \cup R = R$
c. $S \subseteq R$ d. $S \subseteq Q$
e. $S \subset R$ f. $R \cap S = Q$

6. Each pair of numbers below gives the lengths of two legs in a right triangle. Compute the length of the hypotenuse in each case. Then, from the table of square roots give the three-place approximation to the length h of the hypotenuse.

Example   2, 9; $h^2 = 2^2 + 9^2 = 4 + 81 = 85$
$$h = \sqrt{85} \doteq 9.220.$$

[*Note:* We use the symbol "$\doteq$" to mean "is approximately equal to."]

a. 4, 6   b. 2, 3   c. 7, 4   d. 5, 5

7. The following decimal names for rational numbers are repeating decimals. Change to irrational numbers by inserting number symbols after each repeating block.

Example   .281281281 . . .
The repeating block is 281. We choose to insert 3, 4, 5, 6, 7, . . . after each repeating block.
.28132814281528l6 . . .
Thus, .281328142815 . . . is an irrational number.

a. .14141414 . . .   b. .289289 . . .
c. .987198719871 . . .   d. .142857142857142857 . . .

2.10 Subtraction of Real Numbers

Earlier in this chapter addition of integers was presented. The clue to the ways of subtracting real numbers is found in arithmetic. Consider the following problem from arithmetic:

$$7 - 2 = ?$$

This problem can be changed to the following related addition problem:

$$? + 2 = 7.$$

2.10 Subtraction of Real Numbers

Since $5 + 2 = 7$, the answer to the original problem is 5, or $7 - 2 = 5$.

The relation between subtraction and addition of *real* numbers is the same as the relation between subtraction and addition of *arithmetic* numbers.

The following examples show how each subtraction problem involving real numbers can be changed to a related addition problem.

$^+9 - {}^+4 = ?$ is related to $? + {}^+4 = {}^+9$.
(To what number is $^+4$ added to give $^+9$?)   Answer: $^+5$.
Therefore, $^+9 - {}^+4 = {}^+5$.

$^+5 - {}^+8 = ?$ is related to $? + {}^+8 = {}^+5$.
(To what number is $^+8$ added to give $^+5$?)   Answer: $^-3$.
Therefore, $^+5 - {}^+8 = {}^-3$.

$^+3 - {}^-6 = ?$ is related to $? + {}^-6 = {}^+3$.
(To what number is $^-6$ added to give $^+3$?)   Answer: $^+9$.
Therefore, $^+3 - {}^-6 = {}^+9$.

$^+7 - {}^-2 = ?$ is related to $? + {}^-2 = {}^+7$.
(To what number is $^-2$ added to give $^+7$?)   Answer: $^+9$.
Therefore, $^+7 - {}^-2 = {}^+9$.

$^-5 - {}^-2 = ?$ is related to $? + {}^-2 = {}^-5$.
(To what number is $^-2$ added to give $^-5$?)   Answer: $^-3$.
Therefore, $^-5 - {}^-2 = {}^-3$.

$^-6 - {}^-10 = ?$ is related to $? + {}^-10 = {}^-6$.
(To what number is $^-10$ added to give $^-6$?)   Answer: $^+4$.
Therefore, $^-6 - {}^-10 = {}^+4$.

$^-2 - {}^+5 = ?$ is related to $? + {}^+5 = {}^-2$.
(To what number is $^+5$ added to give $^-2$?)   Answer: $^-7$.
Therefore, $^-2 - {}^+5 = {}^-7$.

$^-7 - {}^+3 = ?$ is related to $? + {}^+3 = {}^-7$.
(To what number is $^+3$ added to give $^-7$?)   Answer: $^-10$.
Therefore, $^-7 - {}^+3 = {}^-10$.

There is still another way of relating subtraction and addition. It is illustrated in the following examples:

1a. $^+4 - {}^+11 = {}^-7$ b. $^+4 + {}^-11 = {}^-7$
2a. $^+17 - {}^+9 = {}^+8$ b. $^+17 + {}^-9 = {}^+8$
3a. $^+7 - {}^-10 = {}^+17$ b. $^+7 + {}^+10 = {}^+17$
4a. $^+18 - {}^-12 = {}^+30$ b. $^+18 + {}^+12 = {}^+30$
5a. $^-6 - {}^+17 = {}^-23$ b. $^-6 + {}^-17 = {}^-23$

58 The System of Real Numbers: Addition and Subtraction

6a. $\frac{-1}{2} - \frac{+1}{2} = {}^-1$ b. $\frac{-1}{2} + \frac{-1}{2} = {}^-1$
7a. $\frac{-5}{7} - \frac{-5}{7} = 0$ b. $\frac{-5}{7} + \frac{+5}{7} = 0$
8a. $^+11.3 - {}^-11.2 = {}^+22.5$ b. $^+11.3 + {}^+11.2 = {}^+22.5$
9a. $^-25.1 - {}^-3.4 = {}^-21.7$ b. $^-25.1 + {}^+3.4 = {}^-21.7$
10a. $\frac{-3}{8} - \frac{+5}{8} = {}^-1$ b. $\frac{-3}{8} + \frac{-5}{8} = {}^-1$

Note that in changing a subtraction problem to a related addition problem, we use the additive inverse of the number being subtracted. For example, in the first problem, instead of subtracting $^+11$ from $^+4$ we added $^-11$ to $^+4$, and $^-11$ is the additive inverse of $^+11$. In the third problem, instead of subtracting $^-10$ from $^+7$, we added $^+10$ to $^+7$, and $^+10$ is the additive inverse of $^-10$.

Thus, we have a pattern which relates addition and subtraction of real numbers.

Subtracting a real number is the same as adding its additive inverse.

Using variables, this relation between addition and subtraction can be stated in the form of a definition of subtraction.

DEFINITION 2.1 For all x and y, $x - y = x + (-y)$ where $-y$ is the additive inverse of y.

Because addition and subtraction are related in this manner, they are said to be *inverse operations*.

Now look at the following examples.

$$^+5 + {}^-5 = 0$$
$$\tfrac{+3}{7} + \tfrac{-3}{7} = 0$$
$$^-13.7 + {}^+13.7 = 0$$
$$0 + 0 = 0.$$

These examples suggest the following statement:

PROPERTY OF ADDITIVE INVERSE For each x, $x + (-x) = 0$.

Thus, the sum of a real number and its additive inverse is equal to 0.

EXERCISE 2.6

1. Find the answer to each subtraction problem below by changing it to the related addition problem.

Example $^-3 - {}^-28 = ?$
   $? + {}^-28 = {}^-3.$

2.10 Subtraction of Real Numbers

(To what number is ⁻28 added to give ⁻3?) Answer: ⁺25.
Therefore, ⁻3 − ⁻28 = ⁺25.

a. ⁺3 − ⁺1
b. ⁺17 − ⁺10
c. ⁺2 − ⁺7
d. ⁺3 − ⁺13
e. ⁺27 − ⁺136
f. ⁺132 − ⁺36
g. ⁺216 − ⁺18
h. ⁺78 − ⁺92
i. ⁺320 − ⁺185
j. ⁺6 − ⁻12
k. ⁺3 − ⁻42
l. ⁺7 − ⁻39
m. ⁺100 − ⁻12
n. ⁻17 − ⁺17
o. ⁻85 − ⁺13
p. ⁻199 − ⁺99
q. ⁻85 − ⁻15
r. ⁻17 − ⁻53
s. ⁻55 − ⁻116
t. ⁻168 − ⁻32
u. $^{+}\frac{3}{2} - ^{+}\frac{1}{2}$
v. ⁺1.8 − ⁺1.2
w. $^{+}\frac{1}{3} - ^{+}\frac{1}{2}$
x. $^{+}\frac{2}{5} - ^{+}\frac{2}{7}$
y. ⁻.5 − ⁺.2
z. $^{-}\frac{2}{3} - ^{+}\frac{1}{3}$
a'. ⁻2.14 − ⁺1.14
b'. ⁻115.5 − ⁺105.4
c'. ⁺1.2 − ⁻.8
d'. ⁺3.52 − ⁻7.48
e'. $^{+}\frac{2}{9} - ^{-}\frac{1}{3}$
f'. $^{+}\frac{11}{12} - ^{-}\frac{3}{4}$
g'. $^{-}\frac{1}{2} - ^{-}\frac{3}{2}$
h'. $^{-}\frac{3}{4} - ^{-}\frac{3}{4}$
i'. ⁻1.7 − ⁻3.3
j'. ⁻.333 − ⁻.444

2. Give the additive inverse of each of the following numbers.

   a. ⁻13
   b. ⁻176
   c. ⁺398
   d. $\frac{-9}{4}$
   e. $\frac{+17}{3}$
   f. ⁻1.078
   g. the opposite of ⁺12
   h. the opposite of ⁻17
   i. the opposite of the opposite of ⁺1
   j. the opposite of the opposite of ⁻3

3. For each pair of numbers below, tell whether or not one number in each pair is the additive inverse of the other number.

Example 1 ⁺37.5, ⁻73.5   no.

Example 2 ⁻13½, ⁺13½   yes.

   a. ⁺27, ⁻27
   b. ⁻325, ⁺325
   c. ⁺136, ⁻163
   d. ⁻1221, ⁺2112
   e. $^{+}\frac{3}{7}, ^{-}\frac{7}{3}$
   f. ⁺.4, $^{-}\frac{2}{5}$
   g. $|{-}13\frac{1}{2}|, \frac{-27}{2}$
   h. ⁺1, ⁺3 + ⁻4
   i. |⁺5 − ⁺7|, |⁻2|
   j. $|^{-}\frac{1}{2} + ^{-}\frac{1}{2}|, |^{+}\frac{1}{4}|$
   k. 0, 0
   l. ⁺3 − ⁻7, ⁻7 − ⁺3

4. Change each problem to its equivalent, using addition, and then give the answer.

Example $^+3 - {^-7} = {^+3} + {^+7} = {^+10}$.

a. $^+4 - {^+1}$
b. $^+21 - {^+17}$
c. $^+39 - {^+121}$
d. $^+16 - {^+215}$
e. $^-13 - {^-4}$
f. $^-12 - {^-8}$
g. $^-25 - {^-72}$
h. $^-98 - {^-113}$
i. $^+12 - {^-4}$
j. $^+25 - {^-23}$
k. $^+29 - {^-37}$
l. $^+41 - {^-68}$
m. $^-16 - {^+4}$
n. $^-58 - {^+13}$
o. $^-62 - {^+85}$
p. $^-75 - {^+109}$
q. $0 - {^+156}$
r. $0 - {^-76}$
s. $\frac{-1}{2} - \frac{+3}{4}$
t. $\frac{-3}{5} - \frac{-17}{5}$
u. $\frac{-6}{7} - \frac{-13}{7}$
v. $\frac{+3}{5} - \frac{-1}{5}$
w. $\frac{+6}{11} - \frac{+17}{11}$
x. $^-13.6 - {^-2.3}$
y. $^-19.8 - {^-36.3}$
z. $^-29.2 - {^+13.5}$
a'. $^-40.9 - {^+90.4}$
b'. $^+13.9 - {^-12.8}$
c'. $^+58.45 - {^-67.54}$

5. Compute the sum of each pair of numbers given in Problem 3 above.

Example 1 $^+37.5 + {^-73.5} = {^-36.0}$.

Example 2 $^-13\frac{1}{2} + {^+13\frac{1}{2}} = 0$.

6. In each case, replace the letter with a numeral to result in a true statement.

a. $^+7 - x = {^+2}$
b. $y - {^-6} = {^+9}$
c. $^+16 - z = {^-11}$
d. $t - {^+9} = {^+14}$
e. $^-9 - {^-7} = s$
f. $^+2 - n = {^-5}$
g. $m - {^-5} = {^-16}$
h. $^{+\frac{1}{2}} + p = 0$
i. $\frac{-2}{3} + r = \frac{-4}{3}$
j. $^+1.75 - u = {^+1.15}$
k. $^-3.5 + {^-1.6} = s$
l. $\frac{-1}{7} + x = \frac{-4}{21}$
m. $\frac{+2}{9} - \frac{-3}{4} = v$
n. $w - \frac{-3}{5} = {^+1}$

7. Given the universal set $U = \{^-2, {^-1}, 0, {^+1}, {^+2}\}$. For each case below, give all replacements for x for which true statements result.

a. $x < 0$ b. $x > 0$
c. $x \geq 0$ d. $x \leq {}^-2$
e. $x \neq {}^-1$ f. $x \not< 0$
g. $x \not> 0$ h. $x \not\leq 0$
i. $x \not\geq 0$ j. $x > |{}^-2|$
k. $x \geq |{}^-2|$ l. $x < |{}^-2|$
m. $x + {}^+1 \geq {}^+1$ n. $x - {}^+2 \leq 0$

8. Label each statement T for true and F for false. Before deciding, try a few examples.

a. For each x and for each $y < x$, $x - y > 0$.
b. For all $x > 0$ and $y > 0$, $x - y > 0$.
c. For each $x > 0$ and for each $y < 0$, $x - y > 0$.
d. For all $x < 0$ and $y < 0$, $x - y < 0$.
e. For all $x > 0$ and $y > 0$, $x + y > 0$.
f. For all $x < 0$ and $y < 0$, $x + y < 0$.
g. For each $x < 0$ and for each $y > 0$, $x + y > 0$.
h. For each $x < 0$ and for each $y > 0$, if $|x| > |y|$, then $x + y < 0$.
i. For each $x < 0$ and for each $y > 0$, if $|x| < |y|$, then $x + y > 0$.
j. For each $x < 0$ and for each $y > 0$, if $|x| < |y|$, then $x + y = 0$.
k. For all x and y, $|x + y| = |x| + |y|$.
l. For all x and y, $|x + y| < |x| + |y|$.
m. For all x and y, $|x + y| > |x| + |y|$.
n. For all x and y, $|x + y| \leq |x| + |y|$.
o. For all x and y, $|x + y| \geq |x| + |y|$.
p. For all x and y, $|x - y| = |x| - |y|$.
q. For all x and y, $|x - y| < |x| - |y|$.
r. For all x and y, $|x - y| > |x| - |y|$.
s. For all x and y, $|x - y| \leq |x| - |y|$.
t. For all x and y, $|x - y| \geq |x| - |y|$.

GLOSSARY

Absolute value: For each $x \geq 0$, $|x| = x$ and for each $x < 0$, $|x| = -x$. (The symbol $|\ |$ means absolute value.)

Additive inverse (of a real number): For each real number x, $-x$ is the additive inverse of x.

Directed number: A positive or a negative number.

Integer: The set of integers is $\{\ldots{}^-3, {}^-2, {}^-1, 0, {}^+1, {}^+2, {}^+3, \ldots\}$.

Inverse operations: Addition and subtraction are inverse operations, because for all x and y, $x - y = x + (-y)$.

Irrational number: A real number which has no name of the form $\frac{a}{b}$, where a and b are integers ($b \neq 0$).

Natural number: The set of natural numbers is $\{1, 2, 3, \ldots\}$.

Nonrepeating, nonterminating decimal: Example: $525225222\ldots$.

Number line: A line for which a one-to-one correspondence between its points and the set of real numbers has been established.

Property of Additive Inverses: For each x, $x + (-x) = 0$.

Pythagorean Theorem: In a right triangle, if a and b are the lengths of the two legs and c is the length of the hypotenuse, then $a^2 + b^2 = c^2$.

Rational number: A real number which has a name of the form $\frac{a}{b}$, where a and b are integers ($b \neq 0$).

Repeating decimal: Example: $.373737\ldots$, abbreviated as $.\overline{37}$.

Terminating decimal: Example: $.5603$.

CHAPTER REVIEW PROBLEMS

1. Simplify each of the following.
    a. $4[13(5 + 7)]$
    b. $7\{[6(9) - 8] \div 23\}$
    c. $\frac{1}{2}[(\frac{1}{2} - \frac{1}{3}) \div \frac{1}{6}]$
    d. $(5.1 - 1.3)(5.1 + 1.3)$
    e. $1.0\{[1.1(1.1 + 1.1) + 1.1]1.1\}$
    f. $3 + 7 \times 4$
    g. $12 \div 4 \times 2 \div 6$
    h. $21 - 2 \times 3 \times 3$
    i. $5 \times 0 + 4 - 2 \times 2$
    j. $17 + 2 - 6 - 2 + 4$

2. Add each of the following.
    a. ${}^+4 + {}^+3$
    b. ${}^+27 + {}^+32$
    c. ${}^-2 + {}^+5$
    d. ${}^-7 + {}^+24$
    e. ${}^+3 + {}^-4$
    f. ${}^+12 + {}^-136$
    g. ${}^+17 + {}^-17$
    h. ${}^-219 + {}^+219$
    i. ${}^-4 + {}^-7$
    j. ${}^-216 + {}^-313$

Chapter Review Problems

k. $^{+}\frac{1}{2} + ^{-}\frac{1}{4}$
l. $^{+}\frac{1}{3} + ^{-}\frac{7}{9}$
m. $^{+}\frac{2}{7} + ^{-}\frac{1}{14}$
n. $^{-}\frac{1}{9} + ^{-}\frac{1}{9}$
o. $^{-}\frac{2}{3} + ^{-}\frac{1}{4}$

3. Subtract each of the following.
   a. $^{+}27 - ^{+}5$
   b. $^{+}12 - ^{+}19$
   c. $^{-}2 - ^{+}3$
   d. $^{-}7 - ^{+}2$
   e. $^{+}12 - ^{-}4$
   f. $^{+}5 - ^{-}16$
   g. $^{+}9 - ^{-}9$
   h. $^{-}5 - ^{-}10$
   i. $^{-}17 - ^{-}3$
   j. $^{-}26 - ^{-}26$
   k. $^{+}\frac{1}{3} - ^{-}\frac{2}{3}$
   l. $^{-}\frac{1}{4} - ^{+}\frac{3}{4}$
   m. $^{-}\frac{2}{5} - ^{-}\frac{4}{5}$
   n. $^{-}\frac{1}{3} - ^{-}\frac{1}{2}$
   o. $^{-}\frac{2}{5} - ^{+}\frac{1}{2}$

4. For each true statement write T and for each false statement write F.
   a. $^{-}17 < ^{-}137$
   b. $|^{-}17| < |^{-}137|$
   c. $|^{+}\frac{1}{2}| > |^{-}\frac{1}{3}|$
   d. $|^{-}45| \leq ^{-}45$
   e. $|^{-}\frac{3}{3}| \neq ^{+}1$
   f. $|^{-}7 + ^{-}2| \geq |^{-}7| + |^{-}2|$
   g. $|^{-}2| + |^{-}3| = |^{-}2 + ^{-}3|$
   h. $|^{-}129 + ^{+}129| = |^{-}129| + |^{+}129|$
   i. $|^{-}4 + ^{+}16| \not< |^{-}4| + |^{+}16|$
   j. $|^{-}2| \leq |^{+}2|$
   k. $^{+}2 > ^{-}5000$
   l. $0 < ^{-}3000$
   m. $|^{-}1.5| = |^{+}1.5|$
   n. $|^{+}10| > |^{-}20|$

5. Tell which of the following will result in true statements for all replacements of variables under the specified conditions. Use the set of integers as the universal set.
   a. For all x and y, if $x = y$, then $|x| = |y|$.
   b. For all x and y, if $x > 0$ and $y < 0$, then $x > y$.
   c. For all x and y, if $x < y$, then $x - y < 0$.
   d. For all x and y, if $x < 0$ and $y > 0$, then $|x| < |y|$.
   e. For all x and y, if $x + y > 0$, then $x > 0$ and $y > 0$.
   f. For all x and y, if $x < 0$ and $y < 0$, then $x + y < 0$.

6. For each rational number, give a decimal name. Tell which of the names are terminating decimals and which are repeating decimals.

a. $\frac{1}{2}$ b. $\frac{1}{4}$ c. $\frac{3}{8}$
d. $\frac{1}{3}$ e. $\frac{2}{3}$ f. $\frac{1}{9}$
g. $\frac{3}{7}$ h. $\frac{2}{11}$ i. $\frac{1}{13}$

7. For each number, determine a name of the form $\frac{a}{b}$, where a and b are natural numbers.
   a. .37   b. .506   c. $.\overline{7}$
   d. 1.6   e. 2.3192   f. $.\overline{23}$
   g. $.2\overline{15}$ h. $1.\overline{36}$ i. $2.02\overline{506}$

8. Find the rational number midway between each of the following pairs.
   a. $9\frac{1}{2}$ and $11\frac{1}{3}$ b. $\frac{1}{4}$ and $5\frac{1}{2}$
   c. $^-4$ and $^+3$ d. $^-2$ and $^+7$

9. Compute each of the following to two decimal places.
   a. $\sqrt{12}$ b. $\sqrt{6}$ c. $\sqrt{29}$

10. Using a right triangle, locate by means of a ruler and a compass the point on the number line which corresponds to $\sqrt{7}$.

3

The System of Real Numbers: Multiplication and Division

3.1 Multiplication of Real Numbers

To learn how to multiply real numbers, we shall assume that positive numbers "behave" like the nonzero numbers of arithmetic under multiplication. We can establish a one-to-one correspondence between these numbers of arithmetic and the positive real numbers, as is shown below.

$$
\begin{array}{c}
1 \leftrightarrow {}^+1 \\
\times \begin{bmatrix} \rightarrow 2 \\ \rightarrow 3 \end{bmatrix} \leftrightarrow \begin{matrix} {}^+2 \leftarrow \\ {}^+3 \leftarrow \end{matrix} \begin{bmatrix} \times \\ \rightarrow ? \end{bmatrix} \\
4 \leftrightarrow {}^+4 \\
5 \leftrightarrow {}^+5 \\
\rightarrow 6 \leftrightarrow {}^+6 \leftarrow - \\
7 \leftrightarrow {}^+7 \\
8 \leftrightarrow {}^+8 \\
9 \leftrightarrow {}^+9 \\
10 \leftrightarrow {}^+10 \\
11 \leftrightarrow {}^+11 \\
12 \leftrightarrow {}^+12 \\
13 \leftrightarrow {}^+13 \\
14 \leftrightarrow {}^+14 \\
15 \leftrightarrow {}^+15 \\
\vdots
\end{array}
$$

The dots indicate that this array can be continued on and on. The arrows show how the arithmetic numbers above are associated with positive numbers. Although in the above array we displayed only natural numbers and positive integers, the same pattern holds for all numbers of arithmetic and for all nonnegative real numbers. An array of some of these numbers is displayed below.

$$
\begin{array}{ccc}
0 & \leftrightarrow & 0 \\
\tfrac{1}{3} & \leftrightarrow & {}^+\tfrac{1}{3} \\
\tfrac{1}{2} & \leftrightarrow & {}^+\tfrac{1}{2} \\
1 & \leftrightarrow & {}^+1 \\
\sqrt{2} & \leftrightarrow & {}^+\sqrt{2} \\
1.5 & \leftrightarrow & {}^+1.5 \\
1.97 & \leftrightarrow & {}^+1.97 \\
\vdots & &
\end{array}
$$

These are simply a few selected numbers. It would be impossible to show all of the numbers, say, between 0 and $^+\tfrac{1}{3}$ since there are infinitely many of them.

We shall now explain how the array above is used in multiplying positive real numbers. Suppose we have the following problem: $^+2 \times {}^+3 = ?$ In the array above we find the arithmetic number corresponding to $^+2$. It is 2. Then we find the arithmetic number corresponding to $^+3$. It is 3. We know that $2 \times 3 = 6$, and we find that the positive number $^+6$ corresponds to the arithmetic number 6. Since the positive numbers behave like the arithmetic numbers, it follows that $^+2 \times {}^+3 = {}^+6$. These relationships are indicated in the diagram above by means of arrows. Because of this relationship between the set of the nonzero numbers of arithmetic and the set of the positive real numbers, we say that the two sets are *isomorphic* under multiplication.

From the above example and many others similar to it, it is apparent that:

◄ A positive number multiplied by a positive number gives a positive number for a product.

Next, we make use of a very important characteristic of mathematics. One mathematician has said, "Mathematics is a study of patterns." We can learn a great deal of mathematics by looking for patterns. For example, make up a sequence of examples which begins as follows:

3.1 Multiplication of Real Numbers

$$^+2 \times {}^+4 = {}^+8$$
$$\downarrow \quad \downarrow \quad \downarrow$$
$$^+2 \times {}^+3 = {}^+6$$
$$\downarrow \quad \downarrow \quad \downarrow$$
$$^+2 \times {}^+2 = {}^+4.$$

Three essential features in this sequence are of consequence to us:

(1) $^+2$ is repeated each time at the extreme left.
(2) The second number is decreased by $^+1$ as we move down.
(3) The product decreases by $^+2$ each time.

Having observed these three features, it should be easy to supply the next examples which fit into this pattern. Here are the next six cases:

$$\downarrow \quad \downarrow \quad \downarrow$$
$$^+2 \times {}^+1 = {}^+2$$
$$\downarrow \quad \downarrow \quad \downarrow$$
$$^+2 \times 0 = 0$$
$$\downarrow \quad \downarrow \quad \downarrow$$
$$^+2 \times {}^-1 = {}^-2$$
$$\downarrow \quad \downarrow \quad \downarrow$$
$$^+2 \times {}^-2 = {}^-4$$
$$\downarrow \quad \downarrow \quad \downarrow$$
$$^+2 \times {}^-3 = {}^-6$$
$$\downarrow \quad \downarrow \quad \downarrow$$
$$^+2 \times {}^-4 = {}^-8.$$

Assuming that the pattern continues, we are led to suspect that the product of a positive number and a negative number may be a negative number. Constructing a variety of sequences like the one above, but using different numbers, would convince us that it is indeed so.

Therefore we can agree that:

◄ The product of a positive number and a negative number is a negative number.

Let us now build another pattern.

$$^-2 \times {}^+4 = {}^-8$$
$$\downarrow \quad \downarrow \quad \downarrow$$
$$^-2 \times {}^+3 = {}^-6$$
$$\downarrow \quad \downarrow \quad \downarrow$$
$$^-2 \times {}^+2 = {}^-4.$$

The three essential features here are the following:
(1) $^-2$ is repeated each time at the extreme left.
(2) The second number is decreased by $^+1$ as we move down.
(3) The product increases by $^+2$ each time (note that $^-8 + {^+2} = {^-6}$).

Now it is clear which are the next steps that fit the pattern:

$$\begin{aligned}
^-2 \times {^+1} &= {^-2} \\
^-2 \times 0 &= 0 \\
^-2 \times {^-1} &= {^+2} \\
^-2 \times {^-2} &= {^+4} \\
^-2 \times {^-3} &= {^+6} \\
^-2 \times {^-4} &= {^+8},
\end{aligned}$$

and so on.

By trying similar patterns involving different numbers we will be led to conclude that:

◀ A negative number multiplied by a negative number is a positive number.

We shall approach the multiplication of two negative numbers from a different point of view after examining some properties of the operations of addition and multiplication.

EXERCISE 3.1

1. Determine the following products.

   a. $^+6 \times {^+8}$
   b. $^+12 \times {^-3}$
   c. $^-4 \times {^+15}$
   d. $^-12 \times {^-7}$
   e. $^+\frac{1}{2} \times {^-2}$
   f. $^+\frac{2}{3} \times {^+\frac{2}{7}}$
   g. $^-\frac{4}{5} \times {^+\frac{2}{7}}$
   h. $^-\frac{11}{12} \times {^-\frac{2}{5}}$
   i. $^-.5 \times {^-\frac{2}{5}}$
   j. $^+1.6 \times {^-\frac{1}{3}}$
   k. $^-\frac{4}{5} \times {^+10.5}$
   l. $^+.7 \times {^+\frac{4}{5}}$
   m. $^-1.3 \times {^+.5}$
   n. $^+2.7 \times {^-1.4}$

3.1 Multiplication of Real Numbers

o. $^-.6 \times {}^+36.5$
q. $^+12 \times \frac{-1}{12}$
s. $^-7 \times \frac{+1}{7}$
u. $^+.01 \times {}^-100$
w. $\frac{+4}{9} \times \frac{-9}{4}$
y. $\frac{+3}{2} \times \frac{-2}{3}$
a'. $^-.02 \times {}^-.1$
c'. $^-.07 \times {}^-.07$
e'. $0 \times {}^+7$

p. $^-.5 \times {}^-.12$
r. $^-3 \times \frac{-1}{3}$
t. $^-.1 \times {}^+10$
v. $\frac{-2}{7} \times \frac{-7}{2}$
x. $\frac{-11}{5} \times \frac{-5}{11}$
z. $^-11 \times {}^+111$
b'. $\frac{-1}{4} \times \frac{-1}{4}$
d'. $^-.8 \times {}^+.8$
f'. $^-6 \times 0$

2. Replace the letters with numerals to obtain true statements.

a. $^+2 \times x = {}^+16$
b. $n \times {}^+4 = {}^-24$
c. $^-3 \times y = {}^-6$
d. $^-16 \times \frac{+1}{2} = p$
e. $^+5 \times s = {}^-25$
f. $^+2 \times t = {}^+7$
g. $^-16 \times \frac{+5}{2} = n$
h. $\frac{+2}{3} \times r = \frac{-6}{7}$
i. $m \times \frac{-5}{8} = \frac{+3}{7}$
j. $\frac{+5}{9} \times \frac{-6}{7} = r$
k. $y \times \frac{+4}{3} = \frac{-11}{9}$
l. $z \times \frac{-1}{6} = \frac{+1}{2}$

3. Insert one of the symbols $=$, $<$, $>$, between each pair of expressions below to make true statements.

Example $|{}^+7 \times {}^-5|$? $|{}^+7| \times |{}^-5|$.
$\qquad\qquad |{}^-35|$? $^+7 \times {}^+5$.
$\qquad\qquad {}^+35$ $=$ $^+35$.
Therefore, $|{}^+7 \times {}^-5| = |{}^+7| \times |{}^-5|$.

a. $|{}^+2 \times {}^+3|$? $|{}^+2| \times |{}^+3|$
b. $|\frac{+1}{2} \times \frac{+3}{4}|$? $|\frac{+1}{2}| \times |\frac{+3}{4}|$
c. $|{}^-5 \times {}^-12|$? $|{}^-5| \times |{}^-12|$
d. $|{}^-1.2 \times {}^-2.5|$? $|{}^-1.2| \times |{}^-2.5|$
e. $|{}^+30 \times {}^-6|$? $|{}^+30| \times |{}^-6|$
f. $|{}^-12 \times {}^+12|$? $|{}^-12| \times |{}^+12|$
g. $|\frac{-1}{2} \times \frac{+1}{2}|$? $|\frac{-1}{2}| \times |\frac{+1}{2}|$
h. $|0 \times {}^+4|$? $|0| \times |{}^+4|$
i. $|{}^-17 \times 0|$? $|{}^-17| \times |0|$
j. $|0 \times 0|$? $|0| \times |0|$

4. Examine your answers to Problem 3. What generalization is suggested by your answers? State it using letters.

5. Tell which of the following are true and which are false in the set of real numbers.
   a. For all $x > 0$ and $y > 0$, $xy > 0$.
   b. For all $x \geq 0$ and $y \geq 0$, $xy > 0$.
   c. For all $x < 0$ and $y > 0$, $xy < 0$.
   d. For all $x \leq 0$ and $y \geq 0$, $xy < 0$.
   e. For all $x < 0$ and $y < 0$, $xy < 0$.
   f. For all $x \leq 0$ and $y \leq 0$, $xy \geq 0$.
   g. For each x, $x \times 0 = 0$.
   h. For each x, $x \times {}^+1 = {}^+1$.
   i. For each x, $x \times {}^-1 = x$.
   j. For each x, $x \times {}^-1 = {}^-1 \times x$.

6. In each problem there are two parts in which the computations are to be performed in the order shown by the parentheses or in the order in which the numbers appear. Perform the computations in each part, and for each problem determine whether the answers in both parts are the same.
   a. $({}^-3 + {}^+4) + {}^-2$; ${}^-3 + ({}^+4 + {}^-2)$
   b. ${}^-2 \times ({}^-6 + {}^+5)$; $({}^-2 \times {}^-6) + ({}^-2 \times {}^+5)$
   c. ${}^-3 \times ({}^+2 \times {}^-7)$; $({}^-3 \times {}^+2) \times {}^-7$
   d. $\frac{-1}{2} + \frac{-1}{4}$; $\frac{-1}{4} + \frac{-1}{2}$
   e. ${}^-1.3 \times ({}^+2.9 + {}^-3.2)$; $({}^-1.3 \times {}^+2.9) + ({}^-1.3 \times {}^-3.2)$
   f. ${}^-8 + |{}^-4|$; $|{}^-4| + {}^-8$
   g. $|\frac{-3}{5}| + |\frac{-3}{4}|$; $|\frac{-3}{4}| + |\frac{-3}{5}|$
   h. $\frac{-1}{2} \times \frac{-1}{4}$; $\frac{-1}{4} \times \frac{-1}{2}$
   i. $({}^-5 \times {}^-3.4) + ({}^-5 \times {}^+3.6)$; ${}^-5 \times ({}^-3.4 + {}^+3.6)$
   j. $({}^-6 + {}^+5) \times {}^-1$; ${}^+6 + {}^-5$

3.2 Commutative and Associative Properties

We shall now review some properties, most of which are probably familiar. Assume that the universal set is the set of all real numbers. To say that something is true for all real numbers, we shall use variables and the symbol $\forall$ which is an abbreviation for *for each* or *for every* or *for all*. Then, $\forall_x$ is read: *for each x*. Each property will be given a name. Since we shall use these names on many occasions, abbreviations for these names will be suggested to save writing out long names.

3.2 Commutative and Associative Properties

The following examples suggest a familiar property of addition of real numbers.

$^+2 + {}^-7 = {}^-5$ and $^-7 + {}^+2 = {}^-5$; therefore $^+2 + {}^-7 = {}^-7 + {}^+2$.
$\frac{-1}{2} + {}^-3 = {}^-3\frac{1}{2}$ and $^-3 + \frac{-1}{2} = {}^-3\frac{1}{2}$; therefore $\frac{-1}{2} + {}^-3 = {}^-3 + \frac{-1}{2}$.
$\frac{-3}{2} + \frac{+7}{2} = {}^+2$ and $\frac{+7}{2} + \frac{-3}{2} = {}^+2$; therefore $\frac{-3}{2} + \frac{+7}{2} = \frac{+7}{2} + \frac{-3}{2}$.

The property suggested by the examples above is called

THE COMMUTATIVE PROPERTY OF ADDITION [CPA]
To say, in an abbreviated form, that the commutative property of addition is true for *every* number x and *every* number y in the agreed upon universal set, we write

◀ CPA: $\forall_x \forall_y \; x + y = y + x$.

Here is another group of examples suggesting a property of multiplication of real numbers.

$^+3 \times {}^-24 = {}^-72$ and $^-24 \times {}^+3 = {}^-72$; therefore $^+3 \times {}^-24 = {}^-24 \times {}^+3$.
$\frac{-1}{2} \times {}^+17 = {}^-8\frac{1}{2}$ and $^+17 \times \frac{-1}{2} = {}^-8\frac{1}{2}$; therefore $\frac{-1}{2} \times {}^+17 = {}^+17 \times \frac{-1}{2}$.
$\frac{+2}{3} \times \frac{+4}{5} = \frac{+8}{15}$ and $\frac{+4}{5} \times \frac{+2}{3} = \frac{+8}{15}$; therefore $\frac{+2}{3} \times \frac{+4}{5} = \frac{+4}{5} \times \frac{+2}{3}$.
$^-.36 \times {}^-1.6 = {}^+.576$ and $^-1.6 \times {}^-.36 = {}^+.576$;

therefore $^-.36 \times {}^-1.6 = {}^-1.6 \times {}^-.36$.

The property suggested by the examples above is called

THE COMMUTATIVE PROPERTY OF MULTIPLICATION [CPM]
CPM is stated in an abbreviated form as follows:

◀ CPM: $\forall_x \forall_y \; x \cdot y = y \cdot x$.

We frequently omit the dot in "$x \cdot y$" and write "xy" to indicate the product of x and y. We can show the product of a and b in the following ways: $a \cdot b$, $a \times b$, ab.

The following group of examples suggests another property of addition of real numbers.

$$(^+3 + {}^-13) + {}^+14 = {}^-10 + {}^+14 = {}^+4,$$

and

$$^+3 + ({}^-13 + {}^+14) = {}^+3 + {}^+1 = {}^+4;$$

therefore,
$$(^{+}3 + {}^{-}13) + {}^{+}14 = {}^{+}3 + ({}^{-}13 + {}^{+}14).$$

and
$$(\tfrac{^{+}1}{4} + \tfrac{^{+}3}{4}) + \tfrac{^{-}1}{2} = {}^{+}1 + \tfrac{^{-}1}{2} = \tfrac{^{+}1}{2},$$
$$\tfrac{^{+}1}{4} + (\tfrac{^{+}3}{4} + \tfrac{^{-}1}{2}) = \tfrac{^{+}1}{4} + \tfrac{^{+}1}{4} = \tfrac{^{+}1}{2};$$

therefore,
$$(\tfrac{^{+}1}{4} + \tfrac{^{+}3}{4}) + \tfrac{^{-}1}{2} = \tfrac{^{+}1}{4} + (\tfrac{^{+}3}{4} + \tfrac{^{-}1}{2}).$$

The property illustrated by the examples above is

THE ASSOCIATIVE PROPERTY OF ADDITION [APA]

◄ APA: $\forall_x \forall_y \forall_z (x + y) + z = x + (y + z).$

The choice of the word *associative* is quite sensible here. This can be seen by observing what is involved in this property. Examine, for example, $(^{+}3 + {}^{+}16) + {}^{+}34 = {}^{+}3 + ({}^{+}16 + {}^{+}34)$. On the left side of the equality sign, $^{+}16$ is grouped, or *associated*, with $^{+}3$. On the right side of the equality sign, $^{+}16$ is associated with $^{+}34$. Thus, the associative property of addition tells us that $^{+}16$ can be associated with either $^{+}3$ or $^{+}34$ in these expressions and the answer will be the same. We agree that if parentheses are omitted, as in $^{+}3 + {}^{+}5 + {}^{+}7$, it will mean $(^{+}3 + {}^{+}5) + {}^{+}7$.

An analogous property for multiplication of real numbers is illustrated by the following example:

$$(^{-}2 \times {}^{+}3) \times {}^{-}5 = {}^{-}6 \times {}^{-}5 = {}^{+}30,$$

and
$$^{-}2 \times ({}^{+}3 \times {}^{-}5) = {}^{-}2 \times {}^{-}15 = {}^{+}30;$$

therefore,
$$(^{-}2 \times {}^{+}3) \times {}^{-}5 = {}^{-}2 \times ({}^{+}3 \times {}^{-}5)$$

The name of this property is

THE ASSOCIATIVE PROPERTY OF MULTIPLICATION [APM]

◄ APM: $\forall_x \forall_y \forall_z (xy)z = x(yz).$

Important Agreement: By now, you should have observed that a letter, say x, may occur more than once in an equation. In case we are told to replace x by a numeral, say $^{-}3$, then we are to replace every x in that equation by $^{-}3$.

3.3 Closure

As in the case of commutativity, associativity is a property of an operation on numbers. We have seen that the operations of addition and multiplication are associative. As with addition, we shall agree that if parentheses are omitted, as in $^-2 \times {}^+3 \times {}^+4$, it will mean $(^-2 \times {}^+3) \times {}^+4$.

3.3 Closure

Let us consider the set of natural numbers as our universal set. If we add a pair of natural numbers, the sum is always a natural number. For example, $1 + 21 = 22$, $179 + 382 = 561$, $5 + 5 = 10$, and so on. We would face an impossible task if we should decide to search for a pair of natural numbers whose sum is not a natural number. The name for this property is *closure*. Thus, the set of natural numbers is said to be *closed* under the operation of *addition*.

What about the operation of multiplication? Given a pair of natural numbers, is the product always a natural number? Our experience tells us that the answer to this question is *yes*. Thus, the set of natural numbers is *closed* under the operation of multiplication.

Let us now examine the operation of subtraction. For example, $7 - 3 = 4$; but $3 - 7 = ?$ has no answer among the natural numbers. One example for which an answer does not exist among the natural numbers is sufficient to conclude that the set of natural numbers is *not closed* under the operation of *subtraction*. However, the set of integers *is* closed under subtraction because the difference of any two integers is an integer.

Let us now investigate closure under division. For example, $10 \div 5 = 2$; but $5 \div 10 = ?$ has no answer among natural numbers. Thus, the set of natural numbers is *not closed* under the operation of *division*. The example $5 \div {}^-7 = \frac{-5}{7}$ proves that the set of integers is not closed under division either since $\frac{-5}{7}$ is not an integer.

If we use N for the set of natural numbers, then the closure property of N under addition and multiplication may be stated as follows:

◄ CIPA: $\forall_{a \in N} \forall_{b \in N} (a + b) \in N$.

From our consideration of addition and multiplication of *real numbers*, we can conclude that the set of real numbers is also closed under each of these operations.

EXERCISE 3.2

1. In each exercise below, some property is illustrated. Name the property.

74   The System of Real Numbers: Multiplication and Division

Example 1 $^+3 + {}^-4 = {}^-4 + {}^+3$ CPA is illustrated.

Example 2 $({}^-7 \times {}^+6) \times {}^-2 = {}^-7 \times ({}^+6 \times {}^-2)$ APM is illustrated.

   a. $^+19 + {}^-35 = {}^-35 + {}^+19$
   b. $\frac{-1}{2} + {}^-30 = \frac{-1}{2} + {}^-30$
   c. $({}^+6 + {}^-7) + {}^+9 = {}^+6 + ({}^-7 + {}^+9)$
   d. $^+3 \times \frac{^+1}{3} = \frac{^+1}{3} \times {}^+3$
   e. $^-6 \times {}^+998 = {}^+998 \times {}^-6$
   f. $({}^-1 \times {}^-3) \times {}^+5 = {}^-1 \times ({}^-3 \times {}^+5)$
   g. $(0 \times {}^-2) \times {}^-13 = 0 \times ({}^-2 \times {}^-13)$
   h. $({}^+1 \times {}^+2) + {}^-3 = ({}^+2 \times {}^+1) + {}^-3$

2. The commutative and associative properties of addition and multiplication can be used to good advantage in simplifying numerical expressions. Study the examples below and then find a way of grouping to simplify the computations as much as possible.

Example 1 $({}^+127 + {}^+193) + {}^-93 = {}^+127 + ({}^+193 + {}^-93)$
$= {}^+127 + {}^+100$
$= {}^+227.$

Example 2 $(\frac{^+2}{5} \times {}^-4.7) \times {}^-25 = ({}^-4.7 \times \frac{^+2}{5}) \times {}^-25$
$= {}^-4.7 \times (\frac{^+2}{5} \times {}^-25)$
$= {}^-4.7 \times {}^-10$
$= {}^+47.$

   a. $({}^+56 + {}^+97) + {}^+3$
   b. $({}^+279 + {}^+1013) + {}^-13$
   c. $({}^-159 + {}^-1998) + {}^-2$
   d. $({}^+978 + {}^+316) + {}^+22$
   e. $({}^-901 + {}^-455) + {}^-99$
   f. $({}^+37 \times {}^-300) \times \frac{-1}{3}$
   g. $({}^-200 \times \frac{^+1}{16}) \times {}^-8$
   h. $(\frac{-8}{7} \times \frac{^+6}{11}) \times \frac{-7}{8}$
   i. $({}^-25 \times {}^-73) \times {}^-4$

3. For each set of numbers determine whether the set seems to be closed under the given operations.

3.3 Closure

Example Set: all odd natural numbers; operations: *i.* addition, *ii.* multiplication.

How to Solve. *i.* Is the sum of a pair of odd natural numbers an odd natural number? Let us try.

$$3 + 5 = 8.$$

Answer: No, the set of odd natural numbers is not closed under addition.

ii. The question here is: Is the product of a pair of odd natural numbers an odd natural number? Let us try a few.

$$5 \times 5 = 25$$
$$7 \times 11 = 77$$
$$9 \times 7 = 63.$$

Answer: There seems to be no reason to suspect that the product of a pair of odd numbers is not an odd number (later we will be able to prove it). Therefore we say that the set of odd natural numbers *seems* to be closed under multiplication.

a. Set: all even natural numbers; operations: *i.* addition, *ii.* multiplication.
b. Set: all natural numbers divisible by 3; operations: *i.* addition, *ii.* multiplication.
c. Set: all natural numbers divisible by 4; operations: *i.* addition, *ii.* multiplication.
d. Set: all even natural numbers; operation: take half of the sum of a pair of numbers.
e. Set: all natural numbers divisible by 5; operation: divide by 10.
f. Set: all natural numbers divisible by 3; operation: take half of the product of a pair of numbers.
g. Set: all natural numbers divisible by 9; operation: take one-third of the sum of a pair of numbers.
h. Set: $\{.1, .01, .001, .0001, \ldots\}$; operations: *i.* multiplication, *ii.* division.
i. Set: $\{1, 2\}$; operations: *i.* multiplication, *ii.* addition.
j. Set: $\{\frac{1}{2}, 1, 2\}$; operations: *i.* multiplication, *ii.* division.
k. Set: $\{\frac{1}{2}, \frac{1}{4}, \frac{1}{8}, \frac{1}{16}, \frac{1}{32}, \ldots\}$; operations: *i.* addition, *ii.* multiplication.
l. Set: $\{\frac{1}{3}, \frac{1}{9}, \frac{1}{27}, \frac{1}{81}, \ldots\}$; operation: multiply by $\frac{1}{3}$.
m. Set: $\{0, 1\}$; operations: *i.* addition, *ii.* multiplication, *iii.* subtraction.

76   The System of Real Numbers: Multiplication and Division

n. Set: $\{1, \frac{1}{5}, \frac{1}{25}, \frac{1}{125}, \ldots\}$; operation: multiply by $\frac{1}{5}$.
o. Set: $\{1, 4, 9, 16, 25, 36, 49, 64, \ldots\}$; operation: multiply the number by itself.
p. Set: all prime numbers; operations: *i.* addition, *ii.* multiplication.
q. Set: $\{0, \frac{1}{2}, 1\}$; operation: multiplication.
r. Set: $\{0\}$; operation: addition.

3.4 The Distributive Property

Suppose a man is selling tickets for a benefit at $3 each. If he sells 17 tickets the first day and 4 tickets the next, what are the total receipts?
We could arrange our computation in the following manner:

$$3 \times (17 + 4) = 3 \times 21 = 63.$$

That is, knowing the total number of tickets sold (21) and the price of each ticket ($3), we find the total receipts to be $63.
But we could also write it as follows:

$$(3 \times 17) + (3 \times 4) = 51 + 12 = 63.$$

The product of 3 and 17 indicates the receipts for the first day, and the product of 3 and 4 indicates the receipts for the second day. Their sum is the total intake.
We see, then, that $3 \times (17 + 4) = (3 \times 17) + (3 \times 4)$.
A simple property of "=" permits us to say that, if $3 \times (17 + 4) = (3 \times 17) + (3 \times 4)$, then $(3 \times 17) + (3 \times 4) = 3 \times (17 + 4)$. This property is true for all real numbers and it can be stated in general as follows:

◀ $\forall_x \forall_y$, if $x = y$, then $y = x$.

It is called

THE SYMMETRIC PROPERTY OF EQUALITY [SPE]   The property, illustrated by the statement

$$3 \times (17 + 4) = (3 \times 17) + (3 \times 4),$$

is called

THE DISTRIBUTIVE PROPERTY OF MULTIPLICATION OVER ADDITION [DPMA]   To state this property for all real numbers, we write:

◀ DPMA: $\forall_x \forall_y \forall_z x(y + z) = (xy) + (xz)$.

3.4 The Distributive Property

This property tells us that multiplication is distributive over addition. For example, $^+17$ may replace x, $^-5$ may replace y, and $^+6$ may replace z, to yield

$$^+17 \times (^-5 + {}^+6) = {}^+17 \times {}^+1 = {}^+17$$
$$(^+17 \times {}^-5) + (^+17 \times {}^+6) = {}^-85 + {}^+102 = {}^+17.$$

We now summarize the properties of operations on real numbers which we have observed thus far.

(1) Addition is commutative and associative.
(2) Multiplication is commutative and associative.
(3) Multiplication is distributive over addition.

It is appropriate to ask whether or not some other operation is distributive with respect to addition or any other operation. Let us try some pairs of operations with which you are familiar.

First, let us introduce other symbols to be used in a way similar to the way we have used the symbols x, y, and z. We agreed that we could replace each letter by a numeral. For example, $x + y + z$ becomes $^+6 + \frac{-4}{5} + {}^+.76$ when x is replaced by $^+6$, y is replaced by $\frac{-4}{5}$, and z is replaced by $^+.76$.

The new symbols, let us say $\oplus$ and $\odot$, will be used as replacements for the following symbols for operations: $+, -, \div, \times$.

Now we are in a position to write the equation

$$x \oplus (y \odot z) = (x \oplus y) \odot (x \oplus z),$$

where the letters x, y, and z are replaceable by numerals, and the symbols $\oplus$ and $\odot$ are replaceable by symbols for operations.

We already know that replacing $\oplus$ by $\times$ and $\odot$ by $+$ will always give us true statements no matter what numerals are substituted for x, y and z. Let us now try another replacement. Put $+$ in place of $\oplus$ and $+$ in place of $\odot$. Does the pattern

$$x + (y + z) = (x + y) + (x + z)$$

yield true statements no matter what numerals are put in place of x, y and z? Let us test this.

Is $^-2 + (^-5 + {}^-7) = (^-2 + {}^-5) + (^-2 + {}^-7)$ a true statement?

$$^-2 + (^-5 + {}^-7) = {}^-2 + {}^-12 = {}^-14$$

and

$$(^-2 + {}^-5) + (^-2 + {}^-7) = {}^-7 + {}^-9 = {}^-16.$$

Since $^-14 \neq {}^-16$, the statement is false, and we conclude that the operation of addition is not distributive over itself.

3.5 Special Properties of 1 and 0

Each of the four groups of examples below suggests a certain property.

$$^+165 \times {}^+1 = {}^+165$$
$$0 \times {}^+1 = 0$$
$$\tfrac{^+4}{7} \times {}^+1 = \tfrac{^+4}{7}$$
$$^-1.67 \times {}^+1 = {}^-1.67$$
$$\vdots$$

$$^+79 + 0 = {}^+79$$
$$0 + 0 = 0$$
$$\tfrac{^-12}{19} + 0 = \tfrac{^-12}{19}$$
$$^+17.6 + 0 = {}^+17.6$$
$$\vdots$$

$$^-39 \div {}^+1 = {}^-39$$
$$0 \div {}^+1 = 0$$
$$\tfrac{^+13}{17} \div {}^+1 = \tfrac{^+13}{17}$$
$$^-3.5 \div {}^+1 = {}^-3.5$$
$$\vdots$$

$$^+135 \times 0 = 0$$
$$0 \times 0 = 0$$
$$\tfrac{^+2}{7} \times 0 = 0$$
$$^-6.75 \times 0 = 0$$
$$\vdots$$

The four properties suggested by the examples above are the following:

PROPERTY OF ONE FOR MULTIPLICATION [P1M] $\forall_x \; x \times {}^+1 = x.$

PROPERTY OF ONE FOR DIVISION [P1D] $\forall_x \; x \div {}^+1 = x.$

PROPERTY OF ZERO FOR ADDITION [PZA] $\forall_x \; x + 0 = x.$

PROPERTY OF ZERO FOR MULTIPLICATION [PZM] $\forall_x \; x \times 0 = 0.$

Having DPMA and PZM, we can prove that the product of two negative numbers is a positive number. The argument below shows this for $^-5 \times {}^-7$.

Consider the following:

$$^-5 \times ({}^+7 + {}^-7).$$

According to DPMA,

$$^-5 \times ({}^+7 + {}^-7) = ({}^-5 \times {}^+7) + ({}^-5 \times {}^-7).$$

Since we already know that

$$^-5 \times {}^+7 = {}^-35$$
$$({}^-5 \times {}^+7) + ({}^-5 \times {}^-7) = {}^-35 + ({}^-5 \times {}^-7),$$

we also know that

$$^-5 \times ({}^+7 + {}^-7) = {}^-5 \times 0 = 0.$$

3.5 Special Properties of 1 and 0

Therefore

$$^-35 + (^-5 \times {}^-7) = 0$$

or

$$^-35 + ? = 0.$$

The question is: What number added to $^-35$ gives 0? It is $^+35$. Thus

$$^-5 \times {}^-7 = {}^+35.$$

Later we shall prove that the product of *each* pair of negative numbers is a positive number.

EXERCISE 3.3

1. Give replacements for x, y, and z to obtain true statements.
   a. $^-5 \times ({}^+6 + {}^+9) = (y \times {}^+6) + (y \times {}^+9)$
   b. $^+3 \times (x + {}^-6) = ({}^+3 \times {}^-5) + ({}^+3 \times {}^-6)$
   c. $z \times ({}^+4 + {}^+11) = ({}^-5 \times {}^+4) + ({}^-5 \times {}^+11)$
   d. $({}^{+}\!\tfrac{1}{2} \times {}^{-}\!\tfrac{3}{4}) + ({}^{+}\!\tfrac{1}{2} \times {}^{-}\!\tfrac{3}{10}) = {}^+.5 \times ({}^-.75 + z)$

2. When the symmetric property of equality is applied to the statement ${}^{+}\!\tfrac{1}{2} + {}^{+}\!\tfrac{1}{3} = {}^+1 - {}^{+}\!\tfrac{1}{6}$, the statement ${}^+1 - {}^{+}\!\tfrac{1}{6} = {}^{+}\!\tfrac{1}{2} + {}^{+}\!\tfrac{1}{3}$ is obtained. Write the statement which is obtained from $^-3 \times ({}^+89 + {}^+46) = ({}^-3 \times {}^+89) + ({}^-3 \times {}^+46)$ when the symmetric property of equality is applied to it.

3. For each expression, write the equation which involves the given expression either as an example of the distributive property or as a consequence of the distributive and symmetric properties.

Example 1 $^-5 \times ({}^-7 + {}^+8) = ({}^-5 \times {}^-7) + ({}^-5 \times {}^+8).$

Example 2 $({}^-3 \times {}^-92) + ({}^-3 \times {}^-64) = {}^-3 \times ({}^-92 + {}^-64).$

a. $^-16 \times ({}^+3 + {}^-5)$
b. ${}^{+}\!\tfrac{1}{3} \times ({}^-7 + {}^+3)$
c. $({}^-5 \times {}^+6) + ({}^-5 \times {}^+7)$
d. $({}^+.08 \times {}^{-}\!\tfrac{1}{2}) + ({}^+.08 \times {}^-.3)$
e. ${}^{-}\!\tfrac{1}{2} \times ({}^{+}\!\tfrac{1}{3} + {}^{-}\!\tfrac{1}{4})$
f. ${}^+33 \times ({}^{-}\!\tfrac{1}{3} + {}^{+}\!\tfrac{1}{11})$
g. $({}^-42 \times {}^{+}\!\tfrac{1}{2}) + ({}^-42 \times {}^-7)$
h. ${}^+.1 \times ({}^+.7 + {}^+.5)$
i. $^-.02 \times ({}^-.02 + {}^-.03)$
j. $({}^-5 \times {}^-7) + ({}^-5 \times {}^+13)$
k. $({}^+11 \times {}^+12) + ({}^+11 \times {}^-16)$
l. $({}^{-}\!\tfrac{3}{4})({}^{+}\!\tfrac{5}{6}) + ({}^{-}\!\tfrac{3}{4})({}^{-}\!\tfrac{7}{6})$
m. $({}^+7 \times {}^-20) + ({}^+7 \times {}^-13)$
n. $({}^-37 \times {}^+46) + ({}^-37 \times {}^+4)$
o. ${}^+25 \times ({}^+25 + {}^+25)$
p. $({}^{-}\!\tfrac{1}{2} \times 0) + ({}^{-}\!\tfrac{1}{2} \times {}^-1)$

4. For each case, work out two examples which check whether or not each of the following statements is true.

Statement   Multiplication is distributive over multiplication.

Example 1 $^+3 \times (^-4 \times ^-5) = (^+3 \times ^-4) \times (^+3 \times ^-5)$

$^+3 \times ^+20 \quad\quad | \quad ^-12 \times ^-15$
$^+60 \quad\quad\quad\quad | \quad\, \neq\, ^+180$

Example 2 $\frac{-1}{2} \times (\frac{-2}{3} \times \frac{-4}{7}) = (\frac{-1}{2} \times \frac{-2}{3}) \times (\frac{-1}{2} \times \frac{-4}{7})$

$\frac{-1}{2} \times \frac{^+8}{21} \quad\quad\quad | \quad \frac{^+2}{6} \times \frac{^+4}{14}$
$\frac{-8}{42} \quad\quad\quad\quad | \quad \frac{^+1}{3} \times \frac{^+2}{7}$
$\frac{-4}{21} \quad\quad\quad\quad | \quad\, \neq\, \frac{^+2}{21}$

The statement is false.

   a. Division is distributive with respect to multiplication.
   b. Division is distributive with respect to subtraction.
   c. Multiplication is distributive with respect to subtraction.
   d. Multiplication is distributive with respect to division.
   e. Addition is distributive with respect to multiplication.
   f. Addition is distributive with respect to division.
   g. Addition is distributive with respect to subtraction.
   h. Subtraction is distributive with respect to subtraction.
   i. Subtraction is distributive with respect to multiplication.

5. From what we know already, we can prove that

$$\forall_x\, ^+1 \times x = x.$$

Proof. $^+1 \times x = x \times ^+1$ by reason of CPM
               $= x.$ by reason of PIM

Now, prove each of the following.
   a. $\forall_x\, 0 + x = x$ **b.** $\forall_x\, 0 \times x = 0$ **c.** $\forall_a \forall_b \forall_c\, (a+b)c = ac + bc$

6. In each example below, name the property involved in going from one step to the next.

Example 1 $^-3 \times (^-4 \times ^-7) \stackrel{(1)}{=} {^-3} \times (^-7 \times ^-4)$ CPM
                                $\stackrel{(2)}{=} (^-3 \times ^-7) \times ^-4.$ APM

3.5 Special Properties of 1 and 0

Example 2 $[(^+5 + 0) + (^-6 \times 0)] \times {}^+1 \overset{(1)}{=} [^+5 + (^-6 \times 0)] \times {}^+1$ PZA
$\overset{(2)}{=} [^+5 + 0] \times {}^+1$ PZM
$\overset{(3)}{=} [^+5] \times {}^+1$ PZA
$\overset{(4)}{=} {}^+5.$ P1M

Example 3 $(^-2 + {}^-7) + {}^{+\frac{1}{3}} + {}^-13 \times 0 + {}^-23 \times {}^+1$
$\overset{(1)}{=} (^-7 + {}^-2) + {}^{+\frac{1}{3}} + {}^-13 \times 0 + {}^-23 \times {}^+1$ CPA
$\overset{(2)}{=} {}^-7 + (^-2 + {}^{+\frac{1}{3}}) + {}^-13 \times 0 + {}^-23$ APA and P1M
$\overset{(3)}{=} {}^-7 + ({}^{+\frac{1}{3}} + {}^-2) + 0 + {}^-23.$ CPA and PZM

a. $^-3 + {}^-7 = {}^-7 + {}^-3$
b. $(^-3 + {}^{+\frac{1}{2}}) + {}^{-\frac{1}{3}} = {}^-3 + ({}^{+\frac{1}{2}} + {}^{-\frac{1}{3}})$
c. $^-4 \times {}^+125 = {}^+125 \times {}^-4$
d. $({}^{-\frac{1}{2}} \times {}^{-\frac{1}{4}}) \times {}^{+\frac{1}{7}} = {}^{-\frac{1}{2}} \times ({}^{-\frac{1}{4}} \times {}^{+\frac{1}{7}})$
e. $(^+3 \times {}^-5) \times {}^-7 = (^-5 \times {}^+3) \times {}^-7$
f. $(^-10 + {}^+3) + {}^-27 = (^+3 + {}^-10) + {}^-27$
g. $({}^{-\frac{2}{7}} + {}^{+\frac{3}{4}}) + {}^{-\frac{1}{2}} = {}^{-\frac{2}{7}} + ({}^{+\frac{3}{4}} + {}^{-\frac{1}{2}})$
h. $^-3 \times 0 + ({}^{+\frac{1}{2}} \times {}^{-\frac{1}{3}}) \times {}^{-\frac{1}{4}} \overset{(1)}{=} 0 + ({}^{+\frac{1}{2}} \times {}^{-\frac{1}{3}}) \times {}^{-\frac{1}{4}}$
$\overset{(2)}{=} ({}^{+\frac{1}{2}} \times {}^{-\frac{1}{3}}) \times {}^{-\frac{1}{4}} + 0$
$\overset{(3)}{=} ({}^{+\frac{1}{2}} \times {}^{-\frac{1}{3}}) \times {}^{-\frac{1}{4}}$
$\overset{(4)}{=} {}^{+\frac{1}{2}} \times ({}^{-\frac{1}{3}} \times {}^{-\frac{1}{4}})$
i. $[^+6(^-5 + {}^-8)] + 0 \overset{(1)}{=} [^+6(^-8 + {}^-5)] + 0$
$\overset{(2)}{=} [^+6(^-8) + {}^+6(^-5)] + 0$
$\overset{(3)}{=} {}^+6(^-8) + {}^+6(^-5)$
j. $^+3 \times (^+1 + {}^-4) + {}^-4 \times {}^+7 \overset{(1)}{=} (^+3 \times {}^+1) + (^+3 \times {}^-4) + {}^-4 \times {}^+7$
$\overset{(2)}{=} {}^+3 + (^+3 \times {}^-4) + {}^-4 \times {}^+7$
$\overset{(3)}{=} {}^+3 + (^-4 \times {}^+3) + (^-4 \times {}^+7)$
$\overset{(4)}{=} {}^+3 + {}^-4(^+3 + {}^+7)$
k. $^-5 \times [(^+3 + {}^-9) + {}^-7] \overset{(1)}{=} {}^-5 \times [(^-9 + {}^+3) + {}^-7]$
$\overset{(2)}{=} {}^-5 \times [^-9 + (^+3 + {}^-7)]$
$\overset{(3)}{=} {}^-5 \times (^-9) + {}^-5 \times (^+3 + {}^-7)$
l. $^+5[(^-3 + {}^-4) + {}^-6] \overset{(1)}{=} {}^+5[^-3 + (^-4 + {}^-6)]$
$\overset{(2)}{=} {}^+5[(^-4 + {}^-6) + {}^-3]$
$\overset{(3)}{=} {}^+5(^-4 + {}^-6) + {}^+5(^-3)$
$\overset{(4)}{=} {}^+5(^-4) + {}^+5(^-6) + {}^+5(^-3)$

m. $^+1 \times [^-6 \times ^-3 + ^-6 \times ^-7] \stackrel{(1)}{=} {^+1} \times [^-6(^-3 + ^-7)]$
$\stackrel{(2)}{=} [^-6(^-3 + ^-7)] \times {^+1}$
$\stackrel{(3)}{=} {^-6}(^-3 + ^-7)$

7. Using DPMA, prove that $^-1 \times {^-1} = {^+1}$.

8. Using the result of Problem 7, prove that $^-6 \times {^-4} = {^+24}$.

3.6 Division of Real Numbers

We know that multiplication and division of the numbers of arithmetic are related. Any division problem can be changed to a related multiplication problem. For example,

$$12 \div 6 = 2 \quad \text{and} \quad 2 \times 6 = 12,$$
$$2 \div 6 = \tfrac{1}{3} \quad \text{and} \quad \tfrac{1}{3} \times 6 = 2,$$
$$\tfrac{1}{2} \div \tfrac{2}{3} = \tfrac{3}{4} \quad \text{and} \quad \tfrac{3}{4} \times \tfrac{2}{3} = \tfrac{1}{2}.$$

The same relation exists between division and multiplication of real numbers. On the basis of this relation, we shall learn how to divide real numbers. Study carefully each example below in order to learn how to divide real numbers.

$^+15 \div {^+3} = ?$ is related to $? \times {^+3} = {^+15}$.
(What number multiplied by $^+3$ gives $^+15$?)   Answer: $^+5$.
Therefore, $^+15 \div {^+3} = {^+5}$.

$^+2 \div {^+6} = ?$ is related to $? \times {^+6} = {^+2}$.
(What number multiplied by $^+6$ gives $^+2$?)   Answer: $^{+}\tfrac{1}{3}$.
Therefore, $^+2 \div {^+6} = {^{+}\tfrac{1}{3}}$.

$^+8 \div {^-4} = ?$ is related to $? \times {^-4} = {^+8}$.
(What number multiplied by $^-4$ gives $^+8$?)   Answer: $^-2$.
Therefore, $^+8 \div {^-4} = {^-2}$.

$^+6 \div {^-18} = ?$ is related to $? \times {^-18} = {^+6}$.
(What number multiplied by $^-18$ gives $^+6$?)   Answer: $^{-}\tfrac{1}{3}$.
Therefore, $^+6 \div {^-18} = {^{-}\tfrac{1}{3}}$.

$^-14 \div {^+7} = ?$ is related to $? \times {^+7} = {^-14}$.
(What number multiplied by $^+7$ gives $^-14$?)   Answer: $^-2$.
Therefore, $^-14 \div {^+7} = {^-2}$.

$^-5 \div {^+10} = ?$ is related to $? \times {^+10} = {^-5}$.
(What number multiplied by $^+10$ gives $^-5$?)   Answer: $^{-}\tfrac{1}{2}$.
Therefore, $^-5 \div {^+10} = {^{-}\tfrac{1}{2}}$.

$^-18 \div {^-3} = ?$ is related to $? \times {^-3} = {^-18}$.

3.6 Division of Real Numbers

(What number multiplied by $^-3$ gives $^-18$?) Answer: $^+6$.
Therefore, $^-18 \div {}^-3 = {}^+6$.

$^-5 \div {}^-25 = ?$ is related to $? \times {}^-25 = {}^-5$.
(What number multiplied by $^-25$ gives $^-5$?)   Answer: $^+\frac{1}{5}$.
Therefore, $^-5 \div {}^-25 = {}^+\frac{1}{5}$.

This pattern shows how problems in division are related to problems in multiplication. To state this pattern, we say that

◀ $\forall_x \forall_{y \neq 0} \forall_z$ $(x \div y = z)$ is equivalent to $(z \times y = x)$.

In this definition it is stated that $y \neq 0$. Let us examine closely why this restriction is necessary. Consider the following two cases.

Case 1.   Division of a nonzero number by zero

$$x \div 0 = ? \quad [x \neq 0].$$

Suppose there is a quotient n, such that $x \div 0 = n$. Then $n \times 0 = x$, according to the relation between multiplication and division. But we stated that $x \neq 0$. The last statement means that $n \times 0$ is equal to some number different from 0. This contradicts PZM. Thus, there is *no* number which is the quotient of a nonzero number and 0.

Case 2.   Division of zero by zero

$$0 \div 0 = ?$$

Suppose there is a quotient m, such that $0 \div 0 = m$. Then $m \times 0 = 0$. We know that this is true for *every* number according to PZM. Therefore, $0 \div 0$ can be $^+3$ or $^-7$ or $^-25$ or any number we wish. There is a danger in having such a numeral, because someone may rightly claim that

$$\text{If } \frac{0}{0} = {}^+3 \text{ and } \frac{0}{0} = {}^-7, \text{ then } {}^+3 = {}^-7.$$

This would be a dangerous situation indeed!

$$\text{We, therefore, declare } \frac{0}{0} \text{ meaningless.}$$

The relationship between multiplication and division shown above indicates that these two operations are *inverses* of each other. We can show this relationship for any two operations by using $\odot$ and $\oplus$ as operation variables.

◀ $\odot$ and $\oplus$ are inverses means that $x \odot y = z$ is equivalent to $z \oplus y = x$.

There is another way of linking multiplication and division. This is illustrated by the following examples.

$^+18 \div {}^+3 = {}^+6$ and $^+18 \times {}^{+\frac{1}{3}} = {}^+6$; therefore, $^+18 \div {}^+3 = {}^+18 \times {}^{+\frac{1}{3}}$.

$^+18 \div {}^-3 = {}^-6$ and $^+18 \times {}^{-\frac{1}{3}} = {}^-6$; therefore, $^+18 \div {}^-3 = {}^+18 \times {}^{-\frac{1}{3}}$.

$^-12 \div {}^-3 = {}^+4$ and $^-12 \times {}^{-\frac{1}{3}} = {}^+4$; therefore, $^-12 \div {}^-3 = {}^-12 \times {}^{-\frac{1}{3}}$.

The examples above show that division can be replaced by multiplication in a way different from the one we have already considered.

In the first example we replaced division by $^+3$ with multiplication by $^{+\frac{1}{3}}$. In the second example division by $^-3$ was replaced with multiplication by $^{-\frac{1}{3}}$. In the third example division by $^-3$ was replaced with multiplication by $^{-\frac{1}{3}}$.

The numbers in each pair ($^+3$ and $^{+\frac{1}{3}}$) and ($^-3$ and $^{-\frac{1}{3}}$) are related in such a way that their product is $^+1$,

$$^+3 \times {}^{+\frac{1}{3}} = {}^+1 \qquad {}^-3 \times ({}^{-\frac{1}{3}}) = {}^+1.$$

We thus have the *Property of Reciprocals:*

◄ $\forall_{x \neq 0} \; x \times \dfrac{1}{x} = {}^+1.$

The pattern shown by the three examples above can be shown in general as follows:

◄ $\forall_x \forall_{y \neq 0} \; x \div y = x \times \dfrac{1}{y}.$

We shall call this pattern the *Definition of Division.*

EXERCISE 3.4

1. Find the answer to each division problem below by changing it to the related multiplication problem.

Example $^+12 \div {}^-2 = ?$
The related multiplication problem is $? \times {}^-2 = {}^+12$ or, what number multiplied by $^-2$ gives $^+12$?
The answer is $^-6$; therefore, $^+12 \div {}^-2 = {}^-6$.

a. $^-16 \div {}^-4$ b. $^+100 \div {}^-2$
c. $^-100 \div {}^+50$ d. $^-17 \div {}^+17$
e. $^+35 \div {}^+1$ f. $^+35 \div {}^-1$
g. $^-35 \div {}^+1$ h. $^-35 \div {}^-1$

3.6 Division of Real Numbers

i. $^+99 \div {^-11}$
j. $^-99 \div {^-11}$
k. $^+99 \div {^+11}$
l. $^-99 \div {^+11}$
m. $^-1 \div {^-1}$
n. $^+1 \div {^-1}$
o. $^-1 \div {^+1}$
p. $^+1 \div {^+1}$
q. $\frac{^-1}{2} \div \frac{^-1}{2}$
r. $^-3.673 \div {^-3.673}$
s. $^+4.006 \div {^-4.006}$
t. $^-111.12 \div {^+111.12}$
u. $0 \div {^-25}$

2. The product of a real number and its reciprocal is $^+1$. Give the reciprocal of each of the following numbers.

a. $^+4$
b. $^+3$
c. $^-7$
d. $^+8$
e. $^-12$
f. $^-5$
g. $^+7$
h. $^-6$
i. $\frac{^-1}{2}$
j. $\frac{^+1}{4}$
k. $\frac{^-2}{3}$
l. $\frac{^+4}{7}$
m. $\frac{^-6}{13}$
n. $^+1.6$
o. $^-1.1$
p. $\frac{^-5}{6}$
q. $\frac{^+17}{18}$
r. $\frac{^-3}{4}$
s. $^+3.4$
t. $^-4.23$
u. $^-1.01$
v. $^+2.04$
w. $^-1.05$
x. $^-2.78$

3. In Problem 1, change each division statement $a \div b$ to the related multiplication statement $a \times \frac{1}{b}$.

4. There are two real numbers each of which is its own reciprocal. What are the two numbers?

5. We know that $\forall_{x \neq 0} \frac{1}{x}$ is a reciprocal of x. Label each statement with T for true and F for false.

a. $\forall_{x>0} \frac{1}{x} > 0$
b. $\forall_{x<0} \frac{1}{x} < 0$
c. $\forall_{x>^+1} 0 < \frac{1}{x} < {^+1}$
d. $\forall_{x \geq ^+1} 0 < \frac{1}{x} < {^+1}$
e. $\forall_{x<0} \frac{1}{x} > 0$
f. $\forall_{x>^+10} 0 < \frac{1}{x} < \frac{^+1}{10}$
g. $\forall_x$, if $|x| \not< {^+10}$, then $0 < \frac{1}{x} < \frac{^+1}{10}$
h. $\forall_{x \neq 0}$, the reciprocal of $\frac{1}{x}$ is x

i. $\forall_{x \neq 0} \frac{1}{x}$ is the reciprocal of $\frac{1}{\frac{1}{x}}$

6. Replace the letters with numerals to obtain true statements.

a. $x \div {}^+2 = {}^+6$ b. $y \div {}^-4 = {}^+10$
c. $m \div {}^+10 = {}^-12$ d. $p \div {}^-15 = {}^-4$
e. ${}^+20 \div z = {}^+1$ f. ${}^-10 \div n = {}^+2$
g. ${}^-5 \div p = {}^-2.5$ h. ${}^+10 \div {}^-2.5 = s$
i. ${}^-25 \div {}^-75 = n$ j. $\frac{-1}{2} \div \frac{-1}{4} = r$
k. $\frac{+7}{8} \div \frac{-4}{5} = t$ l. ${}^-7 \div w = \frac{+6}{7}$
m. $z \div {}^+3 = {}^-7$ n. ${}^-9 \div x = \frac{+5}{7}$
o. ${}^+5 \div {}^+7 = w$ p. $v \div {}^-5 = {}^+9$
q. ${}^-13 \div n = {}^+17$ r. ${}^+14 \div \frac{+5}{3} = x$
s. $q \div \frac{+6}{7} = \frac{-5}{9}$ t. $r \div {}^-3.5 = {}^-1$

7. Prove that 0 does not have a reciprocal.

8. Let us invent two new operations: *alphation* and *betation*. All we know about these two operations is that each is the inverse of the other. The symbol for alphation is α and the symbol for betation is β. Change each of the alphation problems below to a related betation problem.

Example $7 \,\alpha\, 3 = 10\frac{1}{2}$.
     The related betation problem is $10\frac{1}{2} \,\beta\, 3 = 7$.

a. $2 \,\alpha\, 2 = 2$ b. $3 \,\alpha\, 4 = 6$
c. $12 \,\alpha\, 5 = 30$ d. $\frac{1}{2} \,\alpha\, \frac{1}{3} = \frac{1}{12}$
e. $.1 \,\alpha\, .02 = .001$ f. $50 \,\alpha\, 10 = 250$
g. $\frac{2}{3} \,\alpha\, \frac{1}{3} = \frac{1}{9}$ h. $10 \,\alpha\, .01 = .05$
i. $500 \,\alpha\, 300 = 75000$ j. $\frac{1}{3} \,\alpha\, \frac{1}{4} = \frac{1}{24}$

9. Look carefully at your answers in Problem 8. Can you tell how the operations of alphation and betation are related to operations with which you are familiar? That is, change the expressions below to expressions involving multiplication and division.

$$x \,\alpha\, y \qquad\qquad x \,\beta\, y$$

3.7 Simpler Numerals for Real Numbers

We shall now introduce some simplifications in writing numerals for real numbers.

Agreement: Omit the raised "+" whenever referring to positive numbers. Thus, instead of ${}^+7$, we write 7 and instead of ${}^+195$, we write 195. With this agreement, we shall keep in mind that whenever we speak

3.8 The Set of Real Numbers as a System

of real numbers and use numerals without the raised "⁺" we mean *positive* real numbers.

Now, we introduce one additional simplification in writing. We shall abbreviate "the additive inverse of" by writing "−". Thus, *the additive inverse of 7* is abbreviated as −7 and *the additive inverse of ⁻5* is abbreviated as −⁻5, and we know that −⁻5 = 5.

Since the additive inverse of 7 is the same as negative 7, that is, −7 = ⁻7, we shall write the lowered "−" instead of the raised "⁻".

Thus, instead of writing ⁻7, we write −7 and instead of writing ⁻2½, we write −2½, and so on.

It follows from the above that:

◀ $\forall_x$ the additive inverse of x is $-x$.

Note that we do not know whether $-x$ refers to a positive number, a negative number, or 0. It depends on what we use as a replacement for x. For example, 5 for x yields -5 for $-x$, and -3 for x yields $-(-3)$ or 3 for $-x$.

We can now simplify the statements of the definition of subtraction, the property of opposites, and the property of reciprocals by making use of the agreements to omit the raised "⁺" and the replace "the additive inverse of" by "−".

DEFINITION OF SUBTRACTION $\quad \forall_x \forall_y \ x - y = x + (-y)$.

PROPERTY OF OPPOSITES $\quad \forall_x \ x + (-x) = 0$.

PROPERTY OF RECIPROCALS $\quad \forall_{x \neq 0} \ x \times \dfrac{1}{x} = 1$.

3.8 The Set of Real Numbers as a System

The set of real numbers possesses eleven basic properties, under the operations of addition and multiplication. In the statements below, keep in mind that the set of real numbers is the universal set. To simplify writing, we shall use R for the set of real numbers.

◀ Any set of elements which has all eleven properties stated below is called a *field*.

R IS CLOSED UNDER ADDITION [CIPA] $\quad \forall_a \forall_b \ (a + b) \in R$.

R IS CLOSED UNDER MULTIPLICATION [CIPM] $\forall_a \forall_b \, (ab) \in R$.

COMMUTATIVE PROPERTY OF ADDITION [CPA] $\forall_a \forall_b \, a + b = b + a$.

COMMUTATIVE PROPERTY OF MULTIPLICATION [CPM]
$\forall_a \forall_b \, ab = ba$.

ASSOCIATIVE PROPERTY OF ADDITION [APA] $\forall_a \forall_b \forall_c \, (a + b) + c = a + (b + c)$.

ASSOCIATIVE PROPERTY OF MULTIPLICATION [APM]
$\forall_a \forall_b \forall_c \, (ab)c = a(bc)$.

DISTRIBUTIVE PROPERTY OF MULTIPLICATION OVER ADDITION
[DPMA] $\forall_a \forall_b \forall_c \, a(b + c) = (ab) + (ac)$.

The real number 1, in addition to possessing the above properties, has an additional interesting property. Any number multiplied by 1 gives that number. We therefore call the number 1 *the multiplicative identity*. Thus,

PROPERTY OF ONE FOR MULTIPLICATION [P1M]   There exists a unique real number 1, such that $\forall_a \, a \times 1 = a$.

There is also a number among the real numbers which "acts" in addition exactly the way the number 1 "acts" in multiplication. That number is 0, because 0 added to any number results in that number. The number 0 is given the name of *additive identity*. Thus,

PROPERTY OF ZERO FOR ADDITION [PZA]   There exists a unique real number 0, such that $\forall_a \, a + 0 = a$.

PROPERTY OF RECIPROCALS $\forall_{a \neq 0}$ there exists a unique real number $\frac{1}{a}$ such that $a \times \frac{1}{a} = 1$.

Note: Since a reciprocal is also called a *multiplicative inverse*, this property is usually referred to as the *Property of Multiplicative Inverses* [PMI].

PROPERTY OF OPPOSITES $\forall_a$ there exists a unique real number $(-a)$ such that $a + (-a) = 0$.

3.8 The Set of Real Numbers as a System

Note: Since an opposite is also called an *additive inverse*, this property is usually referred to as the *Property of Additive Inverses* [PAI].

EXERCISE 3.5

1. Consider the set of natural numbers
$$N = \{1, 2, 3, 4, 5, \ldots\}.$$
   a. Check each of the eleven properties of real numbers to see whether or not they also hold for natural numbers. State the properties, if there are any, which do not hold for N.
   b. Is N a field?

2. Consider the set of integers
$$I = \{\ldots, -5, -4, -3, -2, -1, 0, 1, 2, 3, 4, 5, \ldots\}.$$
   a. Check each of the eleven properties of real numbers to see whether or not they also hold for integers. State the properties, if there are any, which do not hold for I.
   b. Is I a field?

3. Consider the set of rational numbers Q; that is, numbers which have names of the form $\frac{a}{b}$ or $-\frac{a}{b}$ where a and b are natural numbers (a may also be zero).
   a. Check each of the eleven properties of real numbers to see whether or not they also hold for rational numbers. State the properties, if there are any, which do not hold for Q.
   b. Is Q a field?

4. Name the property or definition illustrated by each of the following:
   a. $-7 \times \frac{3}{4} = \frac{3}{4} \times (-7)$
   b. $-5(-1 + 7) = -5 \times (-1) + (-5) \times 7$
   c. $267 + (-267) = 0$
   d. $-4 \div (-2) = -4 \times (-\frac{1}{2})$
   e. $[-\frac{1}{2} + (-3)] \times 1 = -\frac{1}{2} + (-3)$
   f. $[-2 \times (-3)] \times (-4) = -2 \times [-3 \times (-4)]$
   g. $-\frac{2}{7} \times (-\frac{7}{2}) = 1$
   h. $-317 + (-23) = -23 + (-317)$
   i. If $-3 \times 4 = -12$, then $-12 = -3 \times 4$.
   j. $-8 - 3 = -8 + (-3)$

k. $-7 \times 3 + 0 = -7 \times 3$

l. $(-2 \times 4) + (-2 \times 8) = (-2 \times 8) + (-2 \times 4)$

m. $(-2 \times 4) + (-2 \times 8) = (-2 \times 4) + [8 \times (-2)]$

n. $(-8 + 0) + 1 = -8 + (0 + 1)$

5. Consider a finite number system which contains two elements, say 0 and 1. Here are the addition and multiplication tables for this system.

+	0	1		×	0	1
0	0	1		0	0	0
1	1	0		1	0	1

Check to see whether or not this number system is a field; that is, check to see whether or not each of the eleven properties holds for addition and multiplication in this number system.

6. Suppose we had a "five-hour clock." Three hours after two o'clock the time will be zero o'clock $(2 + 3 = 0)$. Three hours after three o'clock will be one o'clock $(3 + 3 = 1)$. Finish supplying the numerals in the addition table for the numbers of the "five-hour clock."

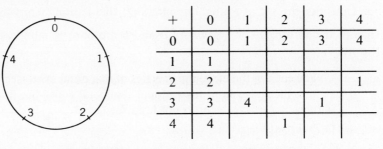

+	0	1	2	3	4
0	0	1	2	3	4
1	1				
2	2				1
3	3	4		1	
4	4		1		

Part of the multiplication table for the numbers of the "five-hour clock" is given below. Supply the missing numerals.

×	0	1	2	3	4
0	0	0	0	0	0
1	0				
2	0	2			
3	0		1	4	2
4	0				1

Check to see whether or not this set of numbers is a field under the operations of addition and multiplication.

7. Repeat the approach described above for a "four-hour clock."

Addition:

+	0	1	2	3
0	0	1	2	3
1	1			0
2	2	3		
3	3		1	2

Multiplication:

×	0	1	2	3
0	0	0	0	0
1	0	1		
2	0		0	
3	0		2	1

Complete the tables and determine whether or not this set of numbers is a field under the operations of addition and multiplication.

3.9 From Word Phrases to Mathematical Phrases

We find many occasions in mathematics to solve problems. Frequently problems are stated in words. Here is a very simple problem.

Three more than twice a number is 4. What is the number?

Sometimes we can obtain an answer to a problem such as this very quickly without having to do any work on paper. At other times the problems may be more involved and we may need to do some writing to get the answer.

Usually, it is best to first write a mathematical phrase which depicts what is stated in words. Let us do that for the problem above.

We use x to refer to the number mentioned in the problem. Then we write

$$2x \quad \text{for} \quad \textit{twice a number}$$

and

$$2x + 3 \quad \text{for} \quad \textit{three more than twice a number.}$$

We obtain the equation

$$2x + 3 = 4$$

which describes what was stated above in words.

EXERCISE 3.6

For each word phrase, write the mathematical phrase. Do not compute the answer.

Example 1   2 less than 3 times n $3n - 2$.

Example 2 four times the sum of a and b $4(a + b)$.

Example 3 The reciprocal of $x + y$ $\dfrac{1}{x+y}$.

Example 4 Jim is n years old now. How old was he 5 years ago? $n - 5$.

1. the sum of 3 and 25
2. the sum of 12 and y
3. 5 more than m
4. the sum of $4x$ and $5y$
5. $9t$ increased by $4s$
6. 5 times the sum of $2x$ and $3y$
7. 9 subtracted from x
8. n subtracted from 12
9. 3 less than 10
10. x less than 12
11. t less than $3x$
12. 15 decreased by 3
13. x decreased by 5
14. 9 decreased by $6t$
15. $7s$ decreased by $2m$
16. 5 times the sum of 6 and x
17. the sum of $2x$ and $3y$, multiplied by 4
18. 20 divided by x
19. $3t$ divided by 7
20. $4x$ divided by the sum of a and b
21. the sum of x and y, divided by n
22. the reciprocal of $2m$
23. the reciprocal of $2x + 3y$
24. the reciprocal of $\dfrac{2}{n}$
25. the reciprocal of $\dfrac{3}{a+b}$
26. the reciprocal of $\dfrac{m+n}{2a+3b}$

27. the reciprocal of $\dfrac{3x - 4y}{5x + 4y}$

28. the reciprocal of $\dfrac{\frac{3}{4}a + b}{a + \frac{1}{2}b}$

29. the sum of the reciprocals of x and y
30. the reciprocal of the sum of x and y
31. the product of the reciprocals of x and y
32. the reciprocal of the product of x and y
33. the difference of the reciprocals of x and y
34. the reciprocal of the difference of x and y
35. the quotient of the reciprocals of x and y
36. the reciprocal of the quotient of x and y
37. If Joe is n years old now, how old will he be m years from now?
38. If Dave is $2n$ years older than Susan and Susan is $(3x + 1)$ years old, how old is Dave?
39. If Dottie is $(2m + 5)$ years old and she is three times as old as Bea, how old is Bea?
40. If Sam is $(x + 2)$ years old, Dave is n years older than Sam, and Pete is 7 years older than Dave, how old is Pete?
41. If Ann will be $(x + 3)$ years old 7 years from now, how old is she now?
42. If Bill will be $(2n - 1)$ years old 3 years from now, how old was he 5 years ago?
43. If Paul was $(3x + 2)$ years old 4 years ago, how old will he be 5 years from now?

GLOSSARY

Associative Property of Addition (APA): $\forall_x \forall_y \forall_z\ (x + y) + z = x + (y + z)$.

Associative Property of Multiplication (APM): $\forall_x \forall_y \forall_z\ (xy)z = x(yz)$.

Closure Property of the Set of Real Numbers under Addition (ClPA): $\forall_{x \epsilon R} \forall_{y \epsilon R}\ (x + y) \epsilon R$.

Closure Property of the Set of Real Numbers under Multiplication (ClPM): $\forall_{x \epsilon R} \forall_{y \epsilon R}\ (xy) \epsilon R$.

Commutative Property of Addition (CPA): $\forall_x \forall_y\ x + y = y + x$.

Commutative Property of Multiplication (CPM): $\forall_x \forall_y\ xy = yz$.

Distributive Property of Multiplication over Addition (DPMA): $\forall_x \forall_y \forall_z \, x(y + z) = (xy) + (xz)$.

Division (definition of): $\forall_x \forall_{y \neq 0} \, x \div y = x \times \dfrac{1}{y}$.

Isomorphism: Two sets are said to be *isomorphic* under a given operation if they "behave" exactly alike under this operation.

Property of One for Multiplication (P1M): There exists a unique real number 1 such that $\forall_x \, x \times 1 = x$.

Property of Reciprocals: $\forall_{x \neq 0}$ there exists a unique real number $\dfrac{1}{x}$ such that $x \times \dfrac{1}{x} = 1$.

Property of Zero for Addition (PZA): There exists a unique real number 0, such that $\forall_x \, x + 0 = x$.

Subtraction (definition of): $\forall_x \forall_y \, x - y = x + (-y)$.

CHAPTER REVIEW PROBLEMS

1. Replace the letters with numerals to obtain true statements. In some cases there is more than one such replacement, and in one case there is none.

   a. $43 + 7 = 7 + x$
   b. $6 \times m = m \times 6$
   c. $7 \times (6 + n) = 7 \times 6 + 7 \times 13$
   d. $y + (3 + z) = (y + 3) + z$
   e. $2 \times 7 + 2 \times p = 2 \times (m + p)$
   f. $(3 \times 10) \times 13 = 3 \times (r \times 13)$
   g. $13 \times (11 + 7) = 13 \times a + 13 \times b$
   h. $u \times (6 + s) = (6 + s) \times u$
   i. $4 \times 5 + 17 = 17 + 4 \times s$
   j. $\frac{4}{3} - \frac{4}{5} = t$
   k. $a - 13 = 39$
   l. $\frac{2}{3} \div \frac{1}{4} = h$
   m. $\frac{1}{5} \div x = \frac{1}{20}$
   n. $\dfrac{5}{0} = v$
   o. $\dfrac{0}{6} = t$
   p. $\dfrac{0}{x} = 0$

2. Name the property or properties used in going from one step to the next.

   a. $[2 \times 0 + 3(4 + 6)] \times 7 \stackrel{(1)}{=} [0 + 3(4 + 6)] \times 7$
   $\stackrel{(2)}{=} [3(4 + 6)] \times 7$
   $\stackrel{(3)}{=} 7 \times [3(4 + 6)]$
   $\stackrel{(4)}{=} 7[3(4) + 3(6)]$

Chapter Review Problems

b. $13 \times [(4 \times 7) \times 11] \stackrel{(1)}{=} 13 \times [4 \times (7 \times 11)]$
$\stackrel{(2)}{=} 13 \times [(7 \times 11) \times 4]$
$\stackrel{(3)}{=} 13 \times [7 \times (11 \times 4)]$
$\stackrel{(4)}{=} [7 \times (11 \times 4)] \times 13$
$\stackrel{(5)}{=} [(7 \times 11) \times 4] \times 13$

c. $7[(5 + 3) + 6] \stackrel{(1)}{=} 7[(3 + 5) + 6]$
$\stackrel{(2)}{=} 7[3 + (5 + 6)]$
$\stackrel{(3)}{=} 7(3) + 7(5 + 6)$

3. For each problem determine whether the given set of numbers is closed under the given operations.
   a. Set: $\{\frac{1}{2}, 0\}$; operations: *i.* addition, *ii.* multiplication.
   b. Set: all real numbers between 0 and 1; operations: *i.* addition, *ii.* subtraction, *iii.* multiplication, *iv.* division.
   c. Set: $\{1, 10, 100, 1000, 10000, \ldots\}$; operations: *i.* multiplication, *ii.* division.
   d. Set: all even natural numbers; operation: average a pair of numbers.
   e. Set: $\{\ldots, 16, 8, 4, 2, 1, \frac{1}{2}, \frac{1}{4}, \frac{1}{8}, \ldots\}$; operations: *i.* addition, *ii.* multiplication.

4. Using the fact that division is the inverse of multiplication, write a related multiplication statement for each division statement.
   a. $51 \div 3 = 17$
   b. $121.8 \div 2.1 = 58$
   c. $\frac{3}{4} \div \frac{7}{5} = \frac{15}{28}$
   d. $x \div y = z \quad [y \neq 0]$

5. Which of the following two statements shows that $\odot$ is an inverse of $\oplus$?
   a. If $a \odot b = c$, then $a \oplus c = b$.
   b. If $a \odot b = c$, then $c \oplus b = a$.

6. Answer the following questions and give an example to illustrate a *yes* answer and give a counterexample to support a *no* answer.
   a. Is multiplication distributive over subtraction?
   b. Is subtraction distributive over division?
   c. Is subtraction associative?

7. Give the reciprocal (multiplicative inverse) of each of the following numbers.
   a. 9
   b. $\frac{3}{7}$
   c. -12
   d. $-\frac{7}{5}$
   e. .7
   f. $-.4$
   g. 3.5
   h. -4.2

8. Tell which of the following statements are true and which are false in the universe of real numbers. You may find it helpful in some cases to try a few replacements before deciding upon your answer.
   a. $\forall_x \forall_y \ x - y = x + (-y)$
   b. $\forall_x \forall_y$, if $|x| = |y|$, then $x + y = 2x$.
   c. $\forall_{x>0} \forall_{y<0}$, if $|x| \geq |y|$, then $x - y \geq 0$.
   d. $\forall_{x>0} \forall_{y<0}$, if $|x| \geq |y|$, then $x - y \leq 0$.
   e. $\forall_{x<0} \forall_{y<0}$, if $|x| \geq |y|$, then $x - y \geq 0$.
   f. $\forall_{x<0} \forall_{y<0}$, if $|x| \geq |y|$, then $x - y \leq 0$.
   g. $\forall_x \forall_y$, if $|x| = |y|$, then $x - y = 0$.
   h. $\forall_{x>0} \forall_{y>0} \ xy > 0$
   i. $\forall_{x<0} \forall_{y<0} \ xy < 0$
   j. $\forall_{x<0} \forall_{y>0} \ \dfrac{x}{y} < 0$

9. Given the set $X = \{-1, 1\}$, label each statement T for true and F for false.
   a. X is closed under multiplication.
   b. X is closed under division.
   c. X is closed under subtraction.
   d. $\forall_{a \in X} \ a^2 \in X$
   e. $\forall_{a \in X} \ a^2 \geq a$
   f. $\forall_{a \in X} \ \dfrac{a}{a} \geq a$
   g. $\forall_{a \in X} \ [a \times (-1)] \in X$
   h. $\forall_{a \in X} \ \left(\dfrac{a}{-1}\right) \in X$

10. For each word phrase, write the mathematical phrase.
    a. 12 increased by n
    b. $3m$ increased by $-4x$
    c. the product of $(6t + 1)$ and $(5s - 3)$
    d. $(5x - 4)$ divided by $(-x + 7)$
    e. the reciprocal of $\dfrac{3n}{4s}$
    f. the reciprocal of $\dfrac{a - x}{y - b}$
    g. the reciprocal of the product of $2n$ and $3t$

h. the product of the reciprocals of $2n$ and $3t$
i. worth of n nickels in cents
j. worth of $(10x + y)$ quarters in cents
k. worth of 1 cent in dollars
l. worth of c cents in dollars
m. worth of $(a + b)$ dollars in cents
n. If Ed is $6x$ years old now, how old will he be 12 years from now?
o. If Jane will be $2y - 1$ years old 12 years from now, how old was she 6 years ago?

4

Exponents

4.1 The Product of Powers

Each of the symbols 5×5 and 5^2 names the same number, that is, $5 \times 5 = 5^2$. The symbol 5^2 is read as *the square of five* or *five squared* or *the second power of five*. Similarly, $5 \times 5 \times 5$ and 5^3 name the same number. We read 5^3 as *the cube of five* or *five cubed* or *the third power of five*.

In 5^3, the 3 is called an *exponent* and the 5 is called a *base*. The entire symbol 5^3 is called a *power*. In this case, 5^3 is the third power of 5. Since $5^3 = 125$, the third power of 5 is 125.

$$5^3 \leftarrow \text{exponent}$$
$$\uparrow$$
$$\text{base}$$

Using a variable and the set of real numbers as the replacement set, we can state that for each real number x the following are true:

$$x^2 = x \times x.$$
$$x^3 = x \times x \times x,$$
$$x^4 = x \times x \times x \times x, \text{ and so on.}$$

We generalize this in the following definition, where R is the set of real numbers and N is the set of natural numbers.

4.1 The Product of Powers

DEFINITION 4.1 $\forall_{x \in R}$ and $\forall_{a \in N(a \geq 2)}$

$$x^a = \underbrace{x \times x \times x \times \ldots \times x}_{x \text{ used } a \text{ times as a factor}}$$

To make this definition complete for all natural numbers, we define x^1 as follows:

$$x^1 = x.$$

This definition helps us to simplify certain multiplication problems. To discover this simplification, consider the following example.

$$2^3 \times 2^4 = (2 \times 2 \times 2) \times (2 \times 2 \times 2 \times 2)$$
$$= 2 \times 2 \times 2 \times 2 \times 2 \times 2 \times 2$$
$$= 2^7.$$

Thus, $2^3 \times 2^4 = 2^7$. Observe also that $3 + 4 = 7$; thus, $2^3 \times 2^4 = 2^{3+4}$, and in general, $x^3 \times x^4 = x^{3+4} = x^7$ for each real number x. This example illustrates the following:

THE PRODUCT OF POWERS PROPERTY $\forall_{a \in N} \forall_{b \in N} \; x^a \times x^b = x^{a+b}$.

In the statements of definitions and properties and in the exercises, we shall agree that if we do not state a replacement set for a variable, then the set of real numbers is the replacement set. Thus, in the property just stated, x has the set of real numbers for its replacement set, whereas a and b have the set of natural numbers for their replacement set.

Observe that the name of the property above reflects the nature of the property. The Product of Powers Property refers to the product of powers, $x^a \times x^b$. Note that there are two powers involved here, x^a and x^b.

The Product of Powers Property can be extended to expressions such as the following:

$$x^3 \times x^4 \times x^5 = x^{3+4+5} = x^{12}$$
$$y^2 \times y \times y^2 \times y^6 = y^{2+1+2+6} = y^{11}.$$

In general,

$$x^a \times x^b \times \ldots \times x^z = x^{a+b+\ldots+z}.$$

Caution: Observe that the product of powers property applies only to cases where the base is the same. For example, it does not apply to $3^2 \times 5^4$ because the bases here are not the same.

When negative real numbers are used as bases in powers, some ambiguity may occur. For example, what is -2^2 equal to? Is it $(-2) \times (-2)$

which equals 4 or is it $-(2 \times 2)$ which equals -4? Since -2^2 is an ambiguous symbol, we must make an agreement concerning its meaning.

Agreement: -2^2 means $-(2 \times 2)$, or -4.

That is, *raising to a power comes first; taking the additive inverse comes second.*

There may also be disagreement in cases where multiplication and raising to a power are involved. For example, what is 2×3^2 equal to? Is it $2 \times (3 \times 3)$, which equals 2×9, or 18; or is it $(2 \times 3)(2 \times 3)$, which equals 6×6, or 36? Since 2×3^2 is an ambiguous symbol, we make the following agreement concerning its meaning.

Agreement: 2×3^2 means $2 \times (3^2)$ which equals 2×9, or 18.

That is, *raising to a power comes first; multiplying by another factor comes second.*

EXERCISE 4.1

1. Give the simplest number name involving no exponents.

Example 1 $(-5)^4 = (-5)(-5)(-5)(-5) = 625.$

Example 2 $-5^4 = -(5 \times 5 \times 5 \times 5) = -625.$

Example 3 $(-\frac{2}{3})^2 = (-\frac{2}{3})(-\frac{2}{3}) = \frac{4}{9}.$

a. 2^3
b. $(-6)^2$
c. $-(6)^2$
d. $(-\frac{1}{2})^3$
e. 3×10^2
f. 10^2
g. 10^5
h. $(.1)^2$
i. $(.1)^3$
j. $(.1)^4$
k. $(-2)^4$
l. $-(2)^4$
m. $(-2)^2(-2)^3$
n. $-(2)^2[-(2)^3]$
o. $(-\frac{1}{2})(-\frac{1}{2})^3$
p. $-(\frac{1}{2})[-(\frac{1}{2})^3]$
q. $(-\frac{1}{3})^2(-\frac{1}{3})^2$

2. Give simpler expressions.

Example 1 $x^4 x^5 = x^{4+5} = x^9.$

Example 2 $x^2 y^3 x^3 y^4 = (x^2 x^3)(y^3 y^4) = x^5 y^7.$

4.1 The Product of Powers

    a. x^2x^5 b. s^3s^4

    c. $m^2m^3m^7$ d. a^3bab^5

    e. $mnm^2n^2mn^5$ f. $abca^2b^2c^2a^4b^4c^4$

3. Before answering each of the following questions, try a few examples.

    a. If a power has a positive-number base and an odd-number exponent, is the power a positive or a negative number?

    b. If a power has a positive-number base and an even-number exponent, is the power a positive or a negative number?

    c. If a power has a negative-number base and an odd-number exponent, is the power a positive or a negative number?

    d. If a power has a negative-number base and an even-number exponent, is the base a positive or a negative number?

4. By merely looking at each of the following, tell whether it names a positive or a negative number. Do not compute the answers.

    a. 2^4 b. $(-1)^3$

    c. $(-1)^{17}$ d. $(-1)^6$

    e. $(-1)^4$ f. $(-17)^4$

    g. $(-17)^{17}$ h. 125^4

    i. 125^{23} j. $(-125)^{37}$

    k. $(-100)^{100}$ l. $(-100)^{1001}$

    m. $(-2)^4(-2)^9$ n. $(-17)^4(-17)^{12}$

    o. $(-100)^{97}(-100)^{98}$

5. If a real number r, which is between 0 and 1, is raised to a natural-number power greater than 1, is the answer greater than, equal to, or less than r?

6. If a real number r, which is greater than 1, is raised to a natural-number power greater than 1, is the answer greater than, equal to, or less than r?

7.   a. If a real number r, which is between 0 and -1, is raised to an even natural-number power, is the answer greater than, equal to, or less than r? Give two examples.

    b. Answer the same question for an odd natural-number power greater than 1. Give two examples.

8.   a. If a real number r, which is less than -1, is raised to an even natural-number power, is the answer greater than, equal to, or less than r? Give two examples.

    b. Answer the same question for an odd natural-number power greater than 1. Give two examples.

4.2 The Power of a Power

The work done in the following examples is based on the definition of a natural-number exponent and on the Product of Powers Property. Study these examples and look for a pattern.

$$(2^4)^3 = 2^4 \times 2^4 \times 2^4 = 2^{12} = 2^{4\times 3}$$
$$(5^2)^4 = 5^2 \times 5^2 \times 5^2 \times 5^2 = 5^8 = 5^{2\times 4}$$
$$(6^8)^2 = 6^8 \times 6^8 = 6^{16} = 6^{8\times 2}.$$

These examples lead us to the following pattern.

THE POWER OF A POWER PROPERTY $\forall_{a \in N} \forall_{b \in N} (x^a)^b = x^{ab}$.

Since $(x^a)^b = x^{ab}$ and $(x^b)^a = x^{ba}$, and since $ab = ba$, it follows that

$$(x^a)^b = (x^b)^a.$$

4.3 The Power of a Product

We shall again use the definition of a natural-number exponent and other basic properties to arrive at another property involving exponents. Study these examples and look for a pattern.

$$\begin{aligned}(3 \times 4)^2 &= (3 \times 4) \times (3 \times 4) \\ &= (3 \times 3) \times (4 \times 4) \\ &= 3^2 \times 4^2.\end{aligned}$$

$$\begin{aligned}(5 \times 7)^3 &= (5 \times 7) \times (5 \times 7) \times (5 \times 7) \\ &= (5 \times 5 \times 5) \times (7 \times 7 \times 7) \\ &= 5^3 \times 7^3.\end{aligned}$$

$$\begin{aligned}(9 \times 6)^4 &= (9 \times 6) \times (9 \times 6) \times (9 \times 6) \times (9 \times 6) \\ &= (9 \times 9 \times 9 \times 9) \times (6 \times 6 \times 6 \times 6) \\ &= 9^4 \times 6^4.\end{aligned}$$

These examples illustrate the following pattern.

THE POWER OF A PRODUCT PROPERTY $\forall_{a \in N} (xy)^a = x^a y^b$.

Examine the last two properties and be sure to notice a difference between them.

EXERCISE 4.2

1. Using the Power of a Power Property, give simpler equivalent expressions which have only one exponent.

4.3 The Power of a Product

Example 1 $(x^3)^5 = x^{3 \times 5} = x^{15}$.

Example 2 $(16^2)^7 = 16^{2 \times 7} = 16^{14}$.

Example 3 $[(-2)^3]^4 = (-2)^{3 \times 4} = (-2)^{12} = 2^{12}$.

Example 4 $[(-2)^3]^3 = (-2)^{3 \times 3} = (-2)^9 = -2^9$.

a. $(r^2)^4$
b. $(a^5)^2$
c. $[(-x)^3]^5$
d. $[(-5)^3]^7$
e. $[(-5)^7]^3$
f. $[(-\frac{1}{3})^2]^4$
g. $[(-\frac{1}{3})^4]^2$
h. $[(-3.5)^5]^{20}$
i. $(x^1)^{17}$

2. By merely looking at each of the following, tell whether it names a positive or a negative number. Do not compute the answers. [*Caution:* $-(3)^4 \neq (-3)^4$.]

a. $[(-1)^3]^5$
b. $[(-1)^{12}]^3$
c. $[(-1)^3]^{12}$
d. $[(-2.65)^{15}]^{15}$
e. $[-(3)^4]^5$
f. $[-(-3)^4]^5$
g. $[-(-3)^4]^6$
h. $[-(-1)^5]^5$

3. Explain why $-(3)^4 \neq (-3)^4$, but $-(3)^3 = (-3)^3$.

4. For each of the following, compute answers in each of two ways:
   a. Multiply first; then raise to a power.
   b. Apply the Power of a Product Property.

Example 1 $(5 \times 3)^2$.
   a. $(5 \times 3)^2 = 15^2 = 225$.
   b. $(5 \times 3)^2 = 5^2 \times 3^2 = 25 \times 9 = 225$.

Example 2 $[(-2)(-3)]^4$.
   a. $[(-2)(-3)]^4 = 6^4 = 1296$.
   b. $[(-2)(-3)]^4 = (-2)^4(-3)^4 = 16 \times 81 = 1296$.

a. $(2 \times 6)^2$
b. $[(-1) \times 4]^2$
c. $[(-1) \times 3]^3$
d. $[(-1)(-1)]^{12}$
e. $[(-1)(-1)]^{13}$
f. $[(-3)(-5)]^2$
g. $[10 \times (-4)]^3$
h. $(\frac{1}{2} \times \frac{1}{3})^4$
i. $[(-1) \times .1]^2$
j. $[(-\frac{3}{4})(-\frac{1}{2})]^3$

k. $[(-1)(-1)(-1)]^{125}$ **l.** $[(-2)(-3)(-4)]^2$

5. Simplify using the appropriate properties of powers.

Example 1 $[(-x)y]^5 = (-x)^5 y^5 = -x^5 y^5$.

Example 2 $(a^2 b^3)^6 = (a^2)^6 (b^3)^6 = a^{12} b^{18}$.

Example 3 $[(-1)x]^4 = (-1)^4 x^4 = 1 \times x^4 = x^4$.

a. $(3r)^2$ **b.** $(4xy)^2$
c. $[(-3)a]^2$ **d.** $[(-3)a]^3$
e. $[(-3)a]^4$ **f.** $[(-3)a]^5$
g. $(5m^3)^4$ **h.** $[(-3)xt]^5$
i. $(3^2 m^3 n)^2$ **j.** $[(-1)^5 x^3 y^2]^4$
k. $[(-.2)^2(-a)^2(-b)]^3$ **l.** $[(-m)(-n)^2(-p)^3]^4$

6. Label each statement with either T for true or F for false.

a. $[5(-3)]^2 = 5(3)^2$ **b.** $[(-5)(-3)]^2 = (-5)^2(-3)^2$
c. $[(-5)(-3)]^2 = (25)(9)$ **d.** $[5(-2)]^3 = 5(-2)^3$
e. $[5(-2)]^3 = 5^3 2^3$ **f.** $[(-5)(-2)]^3 = (-5)^3(-2)^3$
g. $(-5)^3(-2)^3 = -(5^3 \times 2^3)$ **h.** $(-5)^3(-2)^3 = 5^3 \times 2^3$

7. Tell which of the following are true for all replacements of the variables.

a. $3x^2 = 9x^2$ **b.** $(3x)^2 = 9x^2$
c. $(-3x)^2 = 9x^2$ **d.** $-3x^2 = 9x^2$
e. $-(3x)^2 = -9x^2$ **f.** $-(3x)^2 = 9x^2$
g. $(-3x)^3 = -27x^3$ **h.** $-(3x)^3 = -27x^3$
i. $3(xy)^2 = 3x^2 y^2$ **j.** $(3xy)^2 = 3x^2 y^2$
k. $(-3xy)^2 = -9x^2 y^2$ **l.** $-3(xy)^2 = -3x^2 y^2$

4.4 The Power of a Quotient

In raising a quotient to a natural-number power, we again make use of the definition of the natural-number exponent.

$$\left(\frac{x}{y}\right)^n = \underbrace{\frac{x}{y} \times \frac{x}{y} \times \ldots \times \frac{x}{y}}_{n \text{ factors}} = \frac{\overbrace{x \times x \times \ldots \times x}^{n \text{ factors}}}{\underbrace{y \times y \times \ldots \times y}_{n \text{ factors}}} = \frac{x^n}{y^n}$$

4.5 The Quotient of Powers

Thus, we have derived the following:

THE POWER OF A QUOTIENT PROPERTY $\forall_{a \in N} \left(\dfrac{x}{y}\right)^a = \dfrac{x^a}{y^a}$ $[y \neq 0]$.

4.5 The Quotient of Powers

How can $3^5 \div 3^2$ be simplified? Observe the following procedure to see how it is done.

$$3^5 \div 3^2 = \frac{3^5}{3^2} = \frac{3 \times 3 \times 3 \times 3 \times 3}{3 \times 3} = 3 \times 3 \times 3 = 3^3.$$

Observe also that $5 - 2 = 3$.

Each of the following is simplified in a similar fashion.

$$2^6 \div 2^2 = \frac{2^6}{2^2} = \frac{2 \times 2 \times 2 \times 2 \times 2 \times 2}{2 \times 2} = 2^4 \text{ and } 6 - 2 = 4.$$

$$3^4 \div 3 = \frac{3^4}{3} = \frac{3 \times 3 \times 3 \times 3}{3} = 3^3 \text{ and } 4 - 1 = 3.$$

Note: 3 means 3^1.

Instead of working with numbers, we can use a variable.

$$\frac{x^5}{x^2} = \frac{x \times x \times x \times x \times x}{x \times x \times x} = x \times x \times x = x^3 \qquad [x \neq 0].$$

From now on we shall assume that in no case is a divisor equal to 0, even if we do not explicitly state so.

Observe that in all of the examples above, the exponent in the numerator is greater than the exponent in the denominator. These examples lead to the following property:

$$\text{If } b > a, \text{ then } \frac{x^b}{x^a} = x^{b-a}.$$

But what if $b = a$ or if $b < a$? We will now examine these two cases.

First consider the following examples where $b = a$.

$$\frac{2^3}{2^3}, \quad \frac{3^4}{3^4}, \quad \frac{(-3)^2}{(-3)^2}, \quad \frac{4^6}{4^6}.$$

The first example may be computed in the following manner:

$$\frac{2^3}{2^3} = \frac{2 \times 2 \times 2}{2 \times 2 \times 2} = 1.$$

In the same manner it can be shown that

$$\frac{3^4}{3^4} = 1, \qquad \frac{(-3)^2}{(-3)^2} = 1, \qquad \frac{4^6}{4^6} = 1.$$

Also, we may think of each of these examples as a nonzero number divided by itself, which, of course, is equal to 1. Thus,

$$\frac{2^3}{2^3} = 1, \quad \frac{3^4}{3^4} = 1, \quad \frac{(-3)^2}{(-3)^2} = 1, \quad \frac{4^6}{4^6} = 1.$$

In general,

$$\text{if } b = a, \text{ then } \frac{x^b}{x^a} = \frac{x^a}{x^a} = 1.$$

Now consider the following examples where $b < a$:

$$\frac{2^2}{2^5}, \quad \frac{3^5}{3^8}, \quad \frac{(-6)}{(-6)^3}.$$

The first example may be computed as follows:

$$\frac{2^2}{2^5} = \frac{2 \times 2}{2 \times 2 \times 2 \times 2 \times 2} = \frac{1}{2^3}.$$

Using the same procedure, we can show that

$$\frac{3^5}{3^8} = \frac{1}{3^3} \quad \text{and} \quad \frac{(-6)}{(-6)^3} = \frac{1}{(-6)^2}.$$

In examining the examples above, we see that the following pattern holds:

$$\frac{2^2}{2^5} = \frac{1}{2^{5-2}} = \frac{1}{2^3}.$$

In general,

$$\text{if } b < a, \text{ then } \frac{x^b}{x^a} = \frac{1}{x^{a-b}}.$$

Thus, we can summarize the quotient of two powers as follows:

Case I.   If $b > a$, then $\dfrac{x^b}{x^a} = x^{b-a}$.

Case II.   If $b = a$, then $\dfrac{x^b}{x^a} = 1$.

Case III.   If $b < a$, then $\dfrac{x^b}{x^a} = \dfrac{1}{x^{a-b}}$.

Remember that these rules apply to all nonzero real numbers x, and to all natural numbers a and b.

4.5 The Quotient of Powers

EXERCISE 4.3

1. For each natural number n and for each real number $y \neq 0$, what is the value of $\left(\dfrac{0}{y}\right)^n$? of $\dfrac{0^n}{y^n}$?

2. Give equivalent expressions using the Power of a Quotient Property. Simplify wherever possible.

   a. $\left(\dfrac{a}{b}\right)^4$

   b. $\left(\dfrac{-x}{y}\right)^2$

   c. $\left(\dfrac{-x}{y}\right)^3$

   d. $\left[\dfrac{(-2)m}{n}\right]^4$

   e. $\left[\dfrac{(-x)^2(-y)}{-3}\right]^3$

   f. $\left[\dfrac{(-p)^2(-r)^3}{-s}\right]^4$

   g. $\left[\dfrac{(-3c^2d)^2}{2a}\right]^3$

   h. $\left[\dfrac{(-a)(-b)^2}{(-c)(-d)^3}\right]^2$

3. Why is $\left(\dfrac{-2}{3}\right)^4 = \left(\dfrac{2}{3}\right)^4$ true, but $\left(\dfrac{-2}{3}\right)^3 = \left(\dfrac{2}{3}\right)^3$ false?

4. Is $\left(\dfrac{-2}{-3}\right)^3 = \left(\dfrac{2}{3}\right)^3$ true? Why?

5. Discuss the statement $\left(\dfrac{-x}{y}\right)^n = \left(\dfrac{x}{y}\right)^n$ for various replacements; x and y are real numbers and n is a natural number.

6. As in Problem 5, discuss the statement $\left(\dfrac{-x}{-y}\right)^n = \left(\dfrac{x}{y}\right)^n$ for various replacements.

7. For each of the following, give a simplified equivalent expression.

   Example 1 $\dfrac{y^4}{y} = y^{4-1} = y^3$.

   Example 2 $\dfrac{15a^5b^3}{3a^2b} = \dfrac{15}{3} \times \dfrac{a^5}{a^2} \times \dfrac{b^3}{b} = 5 \times a^{5-2} \times b^{3-1} = 5a^3b^2$.

   Example 3 $\dfrac{(x+y)^3}{(x+y)^2} = (x+y)^{3-2} = x+y$.

   Example 4 $\dfrac{x^2y^7}{x^5y^9} = \dfrac{x^2}{x^5} \times \dfrac{y^7}{y^9} = \dfrac{1}{x^{5-2}} \times \dfrac{1}{y^{9-7}} = \dfrac{1}{x^3y^2}$.

   a. $\dfrac{x^5}{x}$

   b. $\dfrac{y^{17}}{y^4}$

c. $\dfrac{a^3}{a}$ d. $\dfrac{c^6}{c^2}$

e. $\dfrac{c^2 m^3}{cm}$ f. $\dfrac{14 y^3}{2y}$

g. $\dfrac{18 x^3 y^2}{3xy^2}$ h. $\dfrac{-15 r^7 s^5}{-3 r^2 s^2}$

i. $\dfrac{-48 r^5 s^7}{-3 r^2 s}$ j. $\dfrac{-3 m^3 s^2}{ms}$

k. $\dfrac{(x+y)^5}{(x+y)^2}$ l. $\dfrac{(a-b)^3(c+d)^{12}}{(a-b)(c+d)^7}$

m. $\dfrac{(2x+y)^6}{(2x+y)^6}$ n. $\dfrac{xy^3}{x^4 y^7}$

o. $\dfrac{27 x(x+y)^4}{9 x^3 (x+y)^8}$ p. $\dfrac{x^2 y^5 z}{y^9 z^{10}}$

q. $\dfrac{(3x-y)^2(z-w)^3}{(3x-y)(z-w)^{12}}$ r. $\dfrac{-12 x(x+3y)}{6 x^2 (x+3y)^4}$

4.6 Zero and Negative Integers as Exponents

When we examined the quotient of two powers, $\dfrac{x^b}{x^a}$, we found that we needed to establish a separate rule for each of the three different possibilities: $b > a$, $b = a$, $b < a$. It would be convenient if we could find a single rule which would cover all of these cases.

Consider again the rule for Case I where $b > a$.

$$\text{If } b > a, \text{ then } \dfrac{x^b}{x^a} = x^{b-a}.$$

If we mechanically apply this rule to Case II where $b = a$, we have

$$\dfrac{x^3}{x^3} = x^{3-3} = x^0.$$

But the symbol x^0 has no meaning since we have defined exponents for natural numbers only. Certainly the symbol cannot mean the result of taking *no* x's and multiplying them together. Since we have not yet given any meaning to the number zero as an exponent, we are free to choose whatever meaning is convenient for us. In referring to the previous example, we have already agreed that

$$\dfrac{x^3}{x^3} = 1.$$

Thus, we see that if we agree upon the following definition, the rule for

4.6 Zero and Negative Integers as Exponents

Case I can be applied to Case II, and we can dispense with the original separate rule for Case II.

DEFINITION 4.2 $x^0 = 1 \quad (x \neq 0)$.

For example,
$$\frac{x^m}{x^m} = x^{m-m} = x^0 = 1.$$

Thus, we conclude that if $b > a$ or $b = a$,
$$\frac{x^b}{x^a} = x^{b-a}.$$

Now let us see if we can also make Case III fit the same rule as Case I. If we apply the rule of Case I to an example where $b < a$, we have
$$\frac{x^2}{x^7} = x^{2-7} = x^{-5}.$$

But the symbol x^{-5} has no meaning since we have defined exponents only for natural numbers and zero (the whole numbers).

We can see that if we agree upon the following definition of a negative integer exponent, then we can dispense with Case III of the quotient of two powers.

DEFINITION 4.3 If $-m$ is a negative integer, then
$$x^{-m} = \frac{1}{x^m}.$$

We can also interpret the definition above to mean that x^{-m} is the multiplicative inverse of x^m.

Now, returning to our previous example, we know that $x^{-5} = \frac{1}{x^5}$. We see that we can simplify $\frac{x^2}{x^7}$ either by using the rule of Case III:
$$\frac{x^2}{x^7} = \frac{1}{x^{7-2}} = \frac{1}{x^5},$$

or by the rule of Case I:
$$\frac{x^2}{x^7} = x^{2-7} = x^{-5} = \frac{1}{x^5}.$$

Finally, we can state a single rule for the quotient of two powers which will apply for all three cases: $b > a$, $b = a$, $b < a$. Let I denote the set of all integers.

THE QUOTIENT OF POWERS PROPERTY $\quad \forall_{a \in I} \forall_{b \in I} \dfrac{x^b}{x^a} = x^{b-a}$
$[x \neq 0]$.

We will find that all of the properties which we have accepted for natural-number exponents also hold for integer exponents. For example, the Product of Powers Property holds:

$$x^5 \times x^{-3} = x^{5+(-3)} = x^2.$$

We can verify this by replacing x^{-3} by $\dfrac{1}{x^3}$:

$$x^5 \times \dfrac{1}{x^3} = \dfrac{x^5}{x^3} = x^{5-3} = x^2.$$

Also, the Power of a Power Property holds for integer exponents. Here is an example which illustrates this point:

$$(x^{-3})^2 = x^{-3 \times 2} = x^{-6}.$$

Replacing x^{-3} with $\dfrac{1}{x^3}$, we have

$$\left(\dfrac{1}{x^3}\right)^2 = \dfrac{1^2}{(x^3)^2} = \dfrac{1}{x^6} = x^{-6}.$$

Now let us illustrate the Power of a Power Property for zero as an exponent:

$$(x^0)^{-6} = x^{0 \times (-6)} = x^0 = 1.$$

Also,

$$(x^0)^{-6} = \dfrac{1}{(x^0)^6} = \dfrac{1}{(1)^6} = \dfrac{1}{1} = 1.$$

4.7 Other Patterns in Exponents

We are ready to derive some other patterns which are based on negative integer exponents. Observe the pattern displayed in the following examples:

$$5 \times 2^{-3} = 5 \times \dfrac{1}{2^3} = \dfrac{5}{2^3}, \qquad 7 \times 3^{-2} = 7 \times \dfrac{1}{3^2} = \dfrac{7}{3^2},$$

$$-3 \times 4^{-5} = -3 \times \dfrac{1}{4^5} = \dfrac{-3}{4^5}.$$

These examples suggest the following pattern:

$$x \times y^{-m} = \dfrac{x}{y^m} \qquad [y \neq 0].$$

4.7 Other Patterns in Exponents

This can be easily derived:

$$x \times y^{-m} = x \times \frac{1}{y^m}$$
$$= \frac{x}{y^m}.$$

Multiplication patterns and the definition of a negative integer exponent also permit us to make the following computations:

$$\frac{5}{2^{-3}} = 5 \times \frac{1}{2^{-3}} = 5 \times 2^3, \quad \frac{7}{3^{-2}} = 7 \times \frac{1}{3^{-2}} = 7 \times 3^2,$$

$$\frac{-3}{4^{-4}} = -3 \times \frac{1}{4^{-4}} = -3 \times 4^4.$$

The pattern shown by these examples is

$$\frac{x}{y^{-m}} = xy^m.$$

This pattern can also be derived:

$$\frac{x}{y^{-m}} = x \times \frac{1}{y^{-m}}$$
$$= x \times y^m.$$

EXERCISE 4.4

1. For each of the following, give a simplified equivalent expression containing only positive exponents.

Example 1 $\quad \dfrac{4^{-3} \times 4^7}{4^2} = \dfrac{4^{-3+7}}{4^2} = \dfrac{4^4}{4^2} = 4^2.$

Example 2 $\quad \dfrac{(-2)^5 \times (-2)^{-2}}{(-2)^{-6}} = \dfrac{(-2)^{5-2}}{(-2)^{-6}} = \dfrac{(-2)^3}{(-2)^{-6}}$
$= (-2)^{3-(-6)} = (-2)^9 = -2^9.$

Example 3 $\quad \dfrac{(m+n)^{-3}(m-n)^5}{(m+n)^{-2}(m-n)^{-1}} = (m+n)^{-3-(-2)}(m-n)^{5-(-1)}$
$= (m+n)^{-1}(m-n)^6 = \dfrac{(m-n)^6}{m+n}.$

a. $\dfrac{5^9 \times 5^{-4}}{5^3}$

b. $\dfrac{10^3 \times 10^{-7}}{10^5}$

c. $\dfrac{(\frac{1}{3})^2(\frac{1}{3})^{-4}}{(\frac{1}{3})^{-2}(\frac{1}{3})^{-5}}$

d. $\dfrac{(1.5)^3(1.5)^4}{(1.5)^{10}}$

e. $\dfrac{x^3 x^4}{x^2}$

f. $\dfrac{m^3 m^2}{m^9}$

g. $\dfrac{m^{-2} n^3}{m^{-7} n^4}$

h. $\dfrac{-2k^{-2} m^{-1}}{4k^3}$

i. $\dfrac{20 r^4 s^{-4}}{5 r^{-1} s^{-3}}$

j. $\dfrac{(a+b)^6}{(a+b)^9}$

k. $\dfrac{(5+m)^3 (3-n)^{-4}}{(3-n)^{-12}}$

l. $\dfrac{12(a+b)(c+d)}{-4(a+b)^3 (c+d)^4}$

m. $\dfrac{3x(z+y)^{-6}(s+t)^{-2}}{y(z+y)(s+t)}$

n. $\dfrac{(x+y+z)^3}{(x+y+z)^6}$

o. $\dfrac{12(a+b)^{-2}(a+b+c)^4}{-6(a+b)^4 (a+b+c)^{-3}}$

p. $\dfrac{(x+y)^{-3}(a+b)^2}{(x+y)^{-5}(a+m)^3}$

q. $\dfrac{(a-b-c)^{-4}(a-b)^5}{(a+b+c)^3 (a-b)^8}$

r. $(x^2 y^3)^{-4}$

s. $[(x+y)^{-2}]^3$

t. $[(x+y)^0]^{-5}$

u. $(x^{-5})^0 (x^3 y^{-4})^{-5}$

2. Verify that $x^{-m} = \dfrac{1}{x^m}$ applies to the case $m = 0$.

3. Verify that $x^{-m} = \dfrac{1}{x^m}$ is true when:

a. x is replaced by 3 and m by -2.

b. x is replaced by -2 and m by -3.

4. For each expression, write the simplest equivalent expression that does not contain negative exponents.

Example 1 $-2(xt)^{-3} = -2 \times \dfrac{1}{(xt)^3} = \dfrac{-2}{(xt)^3}.$

Example 2 $\dfrac{-5}{(x+y)^{-2}} = -5 \times \dfrac{1}{(x+y)^{-2}} = -5(x+y)^2.$

Example 3 $\dfrac{x^{-2} y^{-3}}{a^{-7} b^{-6}} = \dfrac{a^7 b^6}{x^2 y^3}.$

Example 4 $\dfrac{x^2 y^{-3}}{a^{-7} b^6} = \dfrac{x^2 a^7}{y^3 b^6}.$

a. $3x^{-2}$

b. $-2t^{-4}$

4.8 Scientific Notation

c. $5x^2r^{-3}$

d. $-4x^{-2}m$

e. $2(x+y)^{-3}$

f. $-3(t+2)^2(x-4)^{-7}$

g. 8×2^{-3}

h. 8×2^{-4}

i. $3a^{-2}bc^{-4}$

j. $6xyz^{-2}$

k. $6x(yz)^{-2}$

l. $-6(xyz)^{-2}$

m. $\dfrac{3}{x^{-2}}$

n. $\dfrac{-2}{t^{-4}}$

o. $\dfrac{5x^2}{r^{-3}}$

p. $\dfrac{5x^2}{-r^{-3}}$

q. $\dfrac{2}{(x+y)^{-3}}$

r. $\dfrac{-3(t+2)^3}{(x-4)^{-7}}$

s. $\dfrac{8}{2^{-3}}$

t. $\dfrac{8}{-2^{-3}}$

u. $\dfrac{3b}{a^{-2}c^{-4}}$

v. $\dfrac{6xy}{z^{-1}}$

w. $\dfrac{6xy}{-z^{-1}}$

x. $\dfrac{-6}{(xyz)^{-1}}$

y. $\dfrac{3x^{-2}}{a^{-3}b^2}$

z. $\dfrac{-3x^2}{a^3b^{-2}}$

a'. $\dfrac{2x^{-2}}{a^{-3}b^{-2}}$

b'. $\dfrac{12m^2}{4t^{-3}}$

c'. $\dfrac{2x^{-6}}{8y^4}$

d'. $\dfrac{2x^{-6}}{-8y^{-4}}$

e'. $\dfrac{(x+y)^{-3}(m+t)^2}{5r^{-2}}$

f'. $\dfrac{7^{-2}}{x^{-2}}$

g'. $\dfrac{7^{-2}}{-x^{-2}}$

h'. $\dfrac{-7^{-2}}{-x^{-2}}$

i'. $\dfrac{-x^{-2}}{-y^{-3}}$

j'. $\dfrac{-(x-y)^{-3}}{-(a+b)^{-4}}$

4.8 Scientific Notation

What we have learned in this chapter can be used to simplify the problem of dealing with very large or very small numbers. Scientists in particular must constantly deal with such numbers. Consider the following:

(1) A light year is the distance traveled by light in the course of one year: 1 light-year = 6,000,000,000,000 mi.

(2) The most accurate measures of the speed of light show this speed to be 29,979,300,000 cm per sec.

(3) The distance of the earth from the sun is about 93,000,000 mi.

(4) The mass of the sun in tons is

$$2,000,000,000,000,000,000,000,000,000.$$

(5) Every second the sun emits

$$3,900,000,000,000,000,000,000,000,000,000,000$$

ergs of electromagnetic radiation. (One erg is the amount of work required to raise $\frac{1}{981}$ of a gram vertically through 1 cm.)

(6) The wavelength of the longest x-ray is .0000001 cm.

(7) The wavelength of a γ-ray is .0000000002 cm.

(8) Radar waves have photon energy of

.000000000000000000002 erg.

Writing such numerals as those above is extremely unwieldy. We shall now show some of these numbers in a less cumbersome form which we call *scientific notation*.

◄ A number given as a product of the form $k \times 10^n$, where $1 \leq k < 10$ and n is an integer, is said to be given in *scientific notation*.

Example 1 Show a light-year, given in miles, in scientific notation.
$6,000,000,000,000 = 6 \times 1,000,000,000,000 = 6 \times 10^{12}$.
Thus, 1 light-year $= 6 \times 10^{12}$ mi.

Example 2 Show the wavelength of the longest x-rays, given in centimeters, in scientific notation.

$$.0000001 = \frac{1}{10,000,000} = \frac{1}{10^7} = 10^{-7}.$$

Thus, this wavelength is 1×10^{-7} cm.

EXERCISE 4.5

1. For each number, give its name in scientific notation.

Example 1 $352.6 = 3.526 \times 100 = 3.526 \times 10^2$.

Example 2 $.0215 = 2.15 \times \frac{1}{100} = \frac{2.15}{10^2} = 2.15 \times 10^{-2}$.

4.8 Scientific Notation

- a. 65.3
- b. 309.27
- c. 126.302
- d. 20.01
- e. .2
- f. .317
- g. .0021
- h. .000126
- i. .00000091

2. Give each of the following in scientific notation.
   a. the speed of light in cm per sec
   b. the distance of the earth from the sun in miles
   c. the mass of the sun in tons
   d. the number of ergs the sun emits every second in the form of electromagnetic radiation
   e. the wavelength of a γ-ray
   f. the photon energy of radar waves in ergs.

3. Using the appropriate properties of exponents, simplify each of the following.

Example 1 $10^4 \times 10^3 \times 10^{-5} = 10^7 \times 10^{-5} = 10^2$.

Example 2 $\dfrac{5.6 \times 10^3 \times 10^{-5}}{10^4 \times 10} = 5.6 \times \dfrac{10^{-2}}{10^5} = 5.6 \times 10^{-7}$.

- a. $10^3 \times 10^6 \times 10^{-5}$
- b. $10^5 \times 10^{-3} \times 10^3$
- c. $10^2 \times 10^{-5} \times 10^4$
- d. $\dfrac{10^5 \times 10^{-3}}{10^2}$
- e. $\dfrac{10^6 \times 10^{-8}}{10^{-3}}$
- f. $\dfrac{10^3 \times 10^{-4}}{10^2 \times 10^{-5}}$
- g. $(3.13 \times 10^{-3}) \times (2 \times 10^{-2})$
- h. $\dfrac{4.66 \times 10^7}{2 \times 10^3}$
- i. $\dfrac{10^{-3} \times 10^{-5}}{10^{-6} \times 10}$
- j. $\dfrac{10^{-2} \times 10^5 \times 10^{-3}}{10^4 \times 10^{-6}}$
- k. $10^{-2} \div 10^5$
- l. $(4 \times 10^6) \times (2.1 \times 10^{-4})$
- m. $(4.12 \times 10^4) \times (2 \times 10^3)$
- n. $(6.12 \times 10^4) \times (2 \times 10^3)$
- o. $\dfrac{(4 \times 10^5) \times (17 \times 10^2)}{8 \times 10^4}$
- p. $\dfrac{(14.7 \times 10^2) \times (12 \times 10^5)}{(4 \times 10^{-1}) \times (7 \times 10^4)}$

4. Solve each problem.
   a. The wavelength of the visible blue light is 4.5×10^{-5} cm. How many wavelengths should be put together to obtain 1 cm?
   b. If the speed of light is 2.99793×10^{10} cm per sec, what is the speed in km per sec? Give the answer in scientific notation. [*Hint:* 1 cm $= 10^{-5}$ km.]

c. The speed of sound at sea level is approximately 760 mph. Give this speed in ft per sec written in scientific notation. [*Hint:* 1 mi = 5.28×10^3 ft; 1 hr = 3.6×10^3 sec.]

d. It was estimated that among stars the formation of planetary systems similar to ours takes place in 1 out of 100 stars. If there are 10^{11} stars in our galaxy, how many of them possess planetary systems?

e. It is estimated that if all matter were spread out uniformly, the radius of curvature of the universe would be 1.1×10^{10} light-years. How many miles is this? Give the answer in scientific notation.

f. Give, in scientific notation, the number of seconds in a year. [*Hint:* 1 yr = 365 days.]

4.9 Word Phrases to Mathematical Phrases

In solving mathematical problems, we frequently need to write phrases which show relations between various units. For example, to show the number of inches in 7 ft, we write 7×12. To show the worth of x nickels in cents, we write $5x$.

EXERCISE 4.6

Write the mathematical phrase for each word phrase.

Example 1   Worth in cents of $4x$ six-cent stamps $6 \times 4x$, or $24x$.

Example 2   Number of quarts in $3n$ gal $4 \times 3n$, or $12n$.

1. worth in cents of 12 nickels
2. worth in cents of m nickels
3. worth in cents of $3t$ nickels
4. worth in cents of $(a + b)$ dimes
5. worth in cents of $4n$ quarters
6. worth in cents of $(3x + 5)$ dollars
7. worth in cents of $\frac{t}{4}$ dollars
8. worth in cents of 12 six-cent stamps
9. worth in cents of y ten-cent stamps
10. worth in cents of $(3x + 9)$ ten-cent stamps

Chapter Review Problems

11. number of feet in 6 in.
12. number of feet in x in.
13. number of feet in $(a + b)$ in.
14. number of quarts in 5 gal
15. number of quarts in s gal
16. number of quarts in $10n$ gal
17. number of square inches in 1 sq ft
18. number of square inches in 5 sq ft
19. number of square inches in v sq ft
20. worth in cents of $(5x + 6y)$ nickels
21. number of feet in $7t$ yd
22. number of yards in $5x$ ft
23. number of quarts in $(a + 3t)$ gal
24. number of gallons in $(5x + y)$ qt
25. number of feet in n mi
26. number of miles in x ft
27. number of inches in $(a + b)$ ft
28. number of feet in $12m$ in.

GLOSSARY

Base: In 9^5, 9 is the base.

Exponent: In 9^5, 5 is the exponent.

Power: 9^5 is the fifth power of 9.

Scientific notation: A number given as a product of the form $k \times 10^n$, where $1 \leq k < 10$ and k is an integer, is said to be given in scientific notation.

CHAPTER REVIEW PROBLEMS

1. For each of the following, give the simplest number name involving no exponents.
   - a. 4^3
   - b. $(-5)^2$
   - c. 3×6^2
   - d. -5^2
   - e. $(.2)^3$
   - f. $(.03)^2$
   - g. $(-\frac{1}{2})^4$
   - h. $-(\frac{1}{3})^3$

2. For each of the following, give a simpler equivalent expression containing no negative exponents.

a. $n^4 n^7$
b. $(a+b)^3(a+b)^{-7}$
c. $x^4 y^3 x^{-6} y^{-1}$
d. $(x^3)^4$
e. $[(-y)^3]^3$
f. $[(-y)^2]^5$
g. $[(a+b)^2]^4$
h. $[(-a)b]^4$
i. $[(-x)(-y)^2]^3$
j. $(a+b+c)^3(a+b)^{-3}(a+b)^3$
k. $[k^2(a+b)^3]^2$
l. $\dfrac{a^3 b^2}{a^{-2} b^5}$
m. $\dfrac{12(x+y)^2(x-y)^{-3}}{3(x+y)^3(x-y)^{-5}}$
n. $\dfrac{(a+b+c)^2(a+b)^3}{(a+b)^3}$

3. Label each statement with T for true and F for false.

a. $(-5)^{17} > 0$
b. $(-5)^{16} > 0$
c. $(\frac{1}{2})^{12} < 1$
d. $(1.0001)^{10} < 1$
e. $(\frac{2}{5})^7 < \frac{2}{5}$
f. $(\frac{5}{2})^9 < \frac{5}{2}$
g. $(-4)^{17}(-4)^{25} > 0$
h. $(-4)^{16}(-4)^{25} > 0$
i. $(-2)^5 = -2^5$
j. $(-2)^6 = -2^6$
k. $[(-4)^4]^5 > 0$
l. $[(-4)^5]^5 > 0$
m. $-5^3 \neq (-5)^3$
n. $-5^4 \neq (-5)^4$
o. $[(-3)(-3)]^5 = (-3)^5(-3)^5$
p. $[(-3)(-3)]^6 = (-3)^6(-3)^6$
q. $5 \times 4^2 = (5 \times 4)^2$
r. $-2 \times 7^2 = -(2 \times 7)^2$
s. $(-3)^2(-5)^2 = 3^2 \times 5^2$
t. $(-3)^3(-7)^3 = 3^3 \times 7^3$
u. $\left(\dfrac{-2}{5}\right)^2 = \dfrac{4}{25}$
v. $\left(\dfrac{-4}{-5}\right)^3 \neq \dfrac{(-4)^3}{(-5)^3}$
w. $\dfrac{5^3}{5^{-3}} = 5^0$
x. $1^0 = 200^0$
y. $\dfrac{1}{2^{-7}} = 2^7$
z. $(5+2)^2 = 5^2 + 2^2$

4. Tell which of the following are true for all permissible real-number replacements of x and y and for all integer replacements of m and n.

a. $(xy)^m (xy)^n = (xy)^{mn}$
b. $\dfrac{x^m}{y^n} = y^{m-n}$
c. $x^m x^n = x^{n+m}$
d. $(xy)^m = x^m y^m$
e. $(x^m)^n = x^{nm}$
f. $(-x)^m = -(x)^m$
g. $\left(\dfrac{-x}{-y}\right)^m = \left(\dfrac{x}{y}\right)^m$
h. $\left(\dfrac{-x}{-y}\right)^m \neq \dfrac{x^m}{y^m}$

i. $[(-x)^m]^n = [(-x)^n]^m$
j. $(x \times y)^m = [(-x) \times (-y)]^m$
k. $(x \times y)^m \neq (-x)^m \times (-y)^m$
l. $\left(\dfrac{-x}{y}\right)^n = \dfrac{x^n}{y^n}$
m. $\left(-\dfrac{x}{y}\right)^m = \left(\dfrac{x}{y}\right)^m$
n. $\left(\dfrac{-x}{y}\right)^m = \dfrac{(-x)^m}{y^m}$
o. $\dfrac{x^m}{x^{-n}} = x^{m+n}$
p. $x^{-m} = \dfrac{1}{x^m}$
q. $\dfrac{x^{-m}}{x^{-n}} \neq x^{m-n}$
r. $\dfrac{(x+y)^m}{(x-y)^m} = 1$

5. For each expression, write an equivalent expression which contains no negative exponents. Simplify whenever possible.

a. $\dfrac{5}{x^{-3}}$
b. $\dfrac{x^{-4}}{3}$
c. $\dfrac{-2(x+y)^{-3}}{(x+y)^{-5}}$
d. $\dfrac{1}{(xy)^{-2} x^{-5}}$
e. $\dfrac{(x+y)(x+y)^{-3}}{(x+y)^{-2}}$
f. $\dfrac{(x-y)^{-3}}{(x+y)^{-2}}$

6. Tell whether each numeral names a positive number (P) or a negative number (N).

a. $(-1)^{26}$
b. $(-5)^{37}$
c. $[(-2)^3]^{12}$
d. $[(-2)^3]^{13}$
e. $(3-8)^{21}$
f. $(8-3)^{21}$

7. For each number, give its name in scientific notation.

a. 367.25 b. 200,000 c. .0075 d. .00004

8. Using the appropriate properties of exponents, simplify each of the following. State the answers in scientific notation.

a. $(5 \times 10^7) \times (2 \times 10^{-5})$
b. $\dfrac{5.76 \times 10^8}{3 \times 10^9}$
c. $\dfrac{(4 \times 10^3) \times (5 \times 10^{-4})}{2 \times 10^{-5} \times 10^2}$
d. $\dfrac{(2.4 \times 10^5) \times (5 \times 10^{-2})}{(2 \times 10^3) \times (3 \times 10^{-4})}$

5

Open Expressions

5.1 Expressions

Much of the work in algebra consists of handling such expressions as

$$4 \quad -x \quad -7y^3 \quad \sqrt{5}ax^2$$
$$-12mv \quad \tfrac{1}{2}x^2yc \quad -1.675mp \quad y.$$

Each of these expressions is either a real number, a variable, the additive inverse of a variable, or a product of these. Such expressions are called *monomials over the real numbers*. They are also called *terms*.

Now examine each of the following:

$$7 \quad 5x + 7x^2, \quad \tfrac{1}{2}abc - 7.63x + 16u,$$
$$5cy \quad -13y + 16ax^3, \quad 4.6xy - \tfrac{7}{3}a + 1.2mn - 3cd.$$

Expressions containing two terms, such as $3x + 10y$, $-13y + 16ax$, and $-\tfrac{1}{2}y - 3ab$, are called *binomials*. Expressions with three terms, such as $\tfrac{1}{2}abc - 7.63x + 16u$, $x + y - z$, and $-\tfrac{1}{2}a + 3b - 4c$ are called *trinomials*. Each of these belongs to the set of *polynomials over the real numbers*.

DEFINITION 5.1 Each monomial and each sum or difference of monomials over the real numbers is called a *polynomial over the real numbers*.

5.1 Expressions

The examples of monomials given above and Definition 5.1 help us decide that each of the following is not a polynomial over the real numbers.

$$\frac{2}{y} - 5, \quad 5\sqrt{z+5}, \quad \frac{x^2+1}{x^3-1}, \quad \sqrt{-2x^2}.$$

We say that an expression is in *polynomial form* if it is written as a monomial or as a sum of monomials, no two of which have the same non-numerical factors. According to this, each of the following is not a polynomial form.

$$4z + 5z - 1, \quad 3y + x + 2x, \quad \sqrt{2}x(x+3).$$

Whenever an expression contains one or more variables, it is called an *open expression*. One of the important algebraic skills is the ability to tell when two open expressions have the same value as the result of replacing the corresponding variables by the same numerals. For example, do $5x + 7x$ and $12x$ have the same value for every replacement of x? Let us try some replacements.

-2 for x:

$5x + 7x$	$12x$
$5 \times (-2) + 7 \times (-2)$	$12 \times (-2)$
$-10 - 14$	-24
-24	

$\frac{1}{2}$ for x:

$5x + 7x$	$12x$
$5 \times (\frac{1}{2}) + 7 \times (\frac{1}{2})$	$12 \times (\frac{1}{2})$
$\frac{5}{2} + \frac{7}{2}$	6
$\frac{12}{2}$	
6	

Perhaps at this point you have concluded that you need not make any further replacements, since you know $5x + 7x = 12x$ to be true for *all* replacements of x. However, because the replacement set for x is the set of all real numbers, we would never be able to display *all* replacements for x. Thus, we could not verify by means of replacements that $5x + 7x$ and $12x$ have the same values for *all* replacements of x. To establish this, we resort to a *proof*.

The purpose of a proof is to establish, by logical reasoning, something new on the basis of some things which we know to be true. Study the proof below and notice how the known properties are used to support the arguments.

Proof. $\quad 5x + 7x = x \times 5 + x \times 7 \qquad$ CPM
$= x \times (5 + 7) \qquad$ DPMA
$= x \times 12 \qquad\quad\;\,$ Arith. fact: $5 + 7 = 12$
$= 12x. \qquad\qquad\,$ CPM

We started with $5x + 7x$ and concluded with $12x$, proving that $5x + 7x = 12x$ for all real number replacements of x. For this reason we say that $5x + 7x$ and $12x$ are *equivalent expressions*.

We make one more observation about $5x + 7x$ and $12x$: of the two, $12x$ is "simpler looking" than $5x + 7x$. Our work with an expression is often less complicated if we can find an equivalent expression which is simpler than the original. The process of arriving at the simplest equivalent expression is called *simplification*.

EXERCISE 5.1

1. Which of the following are not polynomials over the real numbers?

   a. -3
   b. $\sqrt{2}x^2 + x$
   c. $\dfrac{1}{y}$
   d. $\sqrt{x} + \sqrt{y}$
   e. $\dfrac{4x^2y + 3}{xy}$
   f. $xy^2 + yx^2$

2. Which of the following are in polynomial form?
   a. $3x + 4y - 2$
   b. $9x^3$
   c. $2x(y + x)$
   d. $3x + 2x + (-4y)$
   e. $2x + 3y + 7y$
   f. $2y - (x + 4)$

3. Determine the value of each expression below for the designated replacements of variables.

Example $\quad 6x - 2m; \qquad -2$ for x, $\frac{1}{2}$ for m.
$6 \times (-2) - 2 \times \frac{1}{2} = -12 - 1 = -13.$

   a. $4p - 3; \qquad -7$ for p
   b. $11x - 25y; \qquad -\frac{1}{2}$ for x, 2 for y
   c. $3mv + \frac{1}{2}m; \qquad 1$ for m, -2 for v
   d. $x^2 + x^2; \qquad -5$ for x
   e. $2x^2; \qquad -5$ for x

5.1 Expressions

f. $\dfrac{x^3}{x^2}$; -2 for x

g. $x^2 + y^2$; 3 for x, -1 for y

h. $\dfrac{a+b}{a-b}$; 2 for a, -15 for b

i. $r^2 - s^2$; $\tfrac{1}{2}$ for r, $-\tfrac{3}{4}$ for s

j. $(r+s)(r-s)$; $\tfrac{1}{2}$ for r, $-\tfrac{3}{4}$ for s

k. $\dfrac{m^2 - n^2}{m - n}$; 3 for m, 2 for n

l. $m + n$; 3 for m, 2 for n

m. $xy - xz$; -1 for x, -5 for y, 4 for z

n. $x(y - z)$; -1 for x, -5 for y, 4 for z

4. In each exercise two expressions are given. In each case, make the indicated replacements of the variables to see whether it is reasonable to suspect that the two given expressions may be equivalent expressions.

Example $x^2 - y^2$; $(x-y)(x+y)$.

   i. -2 for x, 5 for y

$x^2 - y^2$	$(x-y)(x+y)$
$(-2)^2 - 5^2$	$(-2 - 5)(-2 + 5)$
$4 - 25$	-7×3
-21	-21

   ii. 0 for x, $-\tfrac{1}{2}$ for y

$x^2 - y^2$	$(x-y)(x+y)$
$0^2 - (-\tfrac{1}{2})^2$	$[0 - (-\tfrac{1}{2})][0 + (-\tfrac{1}{2})]$
$0 - \tfrac{1}{4}$	$\tfrac{1}{2} \times (-\tfrac{1}{2})$
$-\tfrac{1}{4}$	$-\tfrac{1}{4}$

   iii. -4 for x, $.7$ for y

$x^2 - y^2$	$(x-y)(x+y)$
$(-4)^2 - .7^2$	$(-4 - .7)(-4 + .7)$
$16 - .49$	$-4.7 \times (-3.3)$
15.51	15.51

There is reason to believe that $x^2 - y^2$ and $(x-y)(x+y)$ are equivalent expressions.

a. $x + x$; x^2
   i. 0 for x
   ii. 10 for x
   iii. -7 for x

b. $x + x$; $2x$
   i. 4 for x
   ii. -12 for x
   iii. 3.6 for x

c. $x \times x \times x$; $3x$
   i. 0 for x
   ii. 1 for x
   iii. -2 for x

d. $x \times x \times x$; x^3
   i. 1 for x
   ii. -2 for x
   iii. $-\frac{1}{3}$ for x

e. $x^2 \times x^2$; x^4
   i. 0 for x
   ii. -2 for x
   iii. 3 for x

f. $(-n)(-n)$; $-n^2$
   i. 0 for n
   ii. 1 for n
   iii. -2 for n

g. $(-n)(-n)$; $(-n)^2$
   i. 0 for n
   ii. 1 for n
   iii. -2 for n

h. $(-n)(-n)(-n)$; $(-n)^3$
   i. 0 for n
   ii. 1 for n
   iii. -2 for n

i. $\left|\dfrac{c}{c}\right|$ $[c \neq 0]$; $\dfrac{|c|}{|c|}$ $[c \neq 0]$
   i. 5 for c
   ii. -3 for c
   iii. $\frac{1}{4}$ for c

j. $(a+b)^2$; $a^2 + b^2$
   i. 0 for a, 0 for b
   ii. 1 for a, 1 for b
   iii. 1 for a, 2 for b

k. $(a+b)^2$; $a^2 + 2ab + b^2$
   i. 1 for a, 2 for b
   ii. -3 for a, 4 for b
   iii. 5 for a, -1 for b

l. $(a-b)^2$; $a^2 - b^2$
   i. 0 for a, 0 for b
   ii. 1 for a, 1 for b
   iii. 4 for a, 1 for b

m. $(a-b)^2$; $a^2 - 2ab + b^2$
   i. 0 for a, 0 for b
   ii. 1 for a, 1 for b
   iii. 4 for a, 1 for b

n. $x - (3 - y)$; $x - 3 + y$
   i. 0 for x, 0 for y
   ii. 0 for x, 1 for y
   iii. 2 for x, -2 for y

o. $-(x - y)$; $-x - y$
   i. 0 for x, 0 for y
   ii. 1 for x, 2 for y
   iii. -1 for x, -2 for y

p. $-(x - y)$; $-x + y$
   i. 0 for x, 0 for y
   ii. 1 for x, 2 for y
   iii. -1 for x, -2 for y

q. $-(x + y)$; $-x + y$
   i. 0 for x, 0 for y
   ii. 0 for x, 1 for y
   iii. -1 for x, -1 for y

r. $-(x + y)$; $-x - y$
   i. 0 for x, 0 for y
   ii. 0 for x, 1 for y
   iii. -1 for x, -1 for y

5. In each equation replace the variable by several numerals of your own choosing. Then decide if you think resulting statements are true for all real-number replacements.

Example 1 $x^2 + x^2 = x^4$
           1 for x:

$x^2 + x^2$	x^4
$1^2 + 1^2$	1^4
$1 + 1$	1
2	

5.1 Expressions

This single replacement is sufficient to show that $x^2 + x^2$ and x^4 are not equivalent.

Example 2 $|-x| = x$.

Recall that $|-x|$ is the absolute value of the additive inverse of x. Also recall that

$$|x| = x \text{ for every real number } x \geq 0, \text{ and}$$
$$|x| = -x \text{ for every real number } x < 0.$$

0 for x:

$|-x|$	x
$|-0|$	0
0	

2 for x:

$|-x|$	x
$|-2|$	2
2	

Thus, $|-x| = x$ if x is 2.

-1 for x:

$|-x|$	x
$|-(-1)|$	-1
$|1|$	
1	

The last replacement shows that $|-x|$ and x are not equivalent.

a. $x + x + x = 3x$
b. $3m - m = 2m$
c. $3.5y - 2.5y = y$
d. $3\frac{1}{2}a + 7\frac{1}{2}a = 11a$
e. $a - 7 = 7 - a$
f. $\dfrac{4z}{2z} = 2$
g. $\dfrac{20s}{4} = 5s$
h. $|d| = d$
i. $v + 2 = v + 3$
j. $-5n = n$
k. $9f - 3 = 6$
l. $-a + b = b - a$
m. $y = y$
n. $3x - 6 = 6 - 3x$
o. $g = \dfrac{1}{g}$
p. $|x - y| = |y - x|$
q. $(x + 3)^2 = x^2 + 6x + 9$
r. $(p + 5)(p - 5) = p^2 - 25$
s. $|xy| = |x| \times |y|$
t. $-\dfrac{x}{y} = \dfrac{-x}{y}$

u. $\left|\dfrac{a}{t}\right| = \dfrac{|a|}{|t|}$ v. $a(-b) = -(ab)$

6. Choose some replacements for the variables and state your guess as to whether each pair of expressions may be equivalent.

a. $(a+b)^3$; $a^3 + b^3$
b. $(a+b)^3$; $a^3 + 3a^2b + 3ab^2 + b^3$
c. $(a-b)^3$; $a^3 - b^3$
d. $(a-b)^3$; $a^3 - 3a^2b + 3ab^2 - b^3$
e. $x^3 + x^3$; x^6
f. $x^3 + x^3$; $2x^3$
g. $|a+b|$; $|a| + |b|$
h. $|a-b|$; $|a| - |b|$
i. $|ab|$; $|a| \times |b|$

5.2 Proving Expressions Equivalent

Whenever we work with open expressions, we shall assume that the replacement set for the variables is the set of real numbers. Should we need to use another set as the replacement set, we shall say so.

We have already illustrated what is meant by a *proof* by showing that $5x + 7x = 12x$ for all replacements of x. It is rather obvious that there is no exception for this case; *every* replacement of x yields a true statement. There are statements, however, in which there may be some exceptions. Consider, for example, $\dfrac{x}{x} = 1$. Since the meaning of this expression is that any number divided by itself is equal to 1, we know it to be true, with an exception, however. If we replace x by 0, we obtain $\dfrac{0}{0} = 1$, which is meaningless. We shall say that $\dfrac{x}{x} = 1$ is true for all *permissible* real-number replacements of x. We consider 0 to be a nonpermissible replacement for x, since $\dfrac{0}{0}$ is a meaningless expression.

Another proof is presented to illustrate the use of basic properties in writing proofs.

Example   Prove: $x + 4x = 5x$.

Proof. $x + 4x = 1 \times x + 4 \times x$ PIM
$ = x \times 1 + x \times 4$ CPM
$ = x(1 + 4)$ DPMA
$ = x \times 5$ Arith. fact: $1 + 4 = 5$
$ = 5x.$ CPM

Notice how each step in the proof above is supported by a statement whose truth has been established previously.

5.3 Distributivity and Rearrangement

In Chapter 3 we stated the Distributive Property of Multiplication over Addition in the following form:

$$x(y + z) = xy + xz.$$

This property is also called the *Left* Distributive Property of Multiplication over Addition. In proving that $5x + 7x = 12x$, we used the Commutative Property of Multiplication to show that $5x + 7x = x \times 5 + x \times 7$. This enabled us to make use of the left distributivity. However, if we could prove a *theorem* which would apply directly to $5x + 7x$, we would save ourselves some effort. (A *theorem* is a statement which is proved to be true.)

THEOREM 5.1 *Right Distributive Property of Multiplication over Addition* [RDPMA] $(x + y)z = xz + yz.$

Proof. $(x + y)z = z(x + y)$ CPM
$\qquad\qquad = zx + zy$ LDPMA
$\qquad\qquad = xz + yz.$ CPM

Thus, $(x + y)z = xz + yz.$

The left and right distributive properties which we previously considered involved three numbers. Let us now investigate examples involving more than three numbers. In each of the examples below, we will evaluate $w(x + y + z)$ and $wx + wy + wz$ using different replacements for the variables.

$w(x + y + z)$	$wx + wy + wz$
$2(5 + 3 + 4)$	$(2 \times 5) + (2 \times 3) + (2 \times 4)$
$2(12)$	$10 + 6 + 8$
24	24

Therefore, $2(5 + 3 + 4) = (2 \times 5) + (2 \times 3) + (2 \times 4).$

$w(x + y + z)$	$wx + wy + wz$
$(-3)[-4 + 6 + (-7)]$	$(-3)(-4) + (-3)(6) + (-3)(-7)$
$-3(-5)$	$12 - 18 + 21$
15	15

Therefore, $(-3)[-4 + 6 + (-7)] = (-3)(-4) + (-3)(6) + (-3)(-7).$

$w(x+y+z)$	$wx + wy + wz$
$(-\frac{1}{2})[4 + 7 + (-9)]$	$(-\frac{1}{2})(4) + (-\frac{1}{2})(7) + (-\frac{1}{2})(-9)$
$(-\frac{1}{2})(2)$	$-2 - 3\frac{1}{2} + 4\frac{1}{2}$
-1	-1

Therefore, $(-\frac{1}{2})[4 + 7 + (-9)] = (-\frac{1}{2})(4) + (-\frac{1}{2})(7) + (-\frac{1}{2})(-9)$.

It appears that the following is true.

THEOREM 5.2 $w(x+y+z) = wx + wy + wz$.

Proof. $\begin{aligned}w(x+y+z) &= w[(x+y)+z] \\ &= w(x+y) + wz &&\text{LDPMA} \\ &= wx + wy + wz &&\text{LDPMA}\end{aligned}$

Thus, $w(x+y+z) = wx + wy + wz$.

Note that in the first step of the proof, we named $x+y+z$ as $(x+y)+z$. We did this because addition is a *binary* operation, and we can only add two numbers at a time. However, the Associative Property of Addition tells us that we may either add x and y first or we may add y and z first, and the results are the same. Thus, we are justified in choosing to add x and y first.

Thus, we have an extension of left distributivity of multiplication over addition to four numbers. It would be helpful also if we could deal with more than three numbers where the commutative and associative properties apply. For example, in computing the answer to $97 + 65 + 3 + 35$, we would save a lot of work if we knew that

$$97 + 65 + 3 + 35 = (97 + 3) + (35 + 65).$$

Notice that it is much easier to compute the sum by grouping the numbers as is shown in the expression on the right of the equality symbol.

We *can* prove that

$$97 + 65 + 3 + 35 = (97 + 3) + (35 + 65).$$

Proof. $\begin{aligned}97 + 65 + 3 + 35 &= [(97 + 65) + 3] + 35 \\ &= [97 + (65 + 3)] + 35 &&\text{APA} \\ &= [97 + (3 + 65)] + 35 &&\text{CPA} \\ &= [(97 + 3) + 65] + 35 &&\text{APA} \\ &= (97 + 3) + (65 + 35). &&\text{APA}\end{aligned}$

Thus, $97 + 65 + 3 + 35 = (97 + 3) + (35 + 65)$.

5.3 Distributivity and Rearrangement

This is a tedious way to get a result which is easily foreseen as a consequence of repeated applications of commutative and associative properties.

We shall now prove the following:

THEOREM 5.3 $(a + b) + (c + d) = (a + c) + (b + d)$.

Proof.
$(a + b) + (c + d) = [(a + b) + c] + d$ APA
$ = [a + (b + c)] + d$ APA
$ = [a + (c + b)] + d$ CPA
$ = [(a + c) + b] + d$ APA
$ = (a + c) + (b + d)$. APA

Hence, $(a + b) + (c + d) = (a + c) + (b + d)$.

Keep in mind that, although we did not use quantifiers, we agreed that in such cases the statements are true for all real-number replacements of the variables. If there are exceptions, we shall indicate that.

Notice that the only properties used were associative and commutative properties of addition. There are many other rearrangements which we could obtain for the terms a, b, c, and d, but each would be equivalent to $(a + b) + (c + d)$ according to the associative and commutative properties of addition. Therefore, we state the following property.

THE TERM REARRANGEMENT PROPERTY [TR] The terms of a sum may be rearranged in any order and the resulting expression is equivalent to the original expression.

The Term Rearrangement Property may be used to simplify some complicated expressions. Study the following examples to see how it is done.

Example 1 Simplify $4x + 7y + 3 + 8x + 2y + 5$.
$4x + 7y + 3 + 8x + 2y + 5 = (4x + 8x) + (7y + 2y) + (3 + 5)$ TR
$ = (4 + 8)x + (7 + 2)y + (3 + 5)$ RDPMA
$ = 12x + 9y + 8$. Arith. facts

Hence, $4x + 7y + 3 + 8x + 2y + 5 = 12x + 9y + 8$.

Example 2 Simplify $8x + 9x^2 + 6x$.
$8x + 9x^2 + 6x = (8x + 6x) + 9x^2$ TR
$ = (8 + 6)x + 9x^2$ RDPMA
$ = 14x + 9x^2$. Arith. fact: $8 + 6 = 14$

Hence, $8x + 9x^2 + 6x = 14x + 9x^2$.

Example 3 Simplify $5(m + t) + 2(m + 3)$

$$\begin{aligned}
5(m + t) + 2(m + 3) & \\
= 5m + 5t + 2m + 6 & \quad \text{LDPMA} \\
= (5m + 2m) + 5t + 6 & \quad \text{TR} \\
= (5 + 2)m + 5t + 6 & \quad \text{RDPMA} \\
= 7m + 5t + 6. & \quad \text{Arith. fact: } 5 + 2 = 7
\end{aligned}$$

Hence, $5(m + t) + 2(m + 3) = 7m + 5t + 6$.

Next we shall prove a theorem which is analogous to Theorem 5.3 but which involves multiplication rather than addition.

THEOREM 5.4 $(ab)(cd) = (ac)(bd)$.

Proof.
$$\begin{aligned}
(ab)(cd) &= [(ab)c]d & \quad \text{APM} \\
&= [a(bc)]d & \quad \text{APM} \\
&= [a(cb)]d & \quad \text{CPM} \\
&= [(ac)b]d & \quad \text{APM} \\
&= (ac)(bd). & \quad \text{APM}
\end{aligned}$$

Hence, $(ab)(cd) = (ac)(bd)$.

Notice that the only properties used were the associative and commutative properties of multiplication. There are many other rearrangements which we could obtain for the factors a, b, c, and d, but each would be equivalent to $(ab)(cd)$ according to the associative and commutative properties of multiplication. Therefore, we have the following:

THE FACTOR REARRANGEMENT PROPERTY [FR] The factors of a product may be rearranged in any order and the resulting expression is equivalent to the original expression.

Like the Term Rearrangement Property, the Factor Rearrangement Property can also be used to simplify complicated expressions. In this case, however, products rather than sums are involved. Study the following examples to see how it is done.

Example 1 Simplify $(4a)(3ab)(-5a)$.

$$\begin{aligned}
(4a)(3ab)(-5a) &= (-5 \times 4 \times 3)(a \times a \times a) \times b \quad \text{FR} \\
&= -60a^3b.
\end{aligned}$$

Hence, $(4a)(3ab)(-5a) = -60a^3b$.

5.3 Distributivity and Rearrangement

Example 2 Simplify $(-3m)(-7t)(2mt)$.

$$(-3m)(-7t)(2mt) = [(-3)(-7)(2)](mm)(tt) \quad \text{FR}$$
$$= 42m^2t^2.$$

Hence, $(-3m)(-7t)(2mt) = 42m^2t^2$.

Example 3 Simplify $4a[(3a)(2a) + 2a + 1 + (3a)a]$.

$$4a[(3a)(2a) + 2a + 1 + (3a)a] = 4a(6a^2 + 2a + 1 + 3a^2)$$
$$= 4a(6a^2 + 3a^2 + 2a + 1)$$
$$= 4a(9a^2 + 2a + 1)$$
$$= 36a^3 + 8a^2 + 4a.$$

Hence, $4a[(3a)(2a) + 2a + 1 + (3a)a] = 36a^3 + 8a^2 + 4a$.

EXERCISE 5.2

1. Label each statement with T for true and F for false.
   a. $13 + 19 + 27 + 31 = (13 + 27) + (19 + 31)$
   b. $8 + 24 + 16 + 12 = (8 + 12) + (24 + 16)$
   c. $17 + 25 + (-7) + (-5) = [17 + (-7)] + [25 + (-5)]$
   d. $(-9) + 14 + (-24) + (-11) = [(-9) + (-11)] + [14 + (-24)]$
   e. $8 + 15 + 42 + 35 = 8 + 42 + 15 + 35$
   f. $2(4) + 3(7) + 4(5) + 7(6) = 3(7) + 4(5) + 7(6) + 2(4)$
   g. $9 + (-18) + (-19) + (-2) = (-18) + (-2) + 9 + (-19)$
   h. $8 + (-13) + (-8) + (13) = 13 + (-13) + 8 + (-8)$
   i. $(-4) + 7 + (-7) + 4 = 4 + (-4) + 7 + 7$
   j. $2(5) + 5(-2) + 3(4) + 4(-3) = 4(-3) + 3(4) + 5(-2) + 2(5)$
   k. $9 \times 5 \times (-2) \times 6 = 6 \times 5 \times 9 \times (-2)$
   l. $(-1) \times 5 \times (-2) = -2 \times (-5) \times (-1)$
   m. $5 + (-3) = 5 \times (-3)$
   n. $-4 \times 2 = -4 + (-4)$
   o. $6 \times (-2) \times 0 = 25 \times (-10) \times 0$

2. Tell which of the following are true for all replacements of the variables.
   a. $x + y + z + t = x + t + z + y$
   b. $t + y + (-t) + (-y) = y + (-y) + t + (-t)$
   c. $y + t + r + 2(-y) = y + 2(-y) + r + t$

d. $2x + (-3)r + (-3)x + (-4)r = 2x + (-3)x + (-3)r + (-4)r$
e. $9y + 3x + (-2)y + (-5)x = 3x + (-5)x + 9y + (-2)y$
f. $7a + (-3)b = 3b + (-7)a$
g. $x + x = x^2$
h. $3(a + 4) + 2(a + 6) = 5a + 10$
i. $7m + 6 + 3m = 2m + 6 + 8m$
j. $3 + 2c = 5c$
k. $5x + 3 + 2x + 4 = 4x + 5 + 3x + 2$
l. $t^2 + t = 2t^3$
m. $x(y + z)w = wx(y + z)$
n. $(5a)(3b)(2c) = (10bc)(3a)$
o. $-2x(4y)(-z) = 8xyz$
p. $(ab)(ad) = abd$
q. $(-6t)(4t) = (8t)(-3t)$
r. $(5c)(5c)(5c) = 5(c^3)$
s. $3k(8k + 4) = 24k^2 + 4$
t. $5a(6a + 9) = 15a(2a + 3)$
u. $(4fg)(3fg)(-2f) = (6f^2)(2g^2)(-2f)$

3. Simplify.
   a. $6p + 3 + 2p + 5$
   b. $2xy + 3x + 5xy + 4x$
   c. $m + 1 + m + 1$
   d. $2a + 3b + 3a + 2b$
   e. $3k + 4 + 2k$
   f. $7 + 9c + 2 + 6c$
   g. $3x^2 + 5 + 2x^2 + 4x + 7 + 9x$
   h. $5a + 3b + 6 + 9b + 2a + 4$
   i. $\frac{1}{2} + 2y + \frac{1}{4} + 7y$
   j. $1.3 + 4.2t + 7.8v + 2 + 2.3v + 5.4t$
   k. $x + 2 + y + 3y + 5 + x$
   l. $3.2x + 4.3y + 2.1 + 1.8y + 3.4x + 8.7$
   m. $7x^2 + 5x + 2 + 2x^2 + 6x + 8$
   n. $5 + 3xy + 4 + 8xy$
   o. $2(a + b) + 3(a + 7)$
   p. $5(m + n + 2) + 3(6 + n + m)$
   q. $5(3b)$
   r. $5(3 + b)$
   s. $2b(3c)$
   t. $2b(3 + c)$
   u. $-7(6k)$
   v. $-7k(6k)$
   w. $7(-2m)$
   x. $-7(2m)$
   y. $-7(-2m)$
   z. $-3d(-4a)$
   a'. $3d(-4a)$
   b'. $-3d(4a)$

5.3 Distributivity and Rearrangement

c'. $(6c)(8c)$
d'. $(-5t)(-4t)$
e'. $(7a)(3ab)$
f'. $(5k)(3mk)(2mk)$
g'. $-4(3cd)(10cd)$
h'. $2b(5b + 6)$
i'. $3m(7m + 2t)$
j'. $(\frac{1}{2}a)(\frac{1}{3}b)(6ab)$
k'. $3.4(2.6x)(10xy)$
l'. $1.2(3.4m + 1.2)$

4. For each of the following, write the simplest equivalent expression that does not contain grouping symbols.

a. $2(7 + c) + 3(4 + c)$
b. $3(a + b) + 2(b + c) + 5(a + c)$
c. $5 + 3(k + w + 4) + 2(w + 3) + k$
d. $83(9 + t) + 17(9 + t)$
e. $2a(3m + 7)$
f. $3p(4d + 2t)$
g. $16 + 5(3d + 2)$
h. $2m(3m + 7c)$
i. $1.2c(5k + 6)$
j. $\frac{1}{2}b(\frac{1}{3}c + \frac{1}{4}b)$
k. $2a + 3(7 + 4a)$
l. $7x(3y + 2m + 4x)$
m. $100t(3.4a + 2.76b + 5.01c)$
n. $4(3x + 2y) + 5(3y + 4x)$
o. $3(2x + 4y) + 5(3x + y)$
p. $2(4a + 3b + c) + 7(a + b + 2c)$
q. $5(2x + 3) + 7(3x + 2) + 2(5x + 1)$
r. $2m(7 + 4c) + 3c(5m + 2)$
s. $3d(6c + 2r) + 4r(5d + c) + 2c(r + d)$
t. $2(x + m) + 7(x + m) + (x + m)$
u. $3(2a + b) + 2(b + 2a)$
v. $a(x + t) + 3a(t + x)$

5. For each of the following, write the simplest equivalent expression.

a. $2t + 8t + 5t$
b. $2.6m + m$
c. $\frac{1}{2}a + \frac{1}{3}a$
d. $(3a + 9a) + (10b + 7b)$
e. $4s + 2s + 9$
f. $c + c + c$
g. $4x^2 + 9x^2$
h. $2mn + 7mn$

i. $3ab + 7ab + 5$
k. $-6d + 4d$
m. $-5b + 17b$
o. $-2c + (-3c) + (-11c)$
q. $-43g + 4g + (-4g)$
s. $q + (-7q) + 6q$
j. $17m + (-3m)$
l. $-8f + (-4f)$
n. $3a + (-12a)$
p. $-\frac{3}{7}k + \frac{2}{7}k$
r. $d + (-3d)$

6. Prove each of the following. Use the form shown in the examples and be sure to list the reason for each statement. Each sentence applies to all real numbers.

a. $3x + 15x = 18x$
b. $-2y + 7y = 5y$
c. $-3z + (-12z) = -15z$
d. $n + 3n = 4n$
e. $m(-1) + 5m = 4m$
f. $t(-4) + t(-2) = -6t$
g. $5a + 7a + 4 = 12a + 4$
h. $my + ny = (m + n)y$
i. $-cv - dv = [-c + (-d)]v$
j. $u(w + x + y + z) = (uw) + (ux) + (uy) + (uz)$
k. $(w + x + y)z = (wz) + (xz) + (yz)$
l. $(3x + 5) + 4x = 7x + 5$
m. $(6n + 1) + n = 7n + 1$
n. $3a + 2b + 5a + 6b = 8a + 8b$
o. $(5x^2)(3y) = 15x^2y$
p. $(7a)(2b) = 14ab$
q. $x_0(x_1 + x_2 + \ldots + x_n) = (x_0x_1) + (x_0x_2) + \ldots + (x_0x_n)$ [Read "x_0" as "x sub-zero," "x_1" as "x sub-one," "x_n" as "x sub-n."]

5.4 Subtraction and Multiplication by –1

The definition of subtraction,

$$\forall_a \forall_b \ a - b = a + (-b),$$

is useful in simplifying expressions involving subtraction. This is illustrated in the following example:

5.4 Subtraction and Multiplication by −1

$$-3b + a - 2b = -3b + a + (-2b)$$
$$= a + (-3b) + (-2b)$$
$$= a + (-5b)$$
$$= a - 5b.$$

Thus, we can conclude that $-3b + a - 2b = a - 5b$.

Let us look closer at the problem of multiplication by a negative number, in particular, multiplication by −1. Observe the following:

$$-1 \times 5 = -5 \qquad -1 \times \tfrac{1}{2} = -\tfrac{1}{2}$$
$$-1 \times (-3) = 3 \qquad -1 \times (-\tfrac{4}{5}) = \tfrac{4}{5}.$$

The pattern displayed by these examples is

$$-1 \times x = -x.$$

This is read as "negative one times x is equal to the additive inverse of x." We prove this as a theorem.

THEOREM 5.5 *Multiplication by −1* $-1 \times x = -x$.

We know from the property of additive inverses that $x + (-x) = 0$. If we can show that $x + (-1 \times x) = 0$, we would know that $-1 \times x$ is the additive inverse of x, and therefore, $-1 \times x$ is equal to $-x$. We are assuming that each real number has only one additive inverse, which we shall prove later.

Proof. $x + (-1 \times x) = 1 \times x + (-1 \times x)$ PIM
$ = [1 + (-1)] \times x$ RDPMA
$ = 0 \times x$ PAI
$ = 0.$ PZM

Since $x + (-1 \times x) = 0$, then $-1 \times x$ is the additive inverse of x, and therefore, $-1 \times x = -x$. This is what we wished to prove. We thus proved that the product of −1 and any real number is equal to the additive inverse of the number.

Now let us observe another pattern illustrated by the following examples:

$$-(-6) = 6, \qquad -(-\tfrac{1}{2}) = \tfrac{1}{2}, \qquad -(-1.9) = 1.9.$$

It appears that the following is true

$$-(-x) = x.$$

THEOREM 5.6 $-(-x) = x$.

Proof. $\begin{aligned}-(-x) &= -1 \times (-x) & & \text{Theorem 5.5}\\ &= -1 \times (-1 \times x) & & \text{Theorem 5.5}\\ &= [-1 \times (-1)]x & & \text{APM}\\ &= 1 \times x & & \text{Theorem 5.5}\\ &= x. & & \text{P1M}\end{aligned}$

Frequently we shall not state the reasons for the steps in the proofs. Check to see whether you can supply the reasons.

THEOREM 5.7 $x(-z) = -(xz)$.
We prove that $xz + x(-z) = 0$.

Proof. $\begin{aligned}xz + x(-z) &= x[z + (-z)]\\ &= x \times 0\\ &= 0.\end{aligned}$

We proved that the sum of xz and $x(-z)$ is 0; therefore, $x(-z)$ is the additive inverse of xz, or $x(-z) = -(xz)$.

The definition of subtraction also enables us to prove two additional theorems about distributivity of multiplication over subtraction. Study the proofs.

THEOREM 5.8 Left Distributive Property of Multiplication over Subtraction [LDPMS] $x(y - z) = xy - xz$.

Proof. $\begin{aligned}x(y - z) &= x[y + (-z)]\\ &= xy + x(-z)\\ &= xy + [-(xz)]\\ &= xy - (xz).\end{aligned}$

Thus, $x(y - z) = xy - xz$.

Theorem 5.9 Right Distributive Property of Multiplication over Subtraction [RDPMS] $(x - y)z = xz - yz$.

Proof. $\begin{aligned}(x - y)z &= z(x - y)\\ &= zx - zy\\ &= xz - yz.\end{aligned}$

Thus, $(x - y)z = xz - yz$.

5.4 Subtraction and Multiplication by −1

EXERCISE 5.3

1. Label each statement with T for true and F for false.
    a. $(-7)(-1) = -(-7)$
    b. $(-1)[-(-2)] = -2$
    c. $-1 \times (-7) = -7 \times 1$
    d. $-1 \times [-3 + (-4)] = 3 + 4$
    e. $(-2 + 3)(-1) = -3 + 2$
    f. $-[-(-3)] = -3$
    g. $-6 + (-1) \times 5 = -1$
    h. $-(-3) + [-2 \times (-4)] = -11$
    i. $5 - 2 - 7 = -7 - 5 - 2$
    j. $-4 - 3 - 6 = -(6 + 3) - 4$
    k. $5(6 - 7) = 5 \times 6 - 5 \times 7$
    l. $(12 - 67) \times (-3) = 12 \times (-3) - 67 \times (-3)$
    m. $(-2)7 = -(2 \times 7)$
    n. $4(-6) = (-6)(-4)$

2. Label as true those sentences which yield true statements for all replacements of the variables. For others give an example of replacements for the variables which result in a false statement.
    a. $-(-n) = n$
    b. $-1 \times (-x) = -x$
    c. $[3 + (-3)] \times m = m$
    d. $(5 - 7) \times (-x) = -2x$
    e. $(-m)(-n) = -(mn)$
    f. $-(-x) + [-(-y)] = x + y$
    g. $-(-x) + (-y) = y - x$
    h. $-1 \times [-a + (-b)] = -a + (-b)$
    i. $2a - 3b - 5a = -3(a + b)$
    j. $-\{-m + [-(-n)]\} = n - m$
    k. $-(x + y + z) = -y - z - x$
    l. $-(-2a + 3b + 4a) = 2a - 3b$

3. Simplify.
    a. $c(-2b)$
    b. $(-7m)(-b)$

c. $(-4d)(-5d)$
d. $-t(3c)$
e. $-k(5k)$
f. $14m - 8m$
g. $3b - 12b$
h. $5m - 2\frac{1}{3}m$
i. $3x - x$
j. $23ab - 16ab$
k. $0 - 3t$
l. $8x^2 - 5x^2$
m. $(8t - 3t) + (6c - 2c)$
n. $9c - 2c - 3c$
o. $17m - 7m + 4m$
p. $-(-5x)(3y)$
q. $(-x)(-3y)(-4)$
r. $(-6.3x)(-y)(5)$
s. $(-2)(-s)(-6)$
t. $t - 3t$
u. $-3m - 2m$
v. $-7k - 5k$
w. $-d - d$
x. $6m - m$
y. $x^2 - 5x^2$
z. $5v - 6v$

4. Label as true those sentences which will yield true statements for all replacements of the variables.

a. $(6x)(3y) = (-2x)(-9y)$
b. $5(a - 4) = 5a - 4$
c. $-(k \times m) = (-k)(-m)$
d. $3(2t - 7) = 6t - 21$
e. $2(8 - 3f) = 16 - 6f$
f. $8 - 6d = 2d$
g. $2(-ab) = -2ab$
h. $5r(-s) = -5rs$
i. $-7(-m)(-n) = 7mn$
j. $-2p(-q) = -2pq$
k. $-x(-y) = yx$
l. $(3ab)(5abc) = 15abc$

5. We are now familiar with four cases of distributivity: two cases involving multiplication and addition, and two involving multiplication and subtraction. State these four cases.

6. Using Theorem 4.5, prove that $(-2)y = -(2y)$. [*Hint: Use* $-2 = -1 \times 2$ and APM.]

7. Prove: $10x - 4x = 6x$.
8. Prove: $3y - 7y = -4y$.
9. Prove: $\frac{1}{2}m - \frac{3}{4}m = -\frac{1}{4}m$.
10. Prove: $6rs - 2rs = 4rs$.
11. Using Theorem 5.6, prove: $-[-(-y)] = -y$.
12. Prove: $-(-x)y = xy$.
13. Prove: $(-x)(-y) = xy$. [*Hint:* Use Theorem 5.7.]
14. Prove: $x - x = 0$.
15. Prove: $w(x - y - z) = (wx) - (wy) - (wz)$.
16. Prove: $x_0(x_1 - x_2 - \ldots - x_n) = (x_0x_1) - (x_0x_2) - \ldots - (x_0x_n)$.

Chapter Review Problems 139

GLOSSARY

Binomial: A polynomial consisting of two terms.

Equivalent expressions: Expressions which yield the same number for all replacements of the variable(s).

Monomial: A real number, a variable, the additive inverse of a variable, or a product of these.

Open expression: An expression containing one or more variables.

Polynomial: A monomial or a sum (difference) of monomials.

Term: The same as monomial.

Trinomial: A polynomial with three terms.

CHAPTER REVIEW PROBLEMS

1. Label each statement with T for true and F for false.
    a. $(-2)(-5) = -(-10)$
    b. $(-1)[-(-5)] = 5$
    c. $-1 \times (-10) = -10 \times 1$
    d. $-1[-5 + (-6)] = 5 + 6$
    e. $-1(-3 + 2) = 3 + 2$
    f. $-12 + (-1)(-11) = 1$
    g. $3 - 1 - 6 = -6 - 3 - 1$
    h. $-(5 + 6) - 8 = -5 + 6 - 8$
    i. $-(3 + 5) - (2 + 9) = -3 + 5 - 2 + 9$
    j. $-(-2 + 1) - (5 - 3) = 2 - 1 - 5 + 3$

2. Which of the following are not polynomials over the real numbers?
    a. $2x + \dfrac{1}{x}$
    b. $3x^2 - \sqrt{2}x + \sqrt{5}$
    c. $\dfrac{2x^2 + 1}{x}$
    d. $y^2 - 2y + 5$
    e. $\dfrac{4z + 1}{z}$
    f. $\sqrt{-5}x$

3. Which of the following are in polynomial form?
    a. $9x^3 - 2x^2 + 1$
    b. $\sqrt{2} - (x + \sqrt{3})$
    c. $2x(y + 2)$
    d. $3z^2 - x^2 - y^2$

e. $\sqrt{2}y + \sqrt{3}x^2$ f. $(x + 1) + (3y^2 - 1)$

4. Write m for each monomial, b for each binomial, t for each trinomial, and p for each polynomial with more than three terms.

   a. $2x - 3y$
   b. x
   c. $2ab + 3a - 3ac + 4c$
   d. $3x^2 + 2x - y$
   e. -4
   f. $-13abcx$
   g. $\frac{1}{2}x + \frac{1}{3}y + \frac{1}{4}z$
   h. $a - b$
   i. $4a^2 + b^2 + c^3 + d^3$
   j. $3.6xy$

5. Prove each of the following. Give the reason for every statement in your proof.

   a. $5x + 12x = 17x$
   b. $4a + 3b + 6a + 2b = 10a + 5b$
   c. $12m - 3n - 3m - 7n = 9m - 10n$
   d. $3ab + 4a - ac - 3a = a(3b - c) + a$

6. For each of the following, write the simplest equivalent expression.

   a. $9x + 12x + 3x$
   b. $16y + 3a - 12y - 3a$
   c. $\frac{1}{2}x^2 - \frac{1}{3}x + \frac{1}{4}x^2 - \frac{1}{6}x$
   d. $4(2x + y) + 3(x + 2y)$
   e. $-2(3a - b) + 4(b - a)$
   f. $-(5m + n) - (n - 3w)$
   g. $-(-3s - 2t) - (t - s)$
   h. $-(5x^2 + 3x - 2) + (3x^2 - x + 8)$

7. Tell whether each open expression is true for all permissible real-number replacements of the variables. If you are unable to decide, try some replacements.

   a. $x + 2y - 2x = xy - x$
   b. $-(-x + y) = x + y$
   c. $(ax)(ay) = axy$
   d. $y + y^2 = y^3$
   e. $ax + ay + az = a(x + y + z)$
   f. $(2a)(-b)(-c) = 2abc$
   g. $a^2 + a^3 = a^5$
   h. $a^2 + a^3 = a^6$
   i. $5 + 3g = 8g$
   j. $\sqrt{a^2 + b^2} = a + b$
   k. $|a| + |b| = |a + b|$
   l. $|a \times b| = |a| \times |b|$
   m. $\dfrac{|a|}{|b|} = \left|\dfrac{a}{b}\right|$
   n. $|a| - |b| = |a - b|$

8. For each of the following, tell which are not permissible values of the variable or variables involved.

   a. $\dfrac{x + y}{x}$
   b. $\dfrac{x + 3}{x - 1}$
   c. $\dfrac{x}{2x - 3}$
   d. $\dfrac{x + y}{x - y}$

Chapter Review Problems

e. $\dfrac{x-y}{x+y}$

f. $\dfrac{x}{2x+y}$

g. $\dfrac{x}{|x|+|y|}$

h. $\dfrac{x}{|x|-|y|}$

i. $\dfrac{x-5}{2|x|+5}$

6

Solution Sets:
Equations and Inequalities

6.1 What is an Equation?

One of the important objectives in algebra is solving equations. Since there are many kinds of equations, we shall first try to gain some understanding of equations in general. Each of the following is an example of an equation.

(1) $\frac{1}{2} = \frac{4}{8}$ (2) $.5 = 50\%$
(3) $.07 = \frac{14}{100}$ (4) $4x = -28$
(5) $1.7 = 20$ (6) $\frac{1}{x} = 2 + \frac{2}{x}$
(7) $x = x + 1$ (8) $1.7 = 56$
(9) $|x| = -2$ (10) $16x + 2 = 18$
(11) $3 + x = x + 3$

Equations are mathematical sentences which can be classified into three different categories.

(1) *True* sentences: Among the above, Equations (1) and (3) are true sentences.
(2) *False* sentences: Among the above, Equations (2), (5), and (8) are false sentences.

6.2 From Word Sentences to Equations

(3) *Open* sentences: Among the above, Equations (4), (6), (7), (9), (10), and (11) are open sentences. Such sentences are neither true nor false and they can be further classified into three types.
  (a) Sentences in which at least one replacement for the variable, but not all replacements will yield a true sentence; for example, $4x = -28$. In this case -7 in place of x will yield a true sentence.
  (b) Sentences in which no replacement for the variable will yield a true sentence; for example, $x = x + 1$.
  (c) Sentences in which every replacement of the variable will yield a true sentence; for example, $3 + x = x + 3$.

In this chapter we shall be mainly concerned with the study of open sentences. We will develop a method of determining replacements for the variables which will result in true sentences. In each case we will need to know the replacement set for the variables. To simplify the matter, we agree that whenever the replacement set is not specified, it is assumed to be the set of real numbers.

6.2 From Word Sentences to Equations

When solving a problem in mathematics, it is necessary to write an equation or an inequality which fits the conditions stated in the problem. This is a very important skill in problem-solving. The examples and the exercises which follow will provide some practice in this skill. Because we are interested here only in translation, we shall not carry out any computations or simplifications.

Example 1 Sentence: 7 decreased by 2 is 4 increased by 1.

$$\underset{7-2}{7 \text{ decreased by } 2} \quad \underset{=}{\text{is}} \quad \underset{4+1}{4 \text{ increased by } 1}$$

$$7 - 2 = 4 + 1.$$

Example 2 The square of the sum of 6 and 3 is 81.

$$(6 + 3)^2 = 81.$$

Example 3 4 more than t is the product of $2t$ and 7.

$$t + 4 = 2t(7).$$

Example 4 The product of 5 and $x - 2$ is the square of x.

$$5(x - 2) = x^2.$$

Example 5 3 divided by 7 is the reciprocal of $2x$.

$$\frac{3}{7} = \frac{1}{2x}.$$

Example 6 x increased by 5% of x is 123.

$$x + .05x = 123.$$

Example 7 If a certain number is doubled and then decreased by 3, the result is 14.

Let n represent the number: n.
It is doubled: $2n$, and then decreased by 3: $2n - 3$.
The result is 14: $= 14$.
Equation: $2n - 3 = 14$.

Example 8 Mary's age is three times some number. Susan's age is three more than that number. The sum of Mary's and Susan's ages is 23.

Mary's age: $3x$.
Susan's age: $x + 3$.
The sum of Mary's and Susan's ages: $3x + (x + 3)$ is 23: $= 23$.
Equation: $3x + (x + 3) = 23$.

EXERCISE 6.1

1. Classify each sentence into one of the following categories.

   T: true sentence
   F: false sentence
   R: open sentence, in which at least one replacement of the variable, but not all, will result in a true sentence. Find the replacement or several such replacements
   N: open sentence, in which there is no replacement of the variable which will result in a true sentence
   E: open sentence, in which every replacement of the variable will result in a true sentence.

   a. $\sqrt{6.25} = 2.5$ **b.** $|-\frac{6}{2}| = 3$
   c. $.01 = .1\%$ **d.** $10^4 = 10,000$
   e. $(-1)^5 = -1$ **f.** $10x = \frac{1}{2}$
   g. $|y - y| = 0$ **h.** $a^2 = a^3$
   i. $12m + 3 = 0$ **j.** $p(p - 1) = 0$

6.2 From Word Sentences to Equations

k. $75\% = \frac{3}{4}$ l. $z^3 = -1$
m. $11n + 12n = 23$ n. $|2d| = 5$
o. $s^2 = 9$ p. $(-4)^3 = 64$
q. $\frac{2g}{g} = -2$ r. $17 = \frac{b}{2}$
s. $\frac{3}{v} = 5$ t. $w^2 = -16$
u. $x \times x = x^2$ v. $|3a - a| = |a - 3a|$

2. Translate each sentence into an equation. Do not compute or simplify.
   a. 9 decreased by 3 is 2 multiplied by 3.
   b. The cube of the sum of 1 and 2 is 27.
   c. 2 more than m is the product of m and 3.
   d. The sum of 3 and a number is 7.
   e. The sum of 4 and one-half of a number is 9.
   f. A number is subtracted from 6 and the result is -3.
   g. 5 more than two-thirds of a number is 7.
   h. 9 less than twice a number is 2.
   i. The arithmetic mean of a number and 10 is 3.
   j. The arithmetic mean of a number and 3 more than the number is 5.
   k. x is equal to the additive inverse of $2x + 5$.
   l. The absolute value of $x + 2$ is equal to the additive inverse of the reciprocal of $2x$, decreased by 1.
   m. x nickels and $x + 2$ dimes are worth \$1.25.
   n. t three-cent stamps and $5t$ four-cent stamps are worth \$2.30.
   o. Ed is x years old now. Sam is 5 years older than Ed.
      i. Two years ago, the sum of their ages was 15.
      ii. Sam is twice as old as Ed.
   p. A rectangle is x units wide. Its length is 3 more than its width.
      i. The perimeter of this rectangle is 30.
      ii. The area of this rectangle is 54.
   q. $m \angle A = x°$; $m \angle B$ is twice $m \angle A$; $m \angle C$ is 5 more than $m \angle A$. The sum of the 3 measures is 180°. ["$m \angle A$" means "the degree measure of angle A."]
   r. $m \angle T = x°$; $m \angle W = y°$.
      i. $m \angle T$ is twice $m \angle W$.   ii. $\angle T$ and $\angle W$ are supplementary.
      iii. $\angle T$ and $\angle W$ are complementary.

s. x pounds of peanuts worth $.53 per lb mixed with $50 - x$ pounds of pecans worth $.68 per lb form a mix worth $29.50.
t. Ed traveling at 47 mph for x hr goes as far as Carl who travels at 62 mph for $x - 2$ hr.
u. Buffalo Bill carried the mail for x days averaging 180 mi per day. Bill stopped at the end of x days, and then another Pony Express rider carried the mail sack for $x - 2$ days averaging 150 mi per day. The two riders carried the mail for 1350 mi.
v. The sum of three consecutive integers is 33. [*Hint:* x and $x + 1$ are consecutive integers; so are $x - 1$ and x.]
w. The sum of three consecutive odd integers is 15. [*Hint:* x and $x + 2$ may be consecutive odd integers.]
x. The sum of three consecutive even integers is 66.
y. The product of a number, and the number increased by 3, is equal to the quotient of the number, and the number decreased by 1.

6.3 Applications of Equations

Some equations are so simple that no special methods are needed to determine their solutions. We shall illustrate a common-sense way of solving such simple equations. A number which yields a true sentence when its name replaces the variable will be called the *solution* of the given equation. The set consisting of all solutions of a given equation is the *solution set* of that equation.

Example 1 Solve $3x - 11 = 7$.

We cover up $3x$ like this

$$\boxed{3x} - 11 = 7$$

and ask: What number minus 11 is equal to 7?

Answer: 18.

Therefore, $3x = 18$. Now we cover up the x:

$$3 \boxed{x} = 18$$

and ask: 3 multiplied by what number is equal to 18?

Answer: 6.

Now we replace x by 6 in the original equation to see if the result is a true statement.

6.3 Applications of Equations

Check.

$$\begin{array}{c|c} 3x - 11 & 7 \\ \hline 3 \times 6 - 11 & 7 \\ 18 - 11 & \\ 7 & \end{array}$$

Since each side of the equation is 7, we know that 6 is the correct solution. Thus, {6} is the solution set of the equation $3x - 11 = 7$.

Example 2 Solve $3(x + 4) = 9$.

We cover up $x + 4$ like this

$$3 \;\boxed{(x + 4)}\; = 9$$

and ask: 3 times what number is equal to 9?

Answer: 3.

Therefore, $x + 4 = 3$. Now we cover up the x:

$$\boxed{x} + 4 = 3$$

and ask: What number plus 4 is equal to 3?

Answer: -1.

Now we replace x by -1 in the original equation to see if the result is a true statement.

Check.

$$\begin{array}{c|c} 3(x + 4) & 9 \\ \hline 3(-1 + 4) & 9 \\ 3 \times 3 & \\ 9 & \end{array}$$

Thus, $\{-1\}$ is the solution set of $3(x + 4) = 9$.

Example 3 Solve $\dfrac{y + 3}{6} = 2$.

$$\dfrac{\boxed{y + 3}}{6} = 2.$$

Question: What number divided by 6 is equal to 2?

Answer: 12; therefore, $\boxed{y} + 3 = 12$.

Question: What number plus 3 is equal to 12?

Answer: 9.

Check.

$\dfrac{y+3}{6}$	2
$\dfrac{9+3}{6}$	2
$\dfrac{12}{6}$	
2	

Therefore, the solution set of $\dfrac{y+3}{6} = 2$ is $\{9\}$.

Example 4 Solve $5 - (2z + 3) = 4$.

$$5 - \boxed{(2z+3)} = 4.$$

Question: 5 minus what number is equal to 4?

Answer: 1; therefore, $\boxed{2z} + 3 = 1$.

Question: What number plus 3 is equal to 1?

Answer: -2; therefore, $2\,\boxed{z} = -2$.

Question 2 multiplied by what number is equal to -2?

Answer: -1.

Check.

$5 - (2z + 3)$	4
$5 - [2(-1) + 3]$	4
$5 - (-2 + 3)$	
$5 - 1$	
4	

Therefore, the solution set of $5 - (2z + 3) = 4$ is $\{-1\}$.

Frequently equations are written to fit practical problems. Solutions of these equations can then be interpreted as solutions of these practical problems. Let us consider some examples of such problems. Notice that the solutions of equations are also called *roots*.

Example 1 Mr. N thought of a number, multiplied it by 2, added 5 to the product, and obtained 3 as a result. What is the number he thought of?

Since we do not know what this number is, we shall use a letter, say x, in place of a name of this number. Now we shall write expressions corresponding to the things said about this number in the problem.

6.3 Applications of Equations

Multiplied the number by 2: $x \times 2$, or $2x$.
Added 5 to the product: $2x + 5$.
The result is 3: $2x + 5 = 3$.
Equation: $2x + 5 = 3$.

Solving this equation, we find that the root is -1. Thus, Mr. N thought of the number -1.

Check. Multiply the number by 2: $-1 \times 2 = -2$.
Add 5 to the product: $-2 + 5 = 3$.

The result is 3, and therefore, -1 *is* the correct answer.

Example 2 How long is a rectangular plot if its length is 10 ft longer than its width, and its perimeter is 52 ft?

Let w be the number of feet in the width of the rectangle. Then the number of feet in the length is $w + 10$. Let us draw a picture to help us analyze and solve the problem.

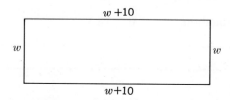

The number of feet in the perimeter is

$$w + (w + 10) + w + (w + 10), \text{ or } 4w + 20.$$

We are now ready to write an equation which fits the problem.

Equation: $4w + 20 = 52$.
Solving the equation: $4w + 20 = 52$
$4w = 32$
$w = 8$.

Check. Since the length is 10 ft longer than the width, it should be 18 ft. These dimensions are correct, since the perimeter is $8 + 18 + 8 + 18$, or 52 ft, as was stated in the problem.

It is important that you check your potential solution in the *original* problem, because the equation you have written may be wrong.

Example 3 The width of a rectangle is given by x and its length by

$2x + 3$. If the perimeter is equal to 18 in., what is the width of the rectangle?

How is the perimeter of any rectangle related to its width and its length?

Answer: It is twice the width plus twice the length.

What is the perimeter in terms of the width and the length of the rectangle given in the problem?

Answer: $2x + 2(2x + 3)$.

Simplifying: $\quad 2x + 2(2x + 3) = 2x + 4x + 6$
$$= 6x + 6.$$

What is the perimeter?

Answer: 18 in.

What is the equation for the problem?

Answer: $6x + 6 = 18$.

Solving the equation: $\quad 6x + 6 = 18$
$$6x = 12$$
$$x = 2.$$

Check. Width: 2.
Length: $2 \times 2 + 3$, or 7.
Perimeter: $2 \times 2 + 2 \times 7$, or 18.

Answer: The width is 2 in.

Example 4 One day Mr. Nickel received t nickels in change. On the following two days he received $3t + 1$ nickels. If he received \$2.65 altogether, how many nickels did he receive the first day?

What is the worth in dollars of the nickels Mr. Nickel received the first day?

Answer: $.05t$.

What is the worth in dollars of nickels Mr. Nickel received during the following two days?

Answer: $.05(3t + 1)$.

What is the equation for the problem?

$$.05t + .05(3t + 1) = 2.65.$$

6.3 Applications of Equations

Solving the equation:

$$.05t + .15t + .05 = 2.65$$
$$.20t + .05 = 2.65$$
$$.20t = 2.60$$
$$t = 13.$$

Check. 13 nickels the first day—$ .65
$3t + 1$ is $3 \times 13 + 1$, or 40 nickels—$2.00
Total $2.65

Answer: Mr. Nickel received 13 nickels the first day.

EXERCISE 6.2

1. Determine the solution set of each equation using the set of integers as the universal set.

   a. $2x - 3 = 5$
   b. $5y + 1 = 8$
   c. $2m - 1 = -3$
   d. $4z - 1 = -3$
   e. $3t + 12 = 0$
   f. $4n - 3 = -3$
   g. $2t + 4 = 4$
   h. $3 + 2t = -1$
   i. $7 - 6u = 1$
   j. $5 + 7n = 5$
   k. $2(x + 1) = 6$
   l. $3(v - 2) = -3$
   m. $4(2m - 3) = 12$
   n. $5(4s - 1) = -5$
   o. $\frac{x}{5} - 5 = -4$
   p. $\frac{2x - 3}{3} = 3$
   q. $\frac{1 + 2m}{-5} = -7$
   r. $\frac{5 - n}{-4} = -1$
   s. $\frac{5s + 1}{3} = -3$
   t. $9 - (3y + 4) = 11$
   u. $4 - (3 - z) = 0$
   v. $10 + (4 - 2t) = 8$

2. Solve each of the following problems.

   a. A number is multiplied by 3; 1 is added to the product, yielding -2. What is the number?
   b. If 2 is added to a number and the sum is multiplied by 3, the result is 0. What is the number?
   c. If 5 is subtracted from a number and the difference is multiplied by 4, the result is -12. What is the number?
   d. The sum of a number and -1 is doubled, yielding 50. What is the number?

e. The sum of a number and 1 is doubled and the result is added to three times the difference of the number and 2. The result is 61. What is the number?

f. How long is a rectangular plot if its length is 9 ft longer than its width, and its perimeter is 94 ft?

g. The width of a rectangle is equal to one-third of its length. What is the width and the length of this rectangle if its perimeter is 32 ft?

h. The difference between the length and the width of a rectangle is 11 in. What is the length and the width of the rectangle if its perimeter is 26 in.?

i. Give all possible pairs of natural numbers which can be used for the length and the width of a rectangle in which the perimeter is 18 in.

j. The length of a rectangle is five times its width. What is the length and the width of this rectangle if its perimeter is 44 ft?

k. At the end of the first week Miss D saved n dimes. At the end of the second week she saved $4n + 1$ dimes, which amounted to $6.10. How many dimes did she have at the end of the first week?

l. Mr. B had saved m quarters during one month and Mr. C had saved $5m - 40$ quarters. Their combined savings amounted to $8.00. How many dollars did each man have?

m. When washing a compact car Jim used x gal of water. When washing a larger car he used $2x + 3$ gal. If he used 35 gal on both cars, how many gallons did he use on the compact car?

n. Square A has an area of s square ft. Square B has an area of $4s + 9$ square ft. If the sum of their areas is 19 square ft, what is the area of square A?

o. Dave Track ran y yd on Monday and $3y + 880$ yd on Tuesday. He ran a total of 6.5 mi in the two days. How many miles did he run on Monday?

p. Square X has a perimeter of p in. Square Y has a perimeter of $3p + 10$ in. If the combined perimeters of the two squares are $5\frac{1}{3}$ ft, how many feet are there in the perimeter of square X?

6.4 Equations Involving Absolute Value

After an equation is written to fit the conditions of a problem, the equation needs to be solved. We have already learned to solve some simple equations. We shall now see how equations involving absolute value are solved.

6.4 Equations Involving Absolute Value

From the study of absolute value, recall that the equation

$$|a| = 3$$

has two roots, -3 and 3, because $|-3| = 3$ and $|3| = 3$. Thus, the equation $|a| = 3$ is equivalent to

$$a = 3 \quad \text{or} \quad a = -3.$$

This observation will help us in solving more complicated equations which involve absolute value.

Example 1 Solve $|z - 5| = 8$.

$$|\boxed{z - 5}| = 8.$$

Question: The absolute value of what numbers is equal to 8?

Answer: 8 and -8.

Therefore, $\qquad z - 5 = 8 \qquad$ or $\qquad z - 5 = -8$.

Thus, $\qquad\qquad z = 13 \qquad$ or $\qquad z = -3$.

Check.

$\|z - 5\|$	8
$\|13 - 5\|$	8
$\|8\|$	8
8	

$\|z - 5\|$	8
$\|-3 - 5\|$	8
$\|-8\|$	
8	

Thus, the solution set of $|z - 5| = 8$ is $\{13, -3\}$.

Example 2 Solve $|2m + 7| = 5$.

$$|\boxed{2m + 7}| = 5.$$

Question: The absolute value of what numbers is equal to 5?

Answer: 5 and -5.

Therefore, $\qquad 2m + 7 = 5 \qquad$ or $\qquad 2m + 7 = -5,$
$\qquad\qquad\qquad 2m = -2 \qquad$ or $\qquad 2m = -12,$
$\qquad\qquad\qquad m = -1 \qquad$ or $\qquad m = -6.$

Check.

$\|2m + 7\|$	5
$\|2(-1) + 7\|$	5
$\|-2 + 7\|$	
$\|5\|$	
5	

$\|2m + 7\|$	5
$\|2(-6) + 7\|$	5
$\|-12 + 7\|$	
$\|-5\|$	
5	

Thus, the solution set of $|2m + 7| = 5$ is $\{-1, -6\}$.

Example 3 Solve $3 - |2t + 1| = 3$.

$$3 - \boxed{|2t+1|} = 3.$$

Question: 3 minus what number is equal to 3?

Answer: 0; therefore, $\boxed{|2t+1|} = 0$.

Question: The absolute value of what number is equal to 0?

Answer: 0; therefore, $2t + 1 = 0$.

$$2t = -1,$$
$$t = -\tfrac{1}{2}.$$

Check.

$$\begin{array}{c|c} 3 - |2t+1| & 3 \\ \hline 3 - |2 \times (-\tfrac{1}{2}) + 1| & 3 \\ 3 - |-1 + 1| & \\ 3 - |0| & \\ 3 - 0 & \\ 3 & \end{array}$$

Thus, the solution set of $3 - |2t + 1| = 3$ is $\{-\tfrac{1}{2}\}$.

Example 4 Solve $|s + 1| - 5 = -3$.

$$\boxed{|s+1|} - 5 = -3.$$

Question: What number minus 5 is equal to -3?

Answer: 2; therefore, $|s + 1| = 2$.

$$\boxed{|s+1|} = 2.$$

Question: The absolute value of what numbers is equal to 2?

Answer: 2 and -2.

Therefore,

$$\begin{array}{lll} s + 1 = 2 & \text{or} & s + 1 = -2. \\ s = 1 & \text{or} & s = -3. \end{array}$$

Check to verify that $\{1, -3\}$ is the solution set of the equation.

Example 5 Solve $|5 - 7v| = -1$.

Note that in this equation it is stated that the absolute value of some number is equal to -1. But we know that the absolute value of any real number is a positive number or 0. Therefore, there is no real number for which the absolute value is -1. Thus, the solution set of $|5 - 7v| = -1$ is ϕ.

6.5 The Addition Properties for Equations

EXERCISE 6.3

Determine the solution sets of the following equations.

1. $|x| = 3$
2. $|x| = 12$
3. $|x| = 0$
4. $|x| = -4$
5. $|x - 1| = 0$
6. $|x - 2| = 0$
7. $|x + 3| = 0$
8. $|x + 10| = 0$
9. $|2x| = 6$
10. $|2x| = -5$
11. $|3n - 2| = 10$
12. $|\frac{1}{3}z + 6| = 15$
13. $|14t - 6| = -12$
14. $|3p - 5| = 4$
15. $|2r| = |-3|$
16. $|y| = 0$
17. $|36z| = 0$
18. $|s + 2| = 0$
19. $|3t + 3| = 0$
20. $|1 - 2u| = 0$
21. $|2b + 5| = 0$
22. $|s + 1| = \frac{1}{2}$
23. $3 - |v| = 2$
24. $|2n| + 1 = 7$
25. $|x + 1| - 2 = 9$
26. $|2y - 3| + 1 = 0$
27. $|\frac{1}{2}x + 4| - 2 = 5$
28. $\left|\frac{y - 1}{4}\right| = 8$
29. $\left|\frac{2 + z}{3}\right| = 2$
30. $\left|\frac{2 - v}{3}\right| = 2$
31. $\left|\frac{x + 1}{3}\right| - 3 = -3$
32. $1 + \left|\frac{2x - 1}{4}\right| = 2$

6.5 The Addition Properties for Equations

The common-sense approach served us well in solving the kinds of equations encountered so far. But consider the equation

$$2x + 6 = 3(x + 1) - 3.$$

This equation involves a variable on the left and on the right of the equality sign. It is not as easy to solve as the previous equations were.

We need to have a more powerful method to handle equations of this kind. Such a method is based on equation properties which we shall consider next.

The first property is suggested by the following arithmetic example:

$$\text{If } 5 = 2 + 3, \text{ then } 5 + 6 = (2 + 3) + 6.$$

We state the general pattern suggested by this example and give it a name.

Solution Sets: Equations and Inequalities

RIGHT-HAND ADDITION PROPERTY FOR EQUATIONS If $x = y$, then $x + z = y + z$.

Using commutativity, we obtain the Left-hand Addition Property.

LEFT-HAND ADDITION PROPERTY FOR EQUATIONS If $x = y$, then $z + x = z + y$.

In essence, these properties tell us that if we add the same number to each member of an equation, then the equation we obtain is equivalent to the original equation. And, of course, two equivalent equations have the same roots. For example, if $x - 7 = 12$,

then $\quad\quad x - 7 + 7 = 12 + 7,\quad$ [Add 7 to each member.]
$\quad\quad\quad\quad\quad x = 19.$

Thus, the solution set of $x - 7 = 12$ is $\{19\}$, which is easily verified.

We have seen that subtraction is the inverse of addition. Therefore, the addition property for equations can be extended to a subtraction property for equations (right-hand only).

SUBTRACTION PROPERTY FOR EQUATIONS If $x = y$, then $x - z = y - z$.

For example, if $x + 3 = 12$,

then $\quad x + 3 - 3 = 12 - 3,\quad$ [Subtract 3 from each member.]
$\quad\quad\quad\quad x = 9.$

It is easy to verify that the solution set of $x + 3 = 12$ is $\{9\}$.

The two equations above are very simple and could have been solved without the use of the Addition Property for Equations. We used them to illustrate the use of this property which will prove a "life-saver" when we are confronted by more complicated equations. We shall first practice, however, the use of these properties in simple equations. The examples below are for your study of the use of equation properties.

Example 1 $\quad\quad x - 2 = -13$
$\quad\quad\quad\quad\quad x - 2 + 2 = -13 + 2\quad$ [Add 2 to each member.]
$\quad\quad\quad\quad\quad\quad\quad x = -11.$

Check.
$$\begin{array}{c|c} x - 2 & -13 \\ \hline -11 - 2 & -13 \\ -13 & \end{array}$$

The solution set of $x - 2 = -13$ is $\{-11\}$.

6.5 The Addition Properties for Equations

Example 2
$$3 - y = 17$$
$$3 - y + y = 17 + y \quad \text{[Add } y \text{ to each member.]}$$
$$3 = 17 + y$$
$$3 - 17 = 17 + y - 17 \quad \text{[Subtract 17 from each member.]}$$
$$-14 = y.$$

Check.

$3 - y$	17
$3 - (-14)$	17
$3 + 14$	
17	

The solution set of $3 - y = 17$ is $\{-14\}$.

Example 3
$$m - 1.7 = -6.3$$
$$m - 1.7 + 1.7 = -6.3 + 1.7$$
$$m = -4.6.$$

Check.

$m - 1.7$	-6.3
$-4.6 - 1.7$	-6.3
-6.3	

The solution set of $m - 1.7 = -6.3$ is $\{-4.6\}$.

Example 4
$$12 + n = 14.5$$
$$n + 12 = 14.5$$
$$n + 12 - 12 = 14.5 - 12$$
$$n = 2.5.$$

Check.

$12 + n$	14.5
$12 + 2.5$	14.5
14.5	

The solution set of $12 + n = 14.5$ is $\{2.5\}$.

Example 5
$$1.5 = 3.2 + s$$
$$1.5 = s + 3.2$$
$$1.5 - 3.2 = s + 3.2 - 3.2$$
$$-1.7 = s.$$

Check to verify that $\{-1.7\}$ is the solution set of $1.5 = 3.2 + s$.

EXERCISE 6.4

1. Below are given pairs of equations. The second equation is obtained from the first by using one of the properties introduced in this section. In

each case, tell what was done to each member of the first equation to obtain the second equation equivalent to it.

a. $x + 3 = 4;$ $\quad x = 1$
b. $x + 8 = 2;$ $\quad x = -6$
c. $x - 4 = 7;$ $\quad x = 11$
d. $x - 15 = 0;$ $\quad x = 15$
e. $x - 5 = -3;$ $\quad x = 2$
f. $x - 6 = 20;$ $\quad x = 26$
g. $x + 9 = 0;$ $\quad x = -9$
h. $2x + 3 = 7;$ $\quad 2x = 4$
i. $3x + 12 = 4;$ $\quad 3x = -8$
j. $4x - 5 = 25;$ $\quad 4x = 30$

2. Use the appropriate equation properties to find the solution set of each equation.

a. $y + 6 = 21$
b. $a + 5 = 17$
c. $c + \frac{1}{3} = \frac{1}{2}$
d. $.7 + x = 0$
e. $d - 3 = 9$
f. $z - 7 = 1$
g. $b - 1 = -3$
h. $m - 15 = -8$
i. $r - \frac{1}{3} = \frac{1}{2}$
j. $s - 16 = 0$
k. $f - \frac{2}{3} = -\frac{1}{3}$
l. $t + \frac{3}{11} = \frac{1}{11}$
m. $u + \frac{4}{7} = -\frac{3}{7}$
n. $w - \frac{3}{2} = -\frac{5}{7}$
o. $-7 + r = -8$
p. $-11 = x + 7$
q. $s - 3.1 = 4.7$
r. $x - (-11) = 14$
s. $(-3.5) - y = 7.5$
t. $14.1 = z - 7.9$
u. $x - \frac{4}{3} = -\frac{5}{3}$
v. $t - 2.2 = -5.3$
w. $-4\frac{1}{4} = s - \frac{3}{4}$
x. $1.4 - x = 7.2$
y. $w + \frac{4}{9} = \frac{1}{9}$
z. $-19 + v = -41.7$
a'. $-4.3 = 7.2 + k$
b'. $y - \frac{5}{3} = -\frac{4}{9}$
c'. $-1.6 - x = 14.1$
d'. $-7.9 = 1.3 - y$
e'. $4\frac{7}{8} + x = -17\frac{1}{8}$
f'. $-\frac{7}{5} + z = \frac{4}{3}$
g'. $v - \frac{5}{3} = -\frac{9}{4}$
h'. $-5.9 = x - 3.4$
i'. $m - 3\frac{1}{2} = 6\frac{3}{4}$
j'. $-x + 7 = 19$
k'. $-4 - y = 13$
l'. $t + \frac{7}{8} = \frac{1}{4}$
m'. $-y + \frac{4}{3} = -\frac{7}{5}$
n'. $-\frac{7}{5} = \frac{7}{10} - z$
o'. $\frac{5}{11} - x = \frac{1}{11}$
p'. $6.23 + y = -7.96$
q'. $6\frac{3}{8} = z - 3\frac{1}{4}$
r'. $-21.8 = x + 7.3$

6.6 The Multiplication Properties for Equations

We now state a fourth equation property, which we call the Right-hand Multiplication Property.

6.6 The Multiplication Properties for Equations

RIGHT-HAND MULTIPLICATION PROPERTY FOR EQUATIONS If $x = y$, then $xz = yz$.

We obtain Left-hand Multiplication Property by using commutativity.

LEFT-HAND MULTIPLICATION PROPERTY FOR EQUATIONS If $x = y$, then $zx = zy$.

These properties tell us that if we multiply each member of an equation by the same number, then the equation we obtain is equivalent to the original equation.

Example
$$\frac{1}{3}x = \frac{1}{2}$$
$$3(\frac{1}{3}x) = 3 \times \frac{1}{2} \quad \text{[Multiply each member by 3.]}$$
$$(3 \times \frac{1}{3})x = \frac{3}{2}$$
$$1 \times x = \frac{3}{2}$$
$$x = \frac{3}{2}.$$

Check.

$\frac{1}{3}x$	$\frac{1}{2}$
$\frac{1}{3} \times \frac{3}{2}$	$\frac{1}{2}$
$\frac{1}{2}$	

Thus, $\{\frac{3}{2}\}$ is the solution set of $\frac{1}{3}x = \frac{1}{2}$.

To extend the Right-hand Multiplication Property to division, we recall the definition of division stated in Chapter 3.

$$\frac{x}{y} = x \times \frac{1}{y}.$$

Applying the Right-hand Multiplication Property to $x = y$ and multiplying each number by $\frac{1}{z}$, $z \neq 0$, we obtain

$$x \times \frac{1}{z} = y \times \frac{1}{z}$$

$$\frac{x}{z} = \frac{y}{z}.$$

Thus, we have the following property:

DIVISION PROPERTY FOR EQUATIONS If $x = y$, then $\frac{x}{z} = \frac{y}{z}$ [$z \neq 0$].

Here is an example of the use of the Division Property.
If $5x = -3$, then

$$\frac{5x}{5} = \frac{-3}{5} \quad \text{[Divide each member by 5.]}$$
$$x = -\frac{3}{5},$$

and the solution set of $5x = -3$ is $\left\{-\frac{3}{5}\right\}$.

Now let us take another look at division in its relation to multiplication. Suppose we are given that

$$\frac{x}{y} = z \quad [y \neq 0].$$

Applying the Right-hand Multiplication Property and multiplying each member of the equation by y, we obtain

$$\frac{x}{y} \times y = z \times y$$
$$\left(x \times \frac{1}{y}\right)y = z \times y$$
$$x \times \left(\frac{1}{y} \times y\right) = z \times y$$
$$x \times 1 = z \times y$$
$$x = z \times y.$$

Thus, we have proved the following theorem.

THEOREM 6.1 If $\frac{x}{y} = z$, then $x = z \times y$ $\quad [y \neq 0]$.

From Theorem 6.1 we see that division is the inverse of multiplication; that is, division "undoes" what multiplication "does." This can be seen from the following development.

Since $z \times y = x$, we can replace x by $z \times y$ in $\frac{x}{y} = z$. This gives

$$\frac{z \times y}{y} = z \quad \text{or} \quad z \times y \div y = z.$$

Thus starting with z, multiplying it by y, then dividing the product by y brings us back to z.

To see how the multiplication and division properties are used in solving equations, study the following examples.

6.6 The Multiplication Properties for Equations

Example 1
$$\tfrac{1}{3}y = -2$$
$$3(\tfrac{1}{3}y) = 3(-2) \quad \text{[Multiply each member by 3.]}$$
$$y = -6.$$

Thus, $\{-6\}$ is the solution set of $\tfrac{1}{3}y = -2$.

Example 2
$$\tfrac{2}{3}u = \tfrac{1}{2}$$
$$\tfrac{3}{2}(\tfrac{2}{3}u) = \tfrac{3}{2} \times \tfrac{1}{2} \quad \text{[Multiply each member by 3.]}$$
$$u = \tfrac{3}{4}.$$

Thus, $\{\tfrac{3}{4}\}$ is the solution set of $\tfrac{2}{3}u = \tfrac{1}{2}$.

We can also solve this equation using division.

$$\tfrac{2}{3}u = \tfrac{1}{2}$$
$$\frac{\tfrac{2}{3}u}{\tfrac{2}{3}} = \frac{\tfrac{1}{2}}{\tfrac{2}{3}} \quad \text{[Divide each member by $\tfrac{2}{3}$.]}$$
$$u = \tfrac{3}{4}.$$

Again we obtain $\{\tfrac{3}{4}\}$ for the solution set.

Example 3
$$-2(3x - 2) = 2$$
$$-6x + 4 = 2$$
$$-6x + 4 - 4 = 2 - 4$$
$$-6x = -2$$
$$\frac{-6x}{-6} = \frac{-2}{-6}$$
$$x = \tfrac{1}{3}.$$

Thus, $\{\tfrac{1}{3}\}$ is the solution set of $-2(3x - 2) = 2$.

EXERCISE 6.5

1. Given below are pairs of equations. The second equation is obtained from the first by using one of the properties. In each case, tell what was done to each member of the first equation to obtain the second equation equivalent to it.

 a. $2x = 8;\quad x = 4$
 b. $-3x = 12;\quad x = -4$
 c. $-5x = -30;\quad x = 6$
 d. $\tfrac{1}{2}x = 5;\quad x = 10$
 e. $-\tfrac{1}{3}x = -2;\quad x = 6$
 f. $.2x = .7;\quad 2x = 7$
 g. $-1.5x = 4.5;\quad 15x = -45$
 h. $-3.6x = -1.2;\quad 36x = 12$

2. Use the appropriate equation properties to find the solution set of each equation.

a. $\frac{1}{2}x = 2$
b. $\frac{1}{3}y = -1$
c. $\frac{1}{4}t = -\frac{1}{2}$
d. $-\frac{1}{2}u = 3$
e. $-\frac{1}{5}s = -1$
f. $4s = 36$
g. $3a = -27$
h. $16x = -48$
i. $-96y = -16$
j. $100z = 1$
k. $10p = -2$
l. $-\frac{10}{3}h = -\frac{1}{3}$
m. $-70 = 3.5r$
n. $\frac{5}{7}x = -\frac{4}{7}$
o. $-4.7y = 9.4$
p. $.7y = 8.4$
q. $-\frac{2}{3}t = \frac{2}{3}$
r. $\frac{1}{4}u = -\frac{3}{2}$
s. $67r = -67$
t. $\frac{9}{2}c = \frac{1}{9}$
u. $-\frac{1}{3}r = \frac{1}{3}$
v. $-\frac{1}{5} = \frac{1}{7}y$
w. $-\frac{2}{3}x = -\frac{2}{3}$
x. $\frac{3}{4}r = \frac{7}{4}$
y. $-\frac{5}{2} = 10y$
z. $\frac{2}{3} = \frac{w}{7}$

a'. $-4.2x = 2.1$
b'. $2\frac{1}{3}y = -\frac{7}{8}$
c'. $-3.9 = \frac{1.3}{x}$
d'. $-2.1 = .7r$
e'. $-\frac{4}{9} = -\frac{1}{3}x$
f'. $-\frac{7}{12}y = -\frac{5}{6}$
g'. $\frac{y}{2.5} = 1.2$
h'. $.36 = 3.6x$

i'. $-10 = -\frac{r}{1.9}$
j'. $\frac{6}{7}t = -1.2$
k'. $1.2 = -3.6w$
l'. $-\frac{5}{7}v = -\frac{3}{14}$
m'. $-\frac{9}{8} = -\frac{1}{4}x$
n'. $-\frac{3}{19}x = \frac{3}{38}$
o'. $\frac{8}{13}y = -\frac{1}{26}$
p'. $-1.3t = -3.9$
q'. $3x + 5 = 2$
r'. $1 - \frac{1}{2}x = -1$
s'. $5 = 2x + 6$
t'. $2 = 1 - 5x$
u'. $3.1x + 2.3 = 8.5$
v'. $10.5 - 2.3x = 3.6$
w'. $2(x + 1) = 8$
x'. $-3(x - 1) = 0$
y'. $3(2x + 5) = 3$
z'. $-5(3x - 1) = 0$
a''. $7 = -7(2 - 3x)$
b''. $-2 = -\frac{1}{2}(4 + 5x)$
c''. $2(x + 1) = 3(2x - 1) + 13$
d''. $-3(-x + 1) + 1 = 4 - (2x + 6)$
e''. $2 - 2(3 - 2x) = 5(x + 4) - 20$
f''. $3(2x + 1) - (4x - 1) = 5(1 - 4x) - 12$
g''. $-2(6x + 2) + 3(1 - 3x) = 5(2 - 3x) - (9x + 10)$

6.7 Using All Equation Properties

We shall now illustrate the use of the equation properties in solving equations. We shall also use other properties which were established earlier.

Example 1
$$2x + 1 = x - 3$$
$$2x + 1 - 1 = x - 3 - 1 \quad \text{[Subtract 1 from each member.]}$$
$$2x = x - 4$$
$$2x - x = x - 4 - x \quad \text{[Subtract } x \text{ from each member.]}$$
$$x = -4.$$

Check.

$2x + 1$	$x - 3$
$2(-4) + 1$	$-4 - 3$
$-8 + 1$	-7
-7	

Thus, $\{-4\}$ is the solution set of $2x + 1 = x - 3$.

Example 2
$$3(y + 2) = 2(2y - 1)$$
$$3y + 6 = 4y - 2$$
$$3y + 6 + 2 = 4y - 2 + 2 \quad \text{[Add 2 to each member.]}$$
$$3y + 8 = 4y$$
$$3y + 8 - 3y = 4y - 3y \quad \text{[Subtract } 3y \text{ from each member.]}$$
$$8 = y.$$

Thus, $\{8\}$ is the solution set of the given equation.

Example 3
$$5a + 1 - 3a = 4a - 2$$
$$5a - 3a + 1 = 4a - 2$$
$$2a + 1 = 4a - 2$$
$$2a + 1 - 2a = 4a - 2 - 2a$$
$$2a - 2a + 1 = 4a - 2a - 2$$
$$1 = 2a - 2$$
$$1 + 2 = 2a - 2 + 2$$
$$3 = 2a$$
$$\tfrac{1}{2} \times 3 = \tfrac{1}{2} \times 2a$$
$$\tfrac{3}{2} = a.$$

Thus, $\{\tfrac{3}{2}\}$ is the solution set of the given equation.

Example 4
$$2(2m - 1) + 3 = 4(m + 4) - 2$$
$$4m - 2 + 3 = 4m + 16 - 2$$
$$4m + 1 = 4m + 14$$
$$4m + 1 - 4m = 4m + 14 - 4m$$
$$1 = 14.$$

We see that the given equation leads to a false statement, no matter what the replacement for m is. Thus, the solution set of

$$2(2m - 1) + 3 = 4(m + 4) - 2$$

is ϕ.

Example 5 $\quad \frac{1}{2}x + \frac{2}{3}x = -7$
$\quad\quad\quad\quad\quad\quad \frac{3}{6}x + \frac{4}{6}x = -7$
$\quad\quad\quad\quad\quad\quad (\frac{3}{6} + \frac{4}{6})x = -7$
$\quad\quad\quad\quad\quad\quad \frac{7}{6}x = -7$
$\quad\quad\quad\quad\quad\quad \dfrac{\frac{7}{6}x}{\frac{7}{6}} = \dfrac{-7}{\frac{7}{6}}$
$\quad\quad\quad\quad\quad\quad x = -6.$

Thus, $\{-6\}$ is the solution set of $\frac{1}{2}x + \frac{2}{3}x = -7$.

6.8 Writing Equivalent Expressions

We have already had some experience in writing an expression equivalent to a given expression. For example, $3x - 5x + 1.5 + 4x$ and $2x + 1.5$ are two equivalent expressions, because if x is replaced throughout by the name of any one number, each expression will name the same number. That this is so can be easily proved.

$$\begin{aligned} 3x - 5x + 1.5 + 4x &= 3x - 5x + 4x + 1.5 \\ &= (3 - 5 + 4)x + 1.5 \\ &= 2x + 1.5. \end{aligned}$$

In the above the Term Rearrangement Property and the Distributivity are the key properties. For many purposes, $2x + 1.5$ is a more convenient expression than $3x - 5x + 1.5 + 4x$.

In solving equations, we frequently found it convenient to write, in place of expressions containing parentheses, equivalent expressions which do not contain parentheses. Let us consider, for example, $2(x+3)$. An expression which is equivalent to this, and which does not contain parentheses, is $2x + 6$. $2x + 6$ is obtained from $2(x + 3)$ by applying the Left Distributive Property of Multiplication over Addition.

Example 1 Write an expression equivalent to $-3(y + 7)$.

$$\begin{aligned} -3(y + 7) &= -3y + (-3)7 \\ &= -3y + (-21) \\ &= -3y - 21. \end{aligned}$$

6.8 Writing Equivalent Expressions

Thus, $-3(y+7)$ and $-3y-21$ are equivalent expressions. One contains parentheses, the other does not.

Example 2 Write an expression equivalent to $-2(m-n-5)$.
$$-2(m-n-5) = -2[m+(-n)+(-5)]$$
$$= -2m+(-2)(-n)+(-2)(-5)$$
$$= -2m+2n+10.$$

Thus, $-2(m-n-5)$ and $-2m+2n+10$ are equivalent expressions.

Example 3 Write an expression equivalent to $(5c-3d)(-\tfrac{1}{2})$.
$$(5c-3d)(-\tfrac{1}{2}) = (5c)(-\tfrac{1}{2}) - (3d)(-\tfrac{1}{2})$$
$$= -\tfrac{5}{2}c + \tfrac{3}{2}d.$$

Thus, $(5c-3d)(-\tfrac{1}{2})$ and $-\tfrac{5}{2}c + \tfrac{3}{2}d$ are equivalent expressions.

You should by now be able to skip the intermediate steps when writing equivalent expressions. In the following examples some steps are omitted. Can you see how the remaining steps are obtained? If not, fill in the missing steps using the appropriate properties and theorems.

Example 1 $\tfrac{1}{3}(a-3b+6c) = \tfrac{1}{3}a - b + 2c.$

Example 2 $(x-y+2z)(-3) = -3x + 3y - 6z.$

EXERCISE 6.6

1. For each expression, write an expression which is equivalent to it and which does not contain parentheses.

a. $3(m+2)$
b. $-1(2x+5)$
c. $-\tfrac{3}{4}(6+3d)$
d. $(x+y)(-5)$
e. $12(a-4)$
f. $-3(b-7)$
g. $9(m-n)$
h. $-\tfrac{2}{5}(3k-5s)$
i. $(1.5u-3.7)5$
j. $4(a+b-c)$
k. $-3(s-t+11)$
l. $(-2x-3y-4z)(-2)$
m. $(u+v+w)a$
n. $-b(2a+3c-4d)$
o. $-\tfrac{3}{4}k(-m-p-r)$
p. $\tfrac{7}{8}(16x+\tfrac{8}{7}y)$
q. $-\tfrac{3}{4}(r-2s+3t)$
r. $(3.5+v)(-6)$
s. $-3(w+3t)$
t. $\tfrac{1}{4}(\tfrac{2}{3}y - \tfrac{3}{4}x)$

u. $(.07 - 1.3r)(-.2)$
v. $-5(x - 2y + 3z)$
w. $(w + 7v - 9t)(-3)$
x. $\frac{3}{4}(3x - \frac{4}{5}y - 6z)$
y. $-\frac{7}{8}(13 - x)$
z. $-\frac{4}{9}(3x - 2w)$
a'. $-7(3x - 4y - 7z)$
b'. $(4w - 7r + 6t)(-\frac{3}{7})$

2. For each expression, write an expression which is equivalent to it and which does not contain parentheses. Simplify wherever possible.

a. $-2(a - b + c) + 3(a + b + 2c)$
b. $-(2m - 3n + t) - 2(m + n - 2t)$
c. $\frac{1}{3}(6u + 3v - 12s) + \frac{2}{3}(12v - 3u)$
d. $-\frac{3}{4}(8a + 12b - 16c) - \frac{1}{4}(-4a + 20c)$
e. $-0.1(10x - 20y + 100z) + 0.5(50x + 200y)$
f. $3[(2x + 1) + (4x - 1)]$
g. $-2[(1 - 4x) - (a + b)]$
h. $[(x + y) - (2x + z)](-5)$
i. $[(2m - 3w) + (-2p - r)](-4)$
j. $4[(a + b + c) - (2a - 2b - 3c)]$
k. $-7[(m - 2n - 3) - (4n - 5 - 2m)]$
l. $-2[(2x + y - 4) - (3x - 2y - 6)]$

3. Determine the solution set of each equation.

a. $6c - 3c = 9$
b. $5t - 7t = -12$
c. $x - 2x = -\frac{1}{2}$
d. $\frac{1}{3}r + \frac{1}{2}r = 5$
e. $1.7d - .3d = -4.2$
f. $4s - \frac{1}{2}s = 2s + 3$
g. $3 + 4z = z - 4$
h. $\frac{1}{4}m - \frac{3}{4}m + \frac{1}{2} = 2m - \frac{1}{2}$
i. $3x - 2 + x = x - 4x$
j. $3.6b - 1.7b = 2b - .5$
k. $\frac{1}{2} - z = 2z + \frac{3}{4}$
l. $8a - .73 = .36 + 7a$
m. $-19 = \frac{3}{2}x - 13$
n. $.7r - 1.2 = 4.7r$
o. $\frac{x}{-5} + 7 = -11$
p. $\frac{4}{3}x - 7 = 6 + \frac{7}{3}x$
q. $7x + 13 = 13x - 16$
r. $7 - 12x = 7x + 17$
s. $\frac{4}{3}x + 5 = \frac{7}{9}x + 7$
t. $\frac{8}{9} + 3w = w + \frac{2}{9}$
u. $\frac{7}{5}s = \frac{3}{5} + 2s$
v. $\frac{5}{3}x + 13 - \frac{7}{3}x = 15$
w. $2(7x + 3) = -22$
x. $(x - 4) \times (-3) = 21$
y. $(4x + 3) \times (-2) = 0$
z. $-5(1 - x) = -5$
a'. $-7(2 - 3x) = 0$
b'. $\frac{1}{2}(x - \frac{1}{2}) = -2$
c'. $-\frac{1}{3}(\frac{1}{6} - 2x) = \frac{1}{6}$
d'. $.3(2x + .1) = -.03$

e′. $2(3x + 1) + 3(2x - 1) = 11$ **f′.** $3(2x - 5) + 5(3x - 4) = -56$
g′. $-2(3 - x) - 6(2 - x) = 50$ **h′.** $-(6 - 3x) - (4 - 5x) = -10$
i′. $2(3x + 1) = -4(2x - 1)$ **j′.** $-4(1 - x) = 3(3 - 2x)$
k′. $5(2 - x) + 1 = -3(x + 2) - 1$ **l′.** $3 - (x + 3) = -1 + (x - 2)$
m′. $4 - (2x + 1) = -3 - (2 - 5x)$ **n′.** $2 - 4(x + 1) = -7(1 - x) - 1$

6.9 Simple Inequalities

So far we have concentrated on the study of equations. Let us now turn our attention to inequalities.

You will recall that an open sentence such as $x > 5$ is called an *inequality*. It is read "x is greater than 5." The sentence "$5 < x$" has the same meaning as the previous sentence. It is read "5 is less than x."

The sentence "$x \not< 5$" is read "x is not less than 5," which, of course, means that x is greater than or equal to 5.

Inequalities, like equations, have *solution sets*. Solution sets of inequalities are sets of those numbers which satisfy the inequalities. For example, in the system of real numbers, the solution set of the inequality $x > 5$ is the set of all real numbers greater than 5. If we call this set A, we write

$$A = \{x | x > 5\}.$$

We have learned to establish a one-to-one correspondence between numbers and points on a number line. Using the number line, we can locate points which correspond to the numbers which make up the solution set $A = \{x | x > 5\}$. On the graph in Figure 6.1 set A is pictured by the part

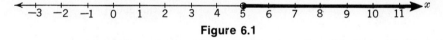

Figure 6.1

of the number line marked with the heavy line. Notice how we have shown that the point corresponding to the number 5 does not belong to the set. To include 5 in the set, we would write $\{x | x \geq 5\}$, which is read "the set of all x such that x is greater than *or* equal to 5." This set can be viewed as the union of the two sets:

$$\{x | x > 5\} \cup \{x | x = 5\}.$$

Its graph is shown in Figure 6.2. Notice that to show that 5 belongs to the set we have a solid dot at 5.

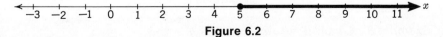

Figure 6.2

Here is another inequality and its graph:

$$-5 \leq m < 3.$$

It is read "negative five is less than or equal to *m and m* is less than three." It can be written as $B = \{m| -5 \leq m < 3\}$ and viewed as the intersection of two sets

$$\{m|m \geq -5\} \cap \{m|m < 3\}.$$

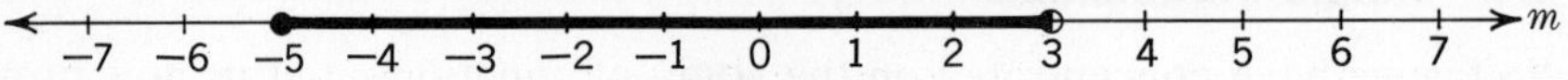

```
◄──┼───┼───●───┼───┼───┼───┼───○───┼───┼───┼──► m
   -7  -6  -5  -4  -3  -2  -1   0   1   2   3   4   5   6   7
```
Figure 6.3

The graph of this set (Figure 6.3) is a set of points making up a line segment with one endpoint missing. Notice that one endpoint belongs to the line segment, namely the point with the coordinate -5. The other endpoint of the segment, namely the point corresponding to 3, does not belong to the segment. We say that the graph of the solution set of $-5 \leq m < 3$ is a *semiopen* (or *semiclosed*) *segment*.

Frequently we do not choose the set of real numbers R as the replacement set. The replacement set may be the set of natural numbers N, the set of integers I, or the set of rational numbers Q.

Example 1 $T = \{x|x \in I \text{ and } -2 < x < 3\}$ is read

"the set of elements x such that x is an integer, and -2 is less than x and x is less than 3."

If we graph on the number line the points which we associate with these numbers, we have Figure 6.4. Note that for integers, the "number line" consists of points which are not connected.

```
◄── •   •   •   ◻   ◻   ◻   ◻   •   •  ──► x
    -4  -3  -2  -1   0   1   2   3   4
```
Figure 6.4

Example 2 If the set $Y = \{w|-2 < w < 3\}$ is graphed on the number line, we have the graph shown in Figure 6.5. A graph such as this is called an *open segment*. Remember that if we do not specify the replacement set, it is assumed to be the set of real numbers.

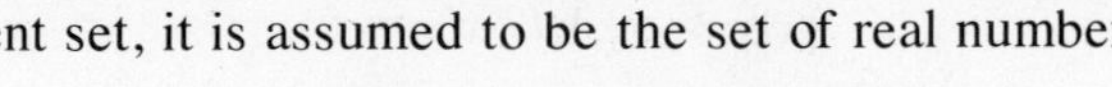

```
◄───┼───┼───○───┼───┼───┼───┼───○───┼──► w
   -4  -3  -2  -1   0   1   2   3   4
```
Figure 6.5

Example 3 Graph $B = \{a|a < -1 \text{ and } a > \frac{1}{2}\}$.

6.9 Simple Inequalities

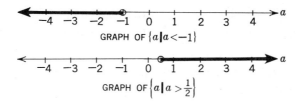

We observe that the graphs of $\{a|a < -1\}$ and $\{a|a > \frac{1}{2}\}$ have no points in common; therefore, the solution set is the empty set. Thus

$$B = \{a|a < -1 \text{ and } a > \tfrac{1}{2}\} = \phi.$$

This can also be written as

$$B = \{a|a < -1\} \cap \{a|a > \tfrac{1}{2}\} = \phi.$$

Example 4 Graph $A = \{s|s \geq 3 \text{ and } s < 4\tfrac{1}{2}\}$.

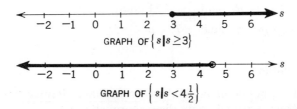

We see that there are points common to both parts of the graph. The graph consists of all points to the right of 3 (including 3) and to the left of $4\tfrac{1}{2}$ (not including $4\tfrac{1}{2}$).

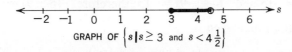

Note that the graph is a semiclosed (or semiopen) segment.

EXERCISE 6.7

1. Graph each sentence in the set of real numbers.
   a. $B = \{p|p > -3\}$
   b. $A = \{a|a < 0\}$
   c. $M = \{s|s \geq 3\}$
   d. $X = \{b|b \leq -1\}$
   e. $Y = \{t|-2 < t < 3\}$
   f. $C = \{r|-\tfrac{1}{2} \leq r < \tfrac{1}{2}\}$
   g. $T = \{u|1 \leq u \leq 2.3\}$
   h. $V = \{w|w \neq 0\}$

i. $G = \{c|c \neq -2\}$ j. $W = \{m|m < 5 \text{ or } m > 7\}$
k. $J = \{x|-1 \leq x < 1\}$ l. $S = \{d|-1 \leq d \leq 1\}$
m. $A = \{x|x > 5 \text{ and } x < -1\}$ n. $M = \{a|a \leq -1\} \cap \{a|a \geq -1\}$
o. $P = \{n|n > -5 \text{ or } n < -5\}$ p. $E = \{c|c \leq 6 \text{ or } c \geq 6\}$
q. $X = \{s|\tfrac{1}{2} < s < 0\}$ r. $C = \{g|-1 < g < -2\}$
s. $D = \{r|r < 2 \text{ or } r < -3\}$

2. Graph the solution set using the replacement set indicated in each case.
   a. $T = \{s|s \in I \text{ and } s > 9 \text{ or } s < 13\}$
   b. $G = \{x|x \in N \text{ and } 6 < x < 9\}$
   c. $H = \{r|r \in R \text{ and } r < 3.5 \text{ and } r > 2.5\}$
   d. $J = \{t|t \in I \text{ and } -3 \leq t \leq 5\}$
   e. $C = \{y|y \in R \text{ and } -2.5 < y < 1 \text{ or } y > 5\}$
   f. $K = \{z|z \in N \text{ and } z \not> 8\}$
   g. $L = \{x|x \in R \text{ and } x > 2\tfrac{1}{2} \text{ and } x \neq 4\}$
   h. $M = \{p|p \in I \text{ and } -7 < p < -5\}$

6.10 Binomials

A binomial is a polynomial with two terms. Since binomials occur in algebra quite often, we shall study them thoroughly.

By carrying out the operations in the examples below, we discover a pattern involving the additive inverse of a binomial.

$-(2 + 5) = -7$ and $-2 + (-5) = -7$; therefore, $-(2 + 5) = -2 + (-5)$.
$-(4 + 1) = -5$ and $-4 + (-1) = -5$; therefore, $-(4 + 1) = -4 + (-1)$.
$-(6 + 9) = -15$ and $-6 + (-9) = -15$; therefore, $-(6 + 9) = -6 + (-9)$.

The pattern displayed is

$$-(x + y) = -x + (-y).$$

We now prove this to be true for all real numbers.

THEOREM 6.2   *Additive Inverse of a Sum* $-(x + y) = -x + (-y)$.

Proof. $-(x + y) = -1 \times (x + y)$ Theorem 5.5
           $= -1 \times x + (-1) \times y$ LDPMA
           $= -x + (-y).$ Theorem 6.2

Here are some examples that suggest another pattern.

$-(7 - 5) = -2$ and $5 - 7 = -2$; therefore, $-(7 - 5) = 5 - 7$.

$-(5 - 2) = -3$ and $2 - 5 = -3$; therefore, $-(5 - 2) = 2 - 5$.
$-(3 - 7) = 4$ and $7 - 3 = 4$; therefore, $-(3 - 7) = 7 - 3$.
$-(1 - 8) = 7$ and $8 - 1 = 7$; therefore, $-(1 - 8) = 8 - 1$.

The pattern displayed is

$$-(x - y) = y - x.$$

We will now prove that this is true for all replacements of x and y.

THEOREM 6.3 *Additive Inverse of a Difference* $-(x - y) = y - x$.

Proof.
$$\begin{aligned}
-(x - y) &= -[x + (-y)] & &\text{Def. subtr.} \\
&= -x + [-(-y)] & &\text{Theorem 6.2} \\
&= -x + y & &\text{Theorem 5.6} \\
&= y + (-x) & &\text{CPA} \\
&= y - x. & &\text{Def. subtr.}
\end{aligned}$$

6.11 Simplification of Expressions

We have already learned how to simplify a variety of expressions. This skill is helpful in solving equations, since expressions are easier to deal with when simplified. For example, when asked to solve the following equation, we find that it is helpful to replace $3(x + 1)$ by its equivalent, $3x + 3$.

$$\begin{aligned}
3(x + 1) - 2 &= x \\
3x + 3 - 2 &= x \\
3x + 1 &= x \\
3x + 1 - x &= x - x \\
2x + 1 &= 0 \\
2x + 1 - 1 &= 0 - 1 \\
2x &= -1 \\
x &= -\tfrac{1}{2}.
\end{aligned}$$

Thus, $\{\tfrac{1}{2}\}$ is the solution set of $3(x + 1) - 2 = x$.

Simplification of expressions, as you have probably noticed, is accomplished by applying theorems and properties which have been established. Study the examples below and be prepared to supply a reason for each step.

Example 1   Simplify $-3(7a)$.

$$\begin{aligned}
-3(7a) &= (-3 \times 7)a \\
&= -21a.
\end{aligned}$$

Thus, $-3(7a) = -21a$.

Example 2 Simplify $2(-10x)$.

$$2(-10x) = [2(-10)]x$$
$$= -20x.$$

Thus, $2(-10x) = -20x$.

Example 3 Simplify $(-4a)(-5b)$.

$$(-4a)(-5b) = (-4)(-5)ab$$
$$= 20ab.$$

Thus, $(-4a)(-5b) = 20ab$.

Example 4 Simplify $4m - m$.

$$4m - m = 4m - 1m$$
$$= (4 - 1)m$$
$$= 3m.$$

Thus, $4m - m = 3m$.

Example 5 Simplify $8n - 4n - 2n$.

$$8n - 4n - 2n = (8 - 4 - 2)n$$
$$= 2n.$$

Thus, $8n - 4n - 2n = 2n$.

Example 6 Simplify $3t - 12t - 5t$.

$$3t - 12t - 5t = (3 - 12 - 5)t$$
$$= -14t.$$

Thus, $3t - 12t - 5t = -14t$.

Example 7 Simplify $9x + 6y - 3x - 4y$.

$$9x + 6y - 3x - 4y = 9x + 6y + [-(3x)] + [-(4y)]$$
$$= 9x + [-(3x)] + 6y + [-(4y)]$$
$$= 9x + [(-3)x] + 6y + [(-4)y]$$
$$= [9 + (-3)]x + [6 + (-4)]y$$
$$= 6x + 2y.$$

Thus, $9x + 6y - 3x - 4y = 6x + 2y$.

As you become proficient in dealing with expressions of this kind, you will write down the minimum number of steps. Would you be able to do this problem by writing down only the following steps?

6.11 Simplification of Expressions

$$9x + 6y - 3x - 4y = (9x - 3x) + (6y - 4y)$$
$$= 6x + 2y.$$

It is important that you be able to fill in and justify each step.

Example 8 Simplify $6.7a + 4.6b - 4.3a - 2.4b$.
Take shortcuts in finding a simpler equivalent expression, and if necessary, fill in the detailed steps.

$$6.7a + 4.6b - 4.3a - 2.4b = (6.7a - 4.3a) + (4.6b - 2.4b)$$
$$= 2.4a + 2.2b.$$

EXERCISE 6.8

1. Label each statement with T for true and F for false.
   a. $-(8 - 12) = -8 + 12$
   b. $-(6 + 4) = -6 - 4$
   c. $-(-3 + 8) = 3 - 8$
   d. $-(-10 - 4) = 10 + 4$
   e. $-[-5 + (-7)] = 5 - 7$
   f. $-(6 - 4 + 8) = -6 + 4 - 8$
   g. $-(12 + 3 - 7) = -12 - 3 + 7$
   h. $-(6 + 10 + 4) = -6 - 10 - 4$
   i. $-(-8 + 3 - 9) = 8 - 3 + 9$
   j. $-(-20 - 5 + 4) = 20 + 5 - 4$
   k. $-6 + 17 - 4 = 17 - 4 - 6$
   l. $-10 - 8 - 2 = -8 - 10 - 2$
   m. $16 + 9 - 2 = 2 + 9 - 16$
   n. $83 + 17 - 40 = 17 - 40 + 83$
   o. $-15 - 6 + 2 = -15 - 2 + 6$
   p. $5 - 2 - (3 + 1) = 5 - 2 - 3 + 1$
   q. $-(3 - 4) - (7 - 2) = 4 - 3 - 2 - 7$
   r. $(-5)(-6) = -(5 \times 6)$
   s. $(-3)(-2)(-23) = -(6)(23)$
   t. $(-1)(-78) = -78$

2. For each expression, write an equivalent expression which does not contain grouping symbols.
   a. $-(x - 5)$
   b. $-(m + 7)$
   c. $-(3x + 4a)$
   d. $-(2d - 6c)$

e. $-(-8 + t)$
f. $-(-c - 6)$
g. $-(-a + 5)$
h. $-(40 + 18pq)$
i. $-(a + m + 14)$
j. $-(a + m - 14)$
k. $-(a - m - 14)$
l. $-(a - m + 14)$
m. $-(9a + 15p - 3)$
n. $-(-r - 2s + t)$
o. $-(-n - 15 - 4q)$
p. $-(y + 20 + 3t)$
q. $-(2t + 4s - 5q - 6z)$
r. $-(-3m + n - 8p - 10c)$
s. $-[-(2c - 4p)]$
t. $-[-(-3t + 5)]$

3. Which pairs of expressions are equivalent?
   a. $3a - 2b + 4c$; $4c - 2b + 3a$
   b. $2p - 3q - r$; $r - 2p - 3q$
   c. $-3x + 7 - x^2$; $7 - 3x - x^2$
   d. $-3m + 2mn - 6n$; $2mn - 6n - 3m$
   e. $-2c - 2d$; $-2d - 2c$
   f. $pk - t$; $-t + pk$
   g. $2x - 9 - 3y$; $3y - 2x - 9$
   h. $5(x - y) + 3(2x + 3y)$; $11x + 4y$
   i. $-(-a + 2b) - (3a - b)$; $4a + b$
   j. $5(x + y - z) - (2x - 2y + z)$; $3x + 7y - 6z$

4. Simplify.
   a. $(-3)(8x)$
   b. $(-6)(-1)(r)$
   c. $5(s)(-3)$
   d. $-6.2(-3x)$
   e. $-(-5x)(3y)$
   f. $(-x)(-3y)(-4)$
   g. $(1.4y)(-3x)$
   h. $(-6.3x)(-y)(5)$
   i. $-1.5(-3r)$
   j. $(-2)(-s)(-5)$

5. Label as true those sentences which yield true statements for all replacements of the variables.
   a. $-(a + b) = -a + b$
   b. $-(a - b) = -a + b$
   c. $-(6x + 3y) = -6x + (-3y)$
   d. $-(3m - 2n) = 2n - 3m$
   e. $-(5 - 6s) = 6s + 5$
   f. $-(7 - a + 3b) = -7 + a + 3b$
   g. $-(2m + 6n - 3) = 2m - 6n + 3$
   h. $-(-8 + 2c - 3d) = -8 - 2c + 3d$

6.11 Simplification of Expressions

6. Simplify.

   a. $7x - 5 + 4x + 3$
   b. $2x - 8 + 5x + 10$
   c. $4c - 5 - 6 - 7c$
   d. $6d - 2f + 3d - 5f$
   e. $6n - 2t + 3n - t$
   f. $x - 2y - x + y$
   g. $3a - 2b + 7 - 5a + 8b - 6$
   h. $5 - 2m - 6 + 7m$
   i. $3r + 5s - 6 - 7r + 4s + 4$
   j. $-u + 3t + 2w - 5 - 3t$
   k. $c - b + a - c + b + a$
   l. $6.7x + 4.6y - 4.3x - 2.4y$
   m. $4xy - 2xy + 6xy$
   n. $10xy - 5rs - 3xy + 9rs - xy$
   o. $6 - 3x + 5y - 9 + 6x - 7y$
   p. $4 + 3y - 2rs + 7rs - 4y$
   q. $4uv + 7uv - 9uv$
   r. $4st + 5xy - 7st - 3xy + 4$
   s. $-4.9r - 6.7r + .2s - 5.8r$
   t. $\frac{4}{3}x - 7 + \frac{2}{5}y - \frac{3}{4}x + 5$

7. Give a reason for each step in the following proofs.

   a. $s - (x - y) = s + [-(x - y)]$
   $ = s + (y - x)$
   $ = s + [y + (-x)]$
   $ = (s + y) + (-x)$
   $ = (s + y) - x$

   b. $x + y - (-x + y) = x + y + [-(-x + y)]$
   $ = x + y + \{-[y + (-x)]\}$
   $ = x + y + [-(y - x)]$
   $ = x + y + (x - y)$
   $ = x + y + [x + (-y)]$
   $ = (x + x) + [y + (-y)]$
   $ = 2x + 0$
   $ = 2x$

8. Prove each of the following.
   a. $-(x + y) = -x - y$
   b. $-(x - y) = -x + y$
   c. $-(-x - y) = x + y$
   d. $-(m + n) = -n - m$
   e. $-(x - y) = -[-(y - x)]$
   f. $-(a - b - c) = c + b - a$

6.12 Theorems on Subtraction of Binomials

We shall now observe some further patterns involving binomials. To do this, carry out the necessary computations to verify that each of the following statements is true. Recall that the operation inside the parentheses is to be performed first. The first illustration is a sample for you to follow.

$$15 - (5 + 3) = 15 - 5 - 3$$
$$\begin{array}{c|c} 15 - 8 & 10 - 3 \\ 7 & 7 \end{array}$$

$$20 - (10 + 4) = 20 - 10 - 4$$
$$17 - (9 + 5) = 17 - 9 - 5$$
$$21 - (6 + 15) = 21 - 6 - 15.$$

The examples above suggest a theorem which we now prove.

THEOREM 6.4 $x - (y + z) = x - y - z$.

Proof. $x - (y + z) = x + [-(y + z)]$ Def. subtr.
$ = x + [-y + (-z)]$ Theorem 6.2
$ = [x + (-y)] + (-z)$ APA
$ = x - y - z.$ Def. subtr.

Now verify that each of the following is true.

$$12 - (4 - 3) = 12 - 4 + 3$$
$$25 - (10 - 5) = 25 - 10 + 5$$
$$32 - (16 - 4) = 32 - 16 + 4$$
$$50 - (20 - 10) = 50 - 20 + 10.$$

The following theorem is suggested by these examples. Supply the reason for each step in its proof.

THEOREM 6.5 $x - (y - z) = x - y + z$.

6.12 Theorems on Subtraction of Binomials

Proof. $x - (y - z) = x + [-(y - z)]$
$= x + (-y + z)$
$= [x + (-y)] + z$
$= x - y + z.$

Now consider the following statements. Perform the necessary computations to verify that each statement is true. The first illustration is already worked out.

$$25 - 2(3 + 4) = 25 - 2 \times 3 - 2 \times 4$$

$25 - 2(7)$	$25 - 6 - 8$
$25 - 14$	$19 - 8$
11	11

$20 - 3(1 + 2) = 20 - 3 \times 1 - 3 \times 2$
$18 - 2(2 + 4) = 18 - 2 \times 2 - 2 \times 4$
$30 - 5(4 + 1) = 30 - 5 \times 4 - 5 \times 1$
$48 - 6(5 + 3) = 48 - 6 \times 5 - 6 \times 3.$

The examples above suggest another theorem.

THEOREM 6.6 $x - y(z + w) = x - yz - yw.$

Proof. $x - y(z + w) = x - (yz + yw)$ LDPMA
$= x - yz - yw.$ Theorem 6.4

The next theorem is analogous to Theorem 6.6, but it deals with subtraction.

THEOREM 6.7 $x - y(z - w) = x - yz + yw.$

Proof. $x - y(z - w) = x - (yz - yw)$ LDPMS
$= x - yz + yw.$ Theorem 6.5

The six theorems we proved are helpful in solving certain kinds of equations, such as the following:

Example 1 Solve: $-(2x + 3) = -(2 - x)$
$-2x - 3 = -2 + x$
$-2x - 3 - x = -2 + x - x$
$-3x - 3 = -2$
$-3x - 3 + 3 = -2 + 3$
$-3x = 1$
$x = -\frac{1}{3}.$

Solution Sets: Equations and Inequalities

Check.

$-(2x + 3)$	$-(2 - x)$
$-[2(-\frac{1}{3}) + 3]$	$-[2 - (-\frac{1}{3})]$
$-(-\frac{2}{3} + 3)$	$-(2 + \frac{1}{3})$
$-2\frac{1}{3}$	$-2\frac{1}{3}$

Thus, $\{-\frac{1}{3}\}$ is the solution set.

Example 2 Solve:
$$3 - (5 - 6x) = 2 - (2 + 3x)$$
$$3 - 5 + 6x = 2 - 2 - 3x$$
$$-2 + 6x + 3x = -3x + 3x$$
$$-2 + 9x = 0$$
$$-2 + 9x + 2 = 0 + 2$$
$$9x = 2$$
$$x = \frac{2}{9}.$$

Check to verify that the solution set is $\{\frac{2}{9}\}$.

EXERCISE 6.9

1. Label each statement with T for true and F for false.
   a. $6 - (9 - 8) = 6 - 9 + 8$
   b. $-5 - (4 + 7) = -5 + 4 - 7$
   c. $14 - (3 - 2) = 14 - 3 + 2$
   d. $-2 - (18 + 6) = -2 - 18 + 6$
   e. $9 - 3(2 + 5) = 9 - 6 + 15$
   f. $-4 - 2[3 + (-7)] = -4 - 6 + 14$
   g. $12 - 4(5 - 2) = 12 - 20 - 8$
   h. $-10 - 3(4 - 7) = -10 + 12 + 21$

2. Which pairs of expressions are equivalent?
   a. $-9 - (2a + 7b);$ $9 - 2a - 7b$
   b. $-6 - (5c - 4d);$ $-6 - 5c - 4d$
   c. $2f + (3 + 4g);$ $2f - 3 - 4g$
   d. $3m + (2n - 9);$ $3m - 2n + 9$
   e. $6 - (-2p + 5q);$ $6 + 2p - 5q$
   f. $a - b(c + d);$ $a - bc - bd$
   g. $9y - m(2n + p);$ $-mp - 2mn + 9y$
   h. $-ax - b(2y + z);$ $-2by - bz - ax$
   i. $-3 - 4(c - d);$ $-3 - 4c + 4d$
   j. $a - b(c - d);$ $bd - bc + a$

6.12 Theorems on Subtraction of Binomials

3. For each expression, write an equivalent expression which does not contain grouping symbols.

a. $7a - (3x - 4t)$
b. $5d - (2a + 7g)$
c. $-3c - (5b - 8p)$
d. $8 - (-16t - 3r)$
e. $-7 - (-5v + pt)$
f. $3 + (a + 5m)$
g. $-3 + (-b + t)$
h. $a - (6 - r + f)$
i. $2 - (a - 3c)$
j. $-\frac{1}{2} - (6t + 4b)$
k. $4.2 - (-2a - 9k)$
l. $x - (-3x^2 + 2x^3)$
m. $5 - 3(x - y)$
n. $5 + 3(x - y)$
o. $7 - 3a(2 + x)$
p. $3x^2 - x(y - 2)$
q. $2x - 3b(4 - x)$
r. $5(3 - x) + 7(a - b)$
s. $-2(6 + 4x) - a(7 + 3b)$
t. $5 + 3(x - t) - 2(a + b)$

4. Solve each equation.

a. $3 - (5x + 7) = 11$
b. $-2 - (2x - 1) = 5$
c. $-(x + 1) = -(2x - 3)$
d. $1 - (1 - x) = -(3x + 1)$
e. $-2 - (2x + 1) = -(-2 - 3x)$
f. $-2 - (2 - 5x) = -5 - (x + 4)$
g. $2(x + 1) + 3(2x - 1) = -25$
h. $-3(2 - x) + 2(3x + 5) = 31$
i. $3 + 2(3 - 4x) - 4(-x - 1) = 2(x + \frac{1}{2})$
j. $1 - (4 - x) - 2(-2x - 3) = 7(x - \frac{1}{7})$

5. Tell which of the following are true for all replacements of the variables.

a. $3a - (2b - 7) = 7 - (2b - 3a)$
b. $6 - (c - d) = 6 + (c + d)$
c. $17 - (t - v) = 17 + (v - t)$
d. $-27 - (3a + 2b) = -27 + (-3a - 2b)$
e. $3 - 5(2x + y) = (2x + y)3 - 5$
f. $4u - 2m(a - b) = 4u - 2[(a - b)m]$
g. $12x - 2(3x + 4) = 6x + 8$
h. $4.3r - 6(1.8r - .3) = 6.5r + 1.8$
i. $5x - 3[2x - 2y - (3x + 4y)] = 8x - 18y$
j. $5(3x - 2y) - 2(2x + y) = 11x - 12y$

6. Tell which are true for all replacements of the variables.
   a. $-2 + x = x - 2$
   b. $(x - y)(r - s) = -(y - x)(r - s)$
   c. $-y = -(-y)$
   d. $(r - s)(x - y) = (s - r)(y - x)$
   e. $-(x - y) = -x - y$
   f. $-[(-x) + (-y)] = x + y$
   g. $-[-(-x)] = -x$
   h. $-[-(x)] + (-x) = 0$
   i. $-[(-x) + x] = x$
   j. $(-x)yz = -[(-x)yz]$
7. Simplify.
   a. $2(a + b) + 4(a - b)$
   b. $-3(x - y) + (x + y)$
   c. $4(t - u) + (-2)(t - u)$
   d. $\frac{1}{2}(2m - 4n) + \frac{1}{3}(9m + 12n)$
   e. $\frac{1}{4}(8y + 12z) - (z + y)$
   f. $(x - y)(-2) + (2x + 3y)(-1)$
   g. $2(x + 2y) - (-2)(x + 2y)$
   h. $0.1(10t + 20n) - 0.2(10t + 20n)$
   i. $3(x + 2y + 3z) - 2(3x + 2y + z)$
   j. $-4(2a + b - 3c) + 3(a + b)$
8. For each expression, write the simplest equivalent expression which does not contain grouping symbols.
   a. $a - [b - (c + d)]$
   b. $a - [b - (c - d)]$
   c. $-3x - [2a - (3b + 4c)]$
   d. $-5t - [-4x - (4y - 3a)]$
   e. $[3a - 4(5x + 4y)] + [-a - 3(-2x - 5y)]$
   f. $[-5x - 4y(-a - b)] - [-2x - y(a + b)]$

GLOSSARY

Division Property for Equations: If $x = y$, then $\frac{x}{z} = \frac{y}{z}$ $[z \neq 0]$.

Left-hand Addition Property for Equations: If $x = y$, then $z + x = z + y$.

Left-hand Multiplication Property for Equations: If $x = y$, then $zx = zy$.
Open segment: A segment without its endpoints.
Right-hand Addition Property for Equations: If $x = y$, then $x + z = y + z$.
Right-hand Multiplication Property for Equations: If $x = y$, then $xz = yz$.
Root of an equation: A number whose name in place of a variable in an equation yields a true sentence.
Semiclosed segment: A segment without one of its endpoints.
Semiopen segment: The same as semiclosed segment.
Solution: The same as *root*.
Solution set: The set of all solutions.
Subtraction Property for Equations: If $x = y$, then $x - z = y - z$.

CHAPTER REVIEW PROBLEMS

1. Classify each of the following sentences into one of the following categories. The replacement set is the set of real numbers.

- T: true sentence
- F: false sentence
- R: open sentence, in which at least one replacement of the variable, but not all, will result in a true sentence. Find the replacement or several such replacements.
- N: open sentence, in which there is no replacement of the variable which will result in a true sentence
- E: open sentence, in which every replacement of the variable will result in a true sentence

a. $x - 1 = 3$
b. $|x| = 2$
c. $|x + 1| = -3$
d. $\sqrt{121} = 11$
e. $\frac{1}{2} = \frac{4}{9}$
f. $.3 = \frac{1}{3}$
g. $|x - y| = |y - x|$
h. $\frac{x}{2} = x$
i. $\frac{2x}{2} = x$
j. $.01\% = .0001$
k. $\sqrt{-1} = -1$
l. $(-1)^{99} = -1$
m. $(-1)^{100} = -1$
n. $2(x - y) = 2x - y$
o. $|x| = -x$
p. $|x| = x$
q. $\frac{1}{x} = 0$
r. $\frac{1}{|x|} = 1$
s. $|x| + |y| = |x + y|$
t. $|x| - |y| = |x - y|$

2. Translate each sentence into an equation. Do not simplify.
   a. 6 increased by 4 is 5 multiplied by 2.
   b. The square of the sum of 2 and -5 is 9.
   c. Subtracting 5 from a number results in -1.
   d. The arithmetic mean of a number and 7 is -1.
   e. The arithmetic mean of a number and 2 more than the number is 3.
   f. n nickels and $n + 3$ dimes are worth $2.00.
   g. Jim is x years old and Larry is twice as old; the sum of their ages is 21 years.
   h. Lora is x years old and Susan is $2x + 3$ years old. The difference of their ages is 33.
   i. The length of a rectangle is k and its width is $\frac{k}{2} - 1$. The perimeter is 88 in.
   j. The product of a number, and the number decreased by 1, is equal to the quotient of the number, and the number increased by 2.

3. For each of the following expressions, write an equivalent expression which does not contain parentheses. Simplify wherever possible.
   a. $3(x + 4)$
   b. $-5(3 - x)$
   c. $-2(7x - \frac{1}{2})$
   d. $-(2y + 3n - 4n)$
   e. $-\frac{1}{2}(10m - 6y - 18w)$
   f. $-\frac{1}{3}(3y - 12x) - \frac{2}{3}(36x + 15y)$
   g. $.4(20y - 16n) - .5(20n - 30y)$
   h. $-.2(50u - 10s) - .3(10s - 20u)$

4. Simplify.
   a. $9y - 5y$
   b. $12x - 4x$
   c. $24a - a$
   d. $2m - 5m$
   e. $\frac{1}{2}n - \frac{3}{4}n$
   f. $\frac{1}{4}p - \frac{1}{8}p$
   g. $3.7h - 1.3h$
   h. $1.2s - 3.9s$
   i. $3t - 5t + 7t$
   j. $3xy + 12xy$
   k. $2x - 5y + 3x$
   l. $a - 3k + 4a + 5k$
   m. $7c - 4d + 2c + d$
   n. $4ay - 2bx - 6ay + 3bx$
   o. $3mv - 2nt - 3mv - 2nt$
   p. $4 - 2ax + 4ay + 6ax - 3ay$
   q. $(-4)(-2x) + \frac{1}{2}(-8x)$
   r. $3(-t)(-5)(t) + t$
   s. $-(a - b) + (a + b)$
   t. $-(3x + 5y) + 2(2y - x)$
   u. $2(u + v + w) - 3(-2u - 3v - 4w)$

Chapter Review Problems

5. Tell which of the following are true for all replacements of the variables.
- a. $1 - 2x = -(2x + 1)$
- b. $-(x + y) = (x + y)(-1)$
- c. $x - 4x = 3x$
- d. $2x - x = x$
- e. $(-1)(x + y) = -x + (-y)$
- f. $x - y = -(y - x)$
- g. $-(x + y) = -x + (-y)$
- h. $3 + x = -(x - 3)$
- i. $-2(4 - x) = 2x + 8$
- j. $-(x + y + z) = -x + y + z$
- k. $(x + y)(z + w) = -(x + y)[-(z + w)]$
- l. $-x = -(-x)$
- m. $-1(-x - y) = x - y$
- n. $(a - b)(c - d) = (b - c)(d - c)$

6. Determine the solution set of each equation.
- a. $x + 1 = 23$
- b. $2x - 4 = 8$
- c. $9 - 4x = 21$
- d. $12 - 7x = 47$
- e. $4(x + 1) = -24$
- f. $-5(2x + 3) = -15$
- g. $-2(3 - 4x) = -6$
- h. $4(2x + 17) = 4$
- i. $|x| = 12$
- j. $|x| = -1$
- k. $|x + 17| = 0$
- l. $|2x + 1| = 9$
- m. $|-3 - 7x| = 4$
- n. $|\frac{1}{3}x + 8| = 11$
- o. $\left|\frac{x + 4}{5}\right| = 1$
- p. $\left|\frac{-2x - 11}{3}\right| = 7$

7. Graph each of the following inequalities. The replacement set is:
  i. the set of integers,
  ii. the set of real numbers.
- a. $A = \{x | -1 \leq x < 2\}$
- b. $B = \{x | x > 1 \text{ or } x \leq -3\}$
- c. $C = \{x | x \geq 5 \text{ and } x \leq 8\}$

8. Prove the following.
- a. $-(x + 2y) = -x + (-2y)$
- b. $x - (y - 5z) = x - y + 5z$
- c. $x - y(2z - 7w) = x - 2yz + 7yw$

9. Solve each problem.
   a. The width of a rectangle is 5 in. shorter than the length. The perimeter is 82 in. What are the width and the length of the rectangle?
   b. The length of the rectangle is 3 more than twice the width. The perimeter is 72 in. What are the length and the width of the rectangle?
   c. A number is multiplied by 4 and 10 is added to the product. The result is 2. What is the number?
   d. The sum of a number and 3 is multiplied by -3. The result is 12. What is the number?
   e. The width of a rectangle is 2 in. more than one-half of its length. The perimeter is 100 in. What are the length and the width of the rectangle?
   f. A number is divided by 3 and the quotient is multiplied by 2. From the result, 24 is subtracted, giving 10. What was the original number?

7

Open Rational Expressions

7.1 Patterns in Multiplication

In this chapter we shall develop rules for operating with expressions which are in the form of quotients of polynomials.

DEFINITION 7.1 A *rational expression* is an expression that can be written in the form $\frac{P}{Q}$, where P and Q are polynomials $\quad [Q \neq 0]$.

To observe what may be true in general, we shall examine some specific cases which fall into general patterns.

$\frac{2 \times 3}{7} = \frac{6}{7}$ and $\frac{2}{7} \times 3 = \frac{6}{7}$; therefore, $\frac{2 \times 3}{7} = \frac{2}{7} \times 3$.

$\frac{3 \times (-4)}{2} = -6$ and $\frac{3}{2} \times (-4) = -6$; therefore, $\frac{3 \times (-4)}{2} = \frac{3}{2} \times (-4)$.

$\frac{-5 \times (-4)}{2} = 10$ and $\frac{-5}{2} \times (-4) = 10$; therefore, $\frac{-5 \times (-4)}{2} = \frac{-5}{2} \times (-4)$.

This leads us to surmise that

$$\frac{xz}{y} \quad \text{and} \quad \frac{x}{y} \times z \quad [y \neq 0].$$

are equivalent expressions. We will now prove this as a theorem. Recall that the replacement set of the variables is the set of real numbers unless otherwise stated.

THEOREM 7.1 $\dfrac{xz}{y} = \dfrac{x}{y} \times z$ $[y \neq 0]$.

Proof. $\dfrac{xz}{y} = xz \times \dfrac{1}{y}$ Def. div.

$= \left(x \times \dfrac{1}{y}\right)(z)$ FR

$= \dfrac{x}{y} \times z.$ Def. div.

Thus, $\dfrac{xz}{y} = \dfrac{x}{y} \times z$ $[y \neq 0]$.

THEOREM 7.2 $\dfrac{xz}{y} = x \times \dfrac{z}{y}$ $[y \neq 0]$.

Proof. $\dfrac{xz}{y} = xz \times \dfrac{1}{y}$ Def. div.

$= x\left(z \times \dfrac{1}{y}\right)$ APM

$= x \times \dfrac{z}{y}.$ Def. div.

We know from work with real numbers that the following statements are true:

$$\dfrac{1}{2} \times \dfrac{1}{3} = \dfrac{1}{2 \times 3}$$

$$\dfrac{1}{-5} \times \dfrac{1}{4} = \dfrac{1}{-5 \times 4}$$

$$\dfrac{1}{-3} \times \dfrac{1}{-5} = \dfrac{1}{-3 \times (-5)}$$

$$\dfrac{1}{6} \times \dfrac{1}{-3} = \dfrac{1}{6 \times (-3)}.$$

It appears that

$$\dfrac{1}{x} \times \dfrac{1}{y} = \dfrac{1}{xy}$$

may be true for all nonzero replacements of x and y. We state this as our next theorem and prove it.

7.1 Patterns in Multiplication

THEOREM 7.3 $\quad \dfrac{1}{x} \times \dfrac{1}{y} = \dfrac{1}{xy} \quad [x \neq 0, y \neq 0]$.

Before writing a proof, let us observe that xy and $\dfrac{1}{xy}$ are multiplicative inverses of each other. We know that the product of multiplicative inverses is equal to 1. Thus, by the Property of Multiplicative Inverse,

$$xy \times \dfrac{1}{xy} = 1.$$

If we could prove that

$$(xy)\left(\dfrac{1}{x} \times \dfrac{1}{y}\right) = 1,$$

we would know that xy and $\left(\dfrac{1}{x} \times \dfrac{1}{y}\right)$ are multiplicative inverses. Since $\dfrac{1}{xy}$ is the multiplicative inverse of xy and since each nonzero real number has exactly *one* multiplicative inverse, we would know that

$$\dfrac{1}{x} \times \dfrac{1}{y} = \dfrac{1}{xy},$$

which is what we want to prove. We will take this approach in our proof.

Proof. $\quad (xy)\left(\dfrac{1}{x} \times \dfrac{1}{y}\right) = \left(x \times \dfrac{1}{x}\right)\left(y \times \dfrac{1}{y}\right) \quad$ FR
$\qquad\qquad\qquad\qquad\;\; = (1)(1) \qquad\qquad\qquad$ Prop. mult. inv.
$\qquad\qquad\qquad\qquad\;\; = 1. \qquad\qquad\qquad\quad\;\;$ P1M

We have thus proved that xy and $\dfrac{1}{x} \times \dfrac{1}{y}$ are multiplicative inverses of each other by showing that their product is 1. Therefore,

$$\dfrac{1}{x} \times \dfrac{1}{y} \text{ is the multiplicative inverse of } xy.$$

THEOREM 7.4 $\quad \dfrac{x}{y} \times \dfrac{z}{w} = \dfrac{xz}{yw} \quad [y \neq 0, w \neq 0]$.

Proof. $\quad \dfrac{x}{y} \times \dfrac{z}{w} = \left(x \times \dfrac{1}{y}\right) \times \left(z \times \dfrac{1}{w}\right) \qquad$ Def. div.
$\qquad\qquad\qquad\; = (xz)\left(\dfrac{1}{y} \times \dfrac{1}{w}\right) \qquad\qquad\;$ FR
$\qquad\qquad\qquad\; = (xz)\left(\dfrac{1}{yw}\right) \qquad\qquad\quad\;\;$ Theorem 7.3
$\qquad\qquad\qquad\; = \dfrac{xz}{yw}. \qquad\qquad\qquad\qquad\;\;$ Def. div.

Thus, $\frac{x}{y} \times \frac{z}{w} = \frac{xz}{yw}$ $[y \neq 0, w \neq 0]$.

Notice that if we replace x and z by 1 in $\frac{x}{y} \times \frac{z}{w} = \frac{xz}{yw}$, we obtain $\frac{1}{y} \times \frac{1}{w} = \frac{1}{yw}$, which is Theorem 7.3. Thus, Theorem 7.3 is a special case of Theorem 7.4.

You have probably observed that the theorems we have proved tell us nothing new about work with real numbers. You have actually been using these theorems with real numbers in your previous study of mathematics. The point is, however, that we are now able to *derive* these theorems from other known facts. We need to develop skills in proving theorems since these skills are very important in the study of mathematics. One mathematician defined mathematics as a study of sentences of the form *if p, then q*. This means that, given a sentence *p*, we attempt to prove that *q* follows logically from *p*.

EXERCISE 7.1

1. Label each statement with T for true and F for false.

   a. $\frac{9 \times 3}{4} = \frac{9}{4} \times 3$

   b. $\frac{-2 \times 5}{11} = \frac{1}{11}(-10)$

   c. $\frac{5 \times 6}{7} = \frac{5}{7} \times \frac{6}{7}$

   d. $\frac{1}{3}(5 \times 7) = \frac{35}{3}$

   e. $\frac{-3}{-11}(-6) = \frac{-6}{-11} \times \frac{-3}{-11}$

   f. $\frac{1}{-9}[(-2)(-9)] = -2$

   g. $3(-8) \times \frac{1}{-3} = -8$

   h. $\frac{4 \times 9}{-4} = -9$

   i. $\frac{(-5)(-11)}{12} = \frac{1}{12} \times \left(-\frac{11}{12}\right) \times \left(-\frac{5}{12}\right)$

   j. $\frac{3}{7}(-4) = \frac{-4 \times 3}{7}$

   k. $\frac{4 \times 9}{-1} = (-4)(-9)$

   l. $\frac{1}{-2} \times [(-3)(-5)] = (-3)\left(\frac{-5}{-2}\right)$

   m. $(-12)\left(\frac{5}{12}\right) = 5$

   n. $\frac{25 \times (-6)}{50} = -12$

   o. $\frac{(-10)(-20)}{-200} = 1$

   p. $\frac{-5 + 3}{4} = -\frac{5}{4} + 3$

   q. $\frac{1}{9} \times \frac{1}{7} = \frac{1}{63}$

   r. $\frac{1}{-3} \times \frac{1}{4} = \frac{1}{-12}$

7.1 Patterns in Multiplication

s. $\dfrac{1}{-6} \times \dfrac{1}{-5} = \dfrac{1}{30}$

t. $\dfrac{2}{7} \times \dfrac{1}{5} = \dfrac{2}{12}$

u. $\dfrac{3}{5} \times \dfrac{2}{-5} = \dfrac{6}{-25}$

v. $\dfrac{-4}{3} \times \dfrac{5}{-9} = \dfrac{-20}{-27}$

w. $\dfrac{-5}{6} \times \dfrac{7}{-3} = \dfrac{35}{-18}$

x. $\dfrac{-3}{-2} \times \dfrac{-5}{-4} = \dfrac{15}{8}$

y. $\dfrac{1}{3} \times \dfrac{-4}{-5} = \dfrac{-4}{-2}$

2. Which pairs of expressions are equivalent? Only permissible real-number values of the variables are considered.

a. $\dfrac{ab}{b}$; a

b. $\dfrac{1}{b}[(-a)(-b)]$; $-a$

c. $\dfrac{(-a)(-b)}{c}$; $\dfrac{a}{c} \times \dfrac{b}{c}$

d. $\dfrac{(a+b)(a-c)}{c}$; $\dfrac{a+b}{c} \times (a-c)$

3. Apply the Commutative Property of Multiplication to the right side of each of the following to obtain another equivalent expression.

a. $\dfrac{xz}{y} = \dfrac{x}{y} \times z$

b. $\dfrac{xz}{y} = x \times \dfrac{z}{y}$

c. $\dfrac{xz}{y} = \dfrac{1}{y} \times xz$

4. As shown in the example, give three different expressions equivalent to the following:

Example $\dfrac{3(6)}{5}$ $\left(\dfrac{3}{5}\right)(6)$; $\dfrac{1}{5}[3(6)]$; $3\left(\dfrac{6}{5}\right)$.

a. $\dfrac{(5)(6)}{7}$

b. $\dfrac{(3)(7)}{5}$

c. $8 \times \left(\dfrac{7}{6}\right)$

d. $\dfrac{1}{3}(4 \times 5)$

e. $\dfrac{4}{7} \times (6)$

f. $9\left(\dfrac{7}{8}\right)$

g. $\dfrac{(4)(9)}{7}$

h. $\dfrac{1}{4}(3 \times 9)$

i. $5\left(\dfrac{7}{8}\right)$

5. For each of the following, write *one* equivalent rational expression. Only permissible real-number values of the variables are considered.

a. $\dfrac{5}{ab} \times (3)$

b. $\dfrac{7}{3c} \times (2d)$

c. $\dfrac{5}{4x} \times (m)$

d. $\dfrac{2}{x-3} \times (a-8)$

e. $\dfrac{10}{3c} \times (2x-5)$

f. $(x-4) \times \dfrac{3}{2c}$

g. $-1 \times \dfrac{x}{2}$　　　**h.** $-3 \times \dfrac{5x}{4c}$

i. $\dfrac{am}{pq} \times (-2)$　　　**j.** $-1 \times \dfrac{2}{t-7}$

k. $-1 \times \dfrac{r}{s}$　　　**l.** $-1 \times \dfrac{5-a}{4}$

6. For each of the following, write a single equivalent rational expression. Only permissible real-number values of the variables are considered.

a. $\dfrac{3-x}{2-y} \times (a+2)$　　　**b.** $(a+b) \times \dfrac{c+d}{x+y}$

c. $\dfrac{1}{x} \times \dfrac{a+b}{2} \times \dfrac{c+d}{3}$　　　**d.** $\dfrac{2}{x+y} \times \dfrac{x^2+y^2}{5} \times \dfrac{3}{x-y}$

e. $\dfrac{a+b}{4} \times \dfrac{-3}{a-b} \times \dfrac{-5}{2a+b}$　　　**f.** $\dfrac{a^3-b^3}{2} \times \dfrac{-7}{a-3b} \times \dfrac{3a-b}{-2}$

7. For each of the following, write a single equivalent rational expression. Only permissible real-number values of the variables are considered.

a. $\dfrac{1}{5} \times \dfrac{1}{2m}$　　　**b.** $\dfrac{1}{3r} \times \dfrac{1}{6}$

c. $\dfrac{1}{-m} \times \dfrac{1}{p}$　　　**d.** $\dfrac{1}{t} \times \dfrac{1}{-v}$

e. $\dfrac{1}{-c} \times \dfrac{1}{-d}$　　　**f.** $\dfrac{1}{5} \times \dfrac{1}{x+y}$

g. $\dfrac{1}{x+2} \times \dfrac{1}{3}$　　　**h.** $\dfrac{1}{4} \times \dfrac{1}{5-m}$

i. $\dfrac{1}{4t} \times \dfrac{1}{x-3}$　　　**j.** $\dfrac{1}{-2} \times \dfrac{1}{x-3}$

k. $\dfrac{1}{-3c} \times \dfrac{1}{x+5}$　　　**l.** $\dfrac{4}{t} \times \dfrac{5}{v}$

m. $\dfrac{x}{7} \times \dfrac{2}{3}$　　　**n.** $\dfrac{2}{5} \times \dfrac{x}{t}$

o. $\dfrac{k}{r} \times \dfrac{5}{7}$　　　**p.** $\dfrac{t}{r} \times \dfrac{y}{7}$

q. $\dfrac{m}{p} \times \dfrac{r}{t}$　　　**r.** $\dfrac{5}{7} \times \dfrac{x-3}{t+2}$

s. $\dfrac{a}{x-c} \times \dfrac{t-d}{b}$　　　**t.** $\dfrac{3x}{4m} \times \dfrac{5xy}{7mp}$

u. $\dfrac{-2}{7} \times \dfrac{x-3}{5m}$　　　**v.** $\dfrac{5}{t-3} \times \dfrac{k}{-4}$

w. $\dfrac{-2a}{5m} \times \dfrac{x-c}{r+3}$ x. $\dfrac{r}{3} \times \dfrac{r}{4}$

7.2 Multiplicative Inverses Theorem

One of the field properties of the set of real numbers which was stated in Chapter 3 was the Property of Multiplicative Inverse. It is the following property:

◂ For every nonzero real number x, there exists a unique real number $\dfrac{1}{x}$, such that $x \times \dfrac{1}{x} = 1$.

We made use of this property in the proof of Theorem 7.3.

The same property, for more general rational expressions, can now be proved using Theorem 7.4. This property is suggested by the following examples. Study them and observe the pattern.

The multiplicative inverse of $\dfrac{7}{3}$ is $\dfrac{3}{7}$, since $\dfrac{7}{3} \times \dfrac{3}{7} = 1$.

The multiplicative inverse of $-\dfrac{2}{11}$ is $-\dfrac{11}{2}$ since, $-\dfrac{2}{11} \times \left(-\dfrac{11}{2}\right) = 1$.

The multiplicative inverse of $-\dfrac{5}{3}$ is $-\dfrac{3}{5}$ since, $-\dfrac{5}{3} \times \left(-\dfrac{3}{5}\right) = 1$.

We are led to suspect that, in general, the multiplicative inverse of $\dfrac{x}{y}$ is $\dfrac{y}{x}$ [$x \neq 0$, $y \neq 0$]. To prove this, we must prove that $\dfrac{x}{y} \times \dfrac{y}{x} = 1$, since the product of a number and its multiplicative inverse is equal to 1. First, however, we shall prove that any nonzero real number divided by itself is equal to 1.

THEOREM 7.5 $\dfrac{r}{r} = 1$ [$r \neq 0$].

Proof. $\dfrac{r}{r}$ names some real number [$r \neq 0$]. Call it x. Then $\dfrac{r}{r} = x$.

$\quad\quad\quad\quad x \times r = r \quad\quad$ Theorem 6.1
$\quad\quad\quad\quad x \times r = 1 \times r \quad$ P1M
$\quad\quad\quad\quad x = 1. \quad\quad\quad\quad$ Mult. prop. for equations

Thus, $\dfrac{r}{r} = 1$ [$r \neq 0$].

Now we are able to prove the following.

THEOREM 7.6 *Multiplicative Inverses Theorem* The multiplicative inverse of $\frac{x}{y}$ is $\frac{y}{x}$ $[x \neq 0, y \neq 0]$.

Proof. $\frac{x}{y} \times \frac{y}{x} = \frac{xy}{yx}$ Theorem 7.4

$\phantom{\frac{x}{y} \times \frac{y}{x}} = \frac{xy}{xy}$ CPM

$\phantom{\frac{x}{y} \times \frac{y}{x}} = 1.$ Theorem 7.5

Hence, $\frac{x}{y} \times \frac{y}{x} = 1$ and $\frac{y}{x}$ is the multiplicative inverse of $\frac{x}{y}$. That is,

$$\frac{y}{x} = \frac{1}{\frac{x}{y}}.$$

7.3 Multiplicative Identity Theorem

We know from work with real numbers that the following statements are true.

$$\frac{2}{4} = \frac{1 \times 2}{2 \times 2} = \frac{1}{2}$$

$$\frac{-12}{-20} = \frac{3 \times (-4)}{5 \times (-4)} = \frac{3}{5}$$

$$\frac{-10}{-14} = \frac{5 \times (-2)}{7 \times (-2)} = \frac{5}{7}.$$

We are led to believe that $\frac{xr}{yr}$ and $\frac{x}{y}$ $[y \neq 0, r \neq 0]$ may be equivalent expressions.

THEOREM 7.7 *Multiplicative Identity Theorem* $\frac{xr}{yr} = \frac{x}{y}$

$[y \neq 0, r \neq 0]$.

Proof. $\frac{xr}{yr} = \left(\frac{x}{y}\right)\left(\frac{r}{r}\right)$ Theorem 7.4

$\phantom{\frac{xr}{yr}} = \frac{x}{y} \times (1)$ Theorem 7.5

$\phantom{\frac{xr}{yr}} = \frac{x}{y}.$ PIM

7.3 Multiplicative Identity Theorem

Thus, $\dfrac{xr}{yr} = \dfrac{x}{y}$ $[y \neq 0, r \neq 0]$.

The Multiplicative Identity Theorem serves as a basis for simplifying such expressions as $\frac{6}{39}$. We find the greatest common factor of 6 and 39, which is 3, and proceed as follows:

$$\frac{6}{39} = \frac{2 \times 3}{13 \times 3} = \frac{2}{13}.$$

Since 2 and 13 have no common factor greater than 1, that is, 2 and 13 are *relatively prime*, we say that $\frac{2}{13}$ is in *simplest form*.

This theorem is also useful when we are simplifying expressions containing variables. For example,

$$\frac{4x + 8}{6} = \frac{2(x + 4)}{2 \times 3} = \frac{x + 4}{3}.$$

EXERCISE 7.2

1. Label each statement with T for true and F for false.

   a. $\dfrac{\frac{1}{7}}{\frac{7}{3}} = \dfrac{3}{7}$ b. $\dfrac{1}{\frac{-3}{4}} = \dfrac{3}{4}$

   c. $\dfrac{1}{-\frac{4}{5}} = \dfrac{5}{4}$ d. $\left(-\dfrac{4}{5}\right) \times \dfrac{5}{4} = 1$

   e. $-10 \times .1 = 1$ f. $-100 \times (-.01) = 1$

   g. -10 is the multiplicative inverse of $.1$.   h. -100 is the multiplicative inverse of $-.01$.

   i. $\dfrac{16}{64} = \dfrac{1}{4}$ j. $\dfrac{26}{65} = \dfrac{2}{5}$

   k. $\dfrac{19}{95} = \dfrac{1}{5}$ l. $\dfrac{310}{7100} = \dfrac{31}{710}$

   m. $\dfrac{4.73}{20} = \dfrac{473}{2000}$ n. $\dfrac{.68}{2.7} = \dfrac{68}{27}$

   o. $\dfrac{-11}{121} = \dfrac{-1}{11}$ p. $\dfrac{-25}{-625} = \dfrac{1}{25}$

   q. $\dfrac{12}{26} = \dfrac{1}{14}$ r. $\dfrac{75}{-3} = -25$

2. Which of the following are true for all permissible replacements of the variables? In each true case, state the replacements of the variable or

variables which yield 0 in the denominator. These replacements, of course, are not permissible.

a. $\dfrac{1}{2t} = \dfrac{3}{2t}$
$\dfrac{}{3}$

b. $\dfrac{1}{\dfrac{1}{x}} = x$

c. $\dfrac{1}{\dfrac{1}{-3y}} = \dfrac{1}{3y}$

d. The multiplicative inverse of $-\dfrac{1}{x+y}$ is $x + y$.

e. $\dfrac{3x}{6x} = \dfrac{1}{2}$

f. $\dfrac{7 - x}{5 - x} = \dfrac{7}{5}$

g. $\dfrac{2(5 - x)}{3(5 - x)} = \dfrac{2}{3}$

h. $\dfrac{10 - 2x}{15 - 3x} = \dfrac{2}{3}$

i. $\dfrac{8x}{x} = 4x$

j. $\dfrac{8x}{2x} = 4$

k. $\dfrac{2x + 7}{7} = 2x$

l. $\dfrac{30 + 7d}{10 + d} = 3 + 7$

m. $\dfrac{6x^2}{x} = 6x$

n. $\dfrac{a^2b}{ab^2} = \dfrac{a}{b}$

o. $\dfrac{a + b}{a - b} = \dfrac{b}{-b}$

p. $\dfrac{x^2 - 1}{x} = x - 1$

q. $\dfrac{x^3}{x^2} = x$

r. $\dfrac{x}{-x} = 1$

s. $\left|\dfrac{x}{-x}\right| = 1$

t. $\dfrac{12 + 18x}{-6} = -2 - 3x$

3. What is the multiplicative inverse of each of the following?

a. -1

b. $-\dfrac{1}{4}$

c. $-.1$

d. $-.01$

e. $.05$

f. $.0001$

g. $\dfrac{5x}{7}$

h. $\dfrac{3a}{2b}$

i. $\dfrac{5 - a}{6}$

j. $\dfrac{3}{7 - c}$

k. $\dfrac{x + 3}{a - 2}$

l. $\dfrac{5x + 9}{3a - 6}$

m. $3m + 7$

n. $\dfrac{1}{2x - 5}$

7.3 Multiplicative Identity Theorem

4. Give equivalent expressions which are in simplest form.

a. $\dfrac{8}{12}$ b. $\dfrac{20}{25}$ c. $\dfrac{14}{21}$

d. $\dfrac{48}{80}$ e. $\dfrac{-13}{169}$ f. $\dfrac{-12}{144}$

g. $\dfrac{-17}{-51}$ h. $\dfrac{-64}{-1024}$ i. $\dfrac{-81}{33}$

j. $\dfrac{121}{-1001}$ k. $\dfrac{117}{26}$ l. $\dfrac{153}{183}$

5. Give equivalent expressions which are in simplest form. Only permissible replacements of the variables are considered.

a. $\dfrac{6x}{10t}$ b. $\dfrac{5x}{5y}$ c. $\dfrac{7m}{9m}$

d. $\dfrac{5ab}{7ac}$ e. $\dfrac{5rs}{10tv}$ f. $\dfrac{7xy}{xy}$

g. $\dfrac{10ab}{25b}$ h. $\dfrac{3a^2b}{5a^2c}$ i. $\dfrac{8x}{14xc}$

j. $\dfrac{7ab^2c}{21ab}$ k. $\dfrac{11xyz}{33axz}$ l. $\dfrac{10ab^2c}{15abc^2}$

m. $\dfrac{57a^2b}{19ab^2}$ n. $\dfrac{24mpx}{20xmy}$ o. $\dfrac{ab}{b^2c^2}$

p. $\dfrac{4(x+7)}{12(t-9)}$ q. $\dfrac{5(x+2)}{9(x+2)}$ r. $\dfrac{5x+10}{5}$

s. $\dfrac{5x+10}{7x+14}$ t. $\dfrac{3x+3}{3}$ u. $\dfrac{4a+8}{40}$

v. $\dfrac{5x+10}{8x+16}$ w. $\dfrac{4c+28}{4c-28}$ x. $\dfrac{3x+3y}{3y-3x}$

y. $\dfrac{7t+21}{14t+42}$ z. $\dfrac{6x^2+4x}{2x}$ a'. $\dfrac{7c-14d}{3c-6d}$

6. What replacements of x, if any, are not permissible in the following expressions?

a. $\dfrac{2x}{3x}$ b. $\dfrac{7x}{x+1}$

c. $\dfrac{5}{x-1}$ d. $\dfrac{5+x}{x^2-4}$

e. $\dfrac{x-3}{(x+7)(x+3)}$ f. $\dfrac{2x+1}{3x+1}$

g. $\dfrac{2-x}{8x-5}$ h. $\dfrac{7+x}{x(x+9)}$

i. $\dfrac{5-x}{8x+x^2}$ j. $\dfrac{7+x}{x^2+3}$

7. Give equivalent expressions which are in simplest form. Only permissible replacements of the variables are considered.

a. $\dfrac{ax+ay}{x+y}$ b. $\dfrac{tm+tn}{t}$

c. $\dfrac{m^2x+m^3y}{m^2}$ d. $\dfrac{ax+by+ay+bx}{a+b}$

e. $\dfrac{ax+by+ay+bx}{x+y}$ f. $\dfrac{ax-by+ay-bx}{a-b}$

7.4 Rational Expressions—Division

From your previous work, you should be able to verify that each of the following statements is true. See Theorem 7.6.

$$\dfrac{1}{\frac{2}{3}}=\dfrac{3}{2} \qquad \dfrac{1}{-\frac{5}{4}}=-\dfrac{4}{5}$$

$$\dfrac{1}{\frac{5}{9}}=\dfrac{9}{5} \qquad \dfrac{1}{-\frac{3}{10}}=-\dfrac{10}{3}.$$

The pattern displayed above suggests another theorem about rational expressions. Before stating and proving this theorem, however, we shall prove that any real number divided by 1 is equal to the number.

THEOREM 7.8 Division by 1 $\quad \dfrac{\frac{x}{y}}{1}=\dfrac{x}{y} \quad [y \neq 0]$.

Proof. $\dfrac{\frac{x}{y}}{1}=\dfrac{x}{y}\times\dfrac{1}{1}$ Def. div.

$\qquad =\dfrac{x\times 1}{y\times 1}$ Theorem 7.4

$\qquad =\dfrac{x}{y}.$ P1M

Thus, $\dfrac{\frac{x}{y}}{1}=\dfrac{x}{y} \quad [y \neq 0]$.

Now we are ready to prove the theorem suggested by the examples

7.4 Rational Expressions—Division

above. Notice that Theorem 7.9 is a case of Theorem 7.6 stated in a different form.

THEOREM 7.9 $\quad \dfrac{1}{\frac{x}{y}} = \dfrac{y}{x} \quad [x \neq 0, y \neq 0]$.

Proof. $\quad \dfrac{1}{\frac{x}{y}} = \dfrac{1 \times \frac{y}{x}}{\frac{x}{y} \times \frac{y}{x}} \quad$ Theorem 7.7

$\qquad\qquad = \dfrac{\frac{y}{x}}{\frac{x}{y} \times \frac{y}{x}} \quad$ P1M

$\qquad\qquad = \dfrac{\frac{y}{x}}{1} \quad$ Theorem 7.6

$\qquad\qquad = \dfrac{y}{x}. \quad$ Theorem 7.8

Thus, $\dfrac{1}{\frac{x}{y}} = \dfrac{y}{x} \quad [x \neq 0, y \neq 0]$.

Now we can prove a more general theorem concerning division.

THEOREM 7.10 Division of Rational Numbers

$$\dfrac{x}{y} \div \dfrac{r}{s} = \dfrac{x}{y} \times \dfrac{s}{r} \quad [y \neq 0, s \neq 0, r \neq 0].$$

Proof. $\quad \dfrac{x}{y} \div \dfrac{r}{s} = \dfrac{x}{y} \times \dfrac{1}{\frac{r}{s}} \quad$ Def. div.

$\qquad\qquad = \dfrac{x}{y} \times \dfrac{s}{r}. \quad$ Theorem 7.9

Thus, $\dfrac{x}{y} \div \dfrac{r}{s} = \dfrac{x}{y} \times \dfrac{s}{r} \quad [y \neq 0, s \neq 0, r \neq 0]$.

This can also be written as $\dfrac{\frac{x}{y}}{\frac{r}{s}} = \dfrac{xs}{yr}$.

Theorem 7.10 serves as a basis for dividing real numbers. For example, according to this theorem and Theorem 7.4,

$$\frac{3}{4} \div \frac{5}{7} = \frac{3}{4} \times \frac{7}{5}$$
$$= \frac{3 \times 7}{4 \times 5}$$
$$= \frac{21}{20}.$$

In the proof of Theorem 7.10, we used the definition of division. We can actually prove this theorem without using this definition.

Alternative Proof. $\dfrac{x}{y} \div \dfrac{r}{s} = \dfrac{\dfrac{x}{y} \times \dfrac{s}{r}}{\dfrac{r}{s} \times \dfrac{s}{r}}$ Theorem 7.7

$\hspace{3.5cm} = \dfrac{\dfrac{x}{y} \times \dfrac{s}{r}}{1}$ Theorem 7.6

$\hspace{3.5cm} = \dfrac{x}{y} \times \dfrac{s}{r}.$ Theorem 7.8

In dealing with rational expressions, the symbol "−" can be used in four different patterns, as shown in the following:

$$\frac{-2}{3} \quad \frac{2}{-3} \quad -\frac{2}{3} \quad \frac{-2}{-3}$$
$$\frac{-x}{y} \quad \frac{x}{-y} \quad -\frac{x}{y} \quad \frac{-x}{-y}.$$

Are these expressions related in any way? We know that they are, because our work with real numbers tells us that

$$\frac{-2}{3} = \frac{2}{-3} = -\frac{2}{3}.$$

We also know that $\dfrac{-2}{-3} = \dfrac{2}{3}.$

Our concern here is with our ability to *prove* the things we know to be true from our previous work. The patterns displayed above are the subject of the next four theorems.

THEOREM 7.11 $\dfrac{-x}{y} = -\dfrac{x}{y}$ $[y \neq 0]$.

7.4 Rational Expressions—Division

Proof. $\dfrac{-x}{y} = \dfrac{-1 \times x}{y}$ Theorem 5.5

$\phantom{\dfrac{-x}{y}} = -1 \times \dfrac{x}{y}$ Theorem 7.2

$\phantom{\dfrac{-x}{y}} = -\dfrac{x}{y}.$ Theorem 5.5

Thus, $\dfrac{-x}{y} = -\dfrac{x}{y}$ for all rational numbers $\dfrac{x}{y}$ $[y \neq 0]$.

Theorem 5.5 was important for the proof above. In this theorem it was asserted that the product of -1 and x is equal to the additive inverse of x, that is, $-x$.

Remember that the definition of division, $\dfrac{x}{y} = x \times \dfrac{1}{y}$, states that dividing x by y is the same as multiplying x by the multiplicative inverse of y. We shall use this definition to prove a theorem concerning division of any real number by -1. We shall assume that $\dfrac{1}{-1} = -1$.

THEOREM 7.12   *Division by* -1 $\dfrac{\frac{x}{y}}{-1} = -\dfrac{x}{y}$ $[y \neq 0]$.

Proof. $\dfrac{\frac{x}{y}}{-1} = \dfrac{x}{y} \times \dfrac{1}{-1}$ Def. div.

$\phantom{\dfrac{\frac{x}{y}}{-1}} = \dfrac{x}{y} \times (-1)$ $\dfrac{1}{-1} = -1$

$\phantom{\dfrac{\frac{x}{y}}{-1}} = -\dfrac{x}{y}.$ CPM; Theorem 5.5

Thus, $\dfrac{\frac{x}{y}}{-1} = -\dfrac{x}{y}$ $[y \neq 0]$.

Thus, any rational number divided by -1 is equal to the additive inverse of the number.

THEOREM 7.13 $\dfrac{x}{-y} = -\dfrac{x}{y}$ $[y \neq 0]$.

Proof. $\dfrac{x}{-y} = \dfrac{1 \times x}{-1 \times y}$ P1M Theorem 5.5

$\phantom{\dfrac{x}{-y}} = \dfrac{1}{-1} \times \dfrac{x}{y}$ Theorem 7.4

$\phantom{\dfrac{x}{-y}} = -1 \times \dfrac{x}{y}$ $\dfrac{1}{-1} = -1$

$\phantom{\dfrac{x}{-y}} = -\dfrac{x}{y}.$ Theorem 5.5

Thus, $\dfrac{x}{-y} = -\dfrac{x}{y}$ $[y \neq 0].$

THEOREM 7.14 $\dfrac{-x}{-y} = \dfrac{x}{y}$ $[y \neq 0].$

Proof. $\dfrac{-x}{-y} = \dfrac{-1 \times x}{-1 \times y}$ Theorem 5.5

$\phantom{\dfrac{-x}{-y}} = \dfrac{x}{y}.$ Theorem 7.7

Thus, $\dfrac{-x}{-y} = \dfrac{x}{y}$ $[y \neq 0].$

THEOREM 7.15 $-\dfrac{x}{-y} = \dfrac{x}{y}$ $[y \neq 0].$

Proof. $-\dfrac{x}{-y} = \dfrac{-x}{-y}$ Theorem 7.11

$\phantom{-\dfrac{x}{-y}} = \dfrac{x}{y}.$ Theorem 7.14

THEOREM 7.16 $-\dfrac{-x}{y} = \dfrac{x}{y}$ $[y \neq 0].$

Proof. $-\dfrac{-x}{y} = \dfrac{-(-x)}{y}$ Theorem 7.11

$\phantom{-\dfrac{-x}{y}} = \dfrac{x}{y}.$ Theorem 5.6

THEOREM 7.17 $\dfrac{x(-y)}{r(-s)} = \dfrac{xy}{rs}$ $[r \neq 0, s \neq 0].$

Proof. $\dfrac{x(-y)}{r(-s)} = \dfrac{-(xy)}{-(rs)}$ Theorem 5.7

$\phantom{\dfrac{x(-y)}{r(-s)}} = \dfrac{xy}{rs}.$ Theorem 7.14

7.4 Rational Expressions—Division

Observe how each subsequent theorem depends on the theorems previously proved. This is an illustration of the *deductive nature* of mathematics.

EXERCISE 7.3

1. Using appropriate theorems, write equivalent expressions in simplest form. Only permissible values of the variables are considered.

a. $\dfrac{2}{3} \div \dfrac{5}{6}$
b. $\dfrac{3}{4} \div \dfrac{9}{8}$

c. $\dfrac{5}{m} \div \dfrac{2}{3}$
d. $\dfrac{4}{t} \div \dfrac{6}{v}$

e. $\dfrac{x}{6} \div \dfrac{2}{3}$
f. $\dfrac{m}{8} \div \dfrac{k}{7}$

g. $\dfrac{2}{5} \div \dfrac{x}{t}$
h. $\dfrac{k}{r} \div \dfrac{5}{7}$

i. $\dfrac{t}{r} \div \dfrac{v}{7}$
j. $\dfrac{m}{p} \div \dfrac{r}{t}$

k. $\dfrac{2}{3} \div \dfrac{5}{x-4}$
l. $\dfrac{5}{3} \div \dfrac{x-4}{t+2}$

m. $\dfrac{r-3}{7} \div \dfrac{t-4}{2}$
n. $\dfrac{a}{x-c} \div \dfrac{b}{t-d}$

o. $\dfrac{-3}{7} \div \dfrac{x-3}{2m}$
p. $\dfrac{5}{t-3} \div \dfrac{k}{-4}$

q. $\dfrac{2a}{5m} \div \dfrac{x-c}{r+3}$
r. $\dfrac{3x}{2} \div \dfrac{6c}{4}$

s. $\dfrac{x}{2} \div \dfrac{x}{-2}$
t. $\dfrac{4}{c-7} \div \dfrac{6}{t+2}$

u. $\dfrac{2x}{3} \div \dfrac{2x}{3}$
v. $\dfrac{f+2}{3} \div \dfrac{2+f}{3}$

w. $\dfrac{2(c-5)}{7} \div \dfrac{2c-10}{14}$
x. $\dfrac{x-4}{8} \div \dfrac{4-x}{8}$

y. $\dfrac{3}{c-9} \div \dfrac{-3}{9-c}$
z. $\dfrac{b^2}{2} \div \dfrac{b}{4}$

2. For each of the following, write an equivalent rational expression in simplest form. Use as few "−" signs as possible.

a. $\dfrac{-6x}{-8y}$
b. $-\dfrac{21b}{-3ab}$

c. $\dfrac{-2wr}{-6w}$ d. $-\dfrac{6+c}{t-3}$

e. $-\dfrac{3}{-x-y}$ f. $-\dfrac{-(f-7)}{g-t}$

g. $\dfrac{(-a)b}{(-c)(-d)}$ h. $-\dfrac{m+n}{-3}$

i. $\dfrac{a(-b)}{(-c)d}$ j. $-\dfrac{-3(x^2)}{12x}$

k. $\dfrac{2x(-3y)}{9r(-8s)}$ l. $\dfrac{3(a-b)}{5(b-a)}$

m. $-\dfrac{x(-y)}{rs}$ n. $-\dfrac{x-y}{r-s}$

o. $\dfrac{-8(-5c)}{6(2t)}$

3. Label each statement with T for true and F for false.

a. $-\dfrac{1}{5-3}=\dfrac{1}{3-5}$ b. $-\dfrac{1}{3-7}=\dfrac{1}{7-3}$

c. $-\dfrac{1}{5-1}=\dfrac{1}{5-1}$ d. $-\dfrac{1}{2+3}=\dfrac{1}{-2-3}$

e. $-\dfrac{1}{4+1}=\dfrac{1}{4+1}$ f. $-\dfrac{1}{2+(-3)}=\dfrac{1}{-2-(-3)}$

g. $-\dfrac{1}{-1+(-1)}=\dfrac{1}{1-(-1)}$ h. $-\dfrac{-(2-1)}{3-2}=\dfrac{2-1}{3-2}$

i. $-\dfrac{-(2-6)}{5-9}=\dfrac{6-2}{5-9}$ j. $-\dfrac{-(5-1)}{3-7}=\dfrac{5-1}{7-3}$

k. $-\dfrac{1}{-(5-3)}=\dfrac{-1}{5-3}$ l. $-\dfrac{1}{-(1-7)}=\dfrac{-1}{7-1}$

4. Tell which of the following are true for all permissible replacements of the variables.

a. $(2a)(3b)=6ab$ b. $(-5x)(7y)=2xy$

c. $\left(\dfrac{1}{2}xy\right)(4a)=2axy$ d. $5xy-3ab+xy+7ab$
$\qquad\qquad\qquad\qquad\qquad\qquad =4ab+5xy$

e. $3mn+2k-mn-6k$ f. $\dfrac{10a}{5b}=\dfrac{2a}{b}$
$\qquad\qquad =-4k+3$

g. $\dfrac{10r}{2rs}=\dfrac{5}{s}$ h. $\dfrac{-3x}{2y}=-\dfrac{3x}{2y}$

i. $-\dfrac{-5x}{-3x}=\dfrac{5x}{3x}$ j. $-\dfrac{-w}{r}=\dfrac{-w}{-r}$

7.5 Theorems about Addition and Subtraction

k. $-\dfrac{s}{-t} = \dfrac{s}{t}$

l. $-\dfrac{(-x) \times 3}{y - 7} = \dfrac{(x)(3)}{(y)(7)}$

m. $\dfrac{t}{-y} = \dfrac{-t}{y}$

n. $-\dfrac{-s}{x} = -\dfrac{s}{x}$

o. $-\dfrac{1}{r - s} = \dfrac{1}{s - r}$

p. $-\dfrac{1}{(x + y)} = \dfrac{1}{-x - y}$

q. $\dfrac{-(r - s)}{(x - y)} = \dfrac{r - s}{x - y}$

r. $-\dfrac{1}{-(y - x)} = \dfrac{-1}{y - x}$

s. $-\dfrac{(x - y)}{(r - s)} = \dfrac{xy}{s - r}$

t. $-\dfrac{-(x - y)}{(r - s)} = \dfrac{y - x}{r - s}$

u. $-\dfrac{-(x - y)}{-(r - s)} = \dfrac{x - y}{r - s}$

v. $-\dfrac{-(x - y)(r - s)}{(s - t)(x + y)} = \dfrac{(x - y)(r - s)}{(s - t)(x + y)}$

w. $-\dfrac{(x - y)(r - s)}{-(s - t)(x + y)} = \dfrac{(x - y)(r - s)}{(t - s)(x + y)}$

5. Prove that each is true for all permissible values of the variables.

a. $\dfrac{x - y}{y - x} = -1$

b. $-\dfrac{x(-y)}{rs} = \dfrac{xy}{rs}$

c. $-\dfrac{(x - y)}{r - s} = \dfrac{y - x}{r - s}$

d. $\dfrac{(-x)(-y)}{rs} = \dfrac{xy}{rs}$

7.5 Theorems about Addition and Subtraction

From our work with real numbers, we know that the following are true.

$$\dfrac{2}{5} + \dfrac{1}{5} = \dfrac{2 + 1}{5}$$

$$\dfrac{2}{9} + \dfrac{5}{9} = \dfrac{2 + 5}{9}$$

$$\dfrac{3}{7} + \dfrac{-2}{7} = \dfrac{3 + (-2)}{7}.$$

The pattern displayed by these examples is stated in the next theorem.

THEOREM 7.18 $\dfrac{x}{z} + \dfrac{y}{z} = \dfrac{x + y}{z}$ $\quad [z \neq 0]$.

Proof. $\dfrac{x}{z} + \dfrac{y}{z} = x\left(\dfrac{1}{z}\right) + y\left(\dfrac{1}{z}\right)$ $\quad$ Def. div.

$\qquad = (x + y)\dfrac{1}{z}$ $\quad$ Theorem 5.1

$\qquad = \dfrac{x + y}{z}.$ $\quad$ Def. div.

Thus, $\dfrac{x}{z}+\dfrac{y}{z}=\dfrac{x+y}{z}\qquad [z \neq 0]$.

Theorem 7.18 deals with rational expressions which have the same denominator. Observe now some examples of addition of rational expressions in which denominators are not the same.

$$\frac{2}{3}+\frac{1}{4}=\frac{2\times 4+1\times 3}{3\times 4}$$
$$\frac{5}{6}+\frac{2}{5}=\frac{5\times 5+2\times 6}{6\times 5}$$
$$\frac{2}{5}+\frac{4}{7}=\frac{2\times 7+4\times 5}{5\times 7}.$$

The preceding examples were selected so that the two denominators in each case are relatively prime.

We state the pattern for relatively prime denominators as a theorem and prove it.

THEOREM 7.19 $\quad \dfrac{x}{y}+\dfrac{z}{w}=\dfrac{xw+zy}{yw}\qquad [y \neq 0, w \neq 0]$.

Proof. $\quad\begin{aligned}\dfrac{x}{y}+\dfrac{z}{w}&=\dfrac{xw}{yw}+\dfrac{zy}{wy}\qquad &\text{Theorem 7.7}\\ &=\dfrac{xw}{yw}+\dfrac{zy}{yw}\qquad &\text{CPM}\\ &=\dfrac{xw+zy}{yw}. \qquad &\text{Theorem 7.18}\end{aligned}$

Thus, $\dfrac{x}{y}+\dfrac{z}{w}=\dfrac{xw+zy}{yw}\qquad [y \neq 0, w \neq 0]$.

In previous work with real numbers we encountered expressions in which the denominators were not relatively prime, that is, they had at least one common factor different from 1. For example, $\frac{3}{8}$ and $\frac{1}{6}$ are such expressions since 8 and 6 have a common factor different from 1, namely the number 2. Addition of such numbers can be carried out as follows:

$$\frac{3}{8}+\frac{1}{6}=\frac{3\times 3}{8\times 3}+\frac{1\times 4}{6\times 4}=\frac{9}{24}+\frac{4}{24}=\frac{13}{24}.$$

Observe that 24 is the least common multiple of 8 and 6.

7.5 Theorems about Addition and Subtraction

Another example of this pattern is the following:

$$\frac{2}{9} + \frac{5}{12} = \frac{2 \times 4}{9 \times 4} + \frac{5 \times 3}{12 \times 3} = \frac{8}{36} + \frac{15}{36} = \frac{23}{36}.$$

Here 36 is the least common multiple of 9 and 12. The last two examples may be generalized:

$$\frac{x}{y} + \frac{z}{w} = \frac{xr}{yr} + \frac{zs}{ws},$$

where y and w are not relatively prime, and $yr = ws$. Furthermore, to avoid dealing with unnecessarily large numbers, we choose r and s so that yr is the least common multiple of y and w.

Notice that since $yr = ws$, we have

$$\frac{x}{y} + \frac{z}{w} = \frac{xr}{yr} + \frac{zs}{ws}$$
$$= \frac{xr}{yr} + \frac{zs}{yr}$$
$$= \frac{xr + zs}{yr}.$$

Theorems 7.18 and 7.19 can easily be extended to subtraction, since addition and subtraction are related in the following manner:

$$x - y = x + (-y).$$

Thus, we have the following two theorems.

THEOREM 7.20 $\dfrac{x}{z} - \dfrac{y}{z} = \dfrac{x - y}{z}$ $[z \neq 0]$.

Proof. $\dfrac{x}{z} - \dfrac{y}{z} = x\left(\dfrac{1}{z}\right) - y\left(\dfrac{1}{z}\right)$ Def. div.

$\phantom{Proof.\ \dfrac{x}{z} - \dfrac{y}{z}} = (x - y)\left(\dfrac{1}{z}\right)$ Theorem 5.1

$\phantom{Proof.\ \dfrac{x}{z} - \dfrac{y}{z}} = \dfrac{x - y}{z}.$ Def. div.

THEOREM 7.21 $\dfrac{x}{y} - \dfrac{z}{w} = \dfrac{xw - zy}{yw}$ $[y \neq 0, w \neq 0]$.

Proof. $\dfrac{x}{y} - \dfrac{z}{w} = \dfrac{xw}{yw} - \dfrac{zy}{wy}$ Theorem 7.7

$\phantom{\dfrac{x}{y}} = \dfrac{xw}{yw} - \dfrac{zy}{yw}$ CPM

$\phantom{\dfrac{x}{y}} = \dfrac{xw - zy}{yw}$ Theorem 7.20

EXERCISE 7.4

1. Label each statement with T for true and F for false.
   a. $\dfrac{1}{2} + \dfrac{1}{3} = \dfrac{1}{5}$ b. $\dfrac{2}{3} + \dfrac{7}{7} = \dfrac{7}{10}$
   c. $\dfrac{4}{5} - \dfrac{2}{5} = \dfrac{2}{5}$ d. $\dfrac{5}{2} - \dfrac{1}{2} = 2$
   e. $\dfrac{2}{3} + \dfrac{6}{3} = \dfrac{8}{3}$ f. $\dfrac{4}{7} - \dfrac{1}{6} = 3$

2. Which of the following are true for all permissible replacements of the variables?

   a. $\dfrac{a}{4} + \dfrac{a}{4} = \dfrac{2a}{8}$ b. $\dfrac{d}{6} + \dfrac{d}{6} = \dfrac{d}{3}$

   c. $\dfrac{c-3}{7} + \dfrac{3-c}{7} = 0$ d. $\dfrac{3m}{7m} + \dfrac{4m}{7m} = 1$

   e. $\dfrac{3t-5}{k-2} + \dfrac{4}{3} = \dfrac{3t-1}{3k-6}$ f. $\dfrac{2t-5}{4} + \dfrac{5-2t}{5} = 0$

   g. $\dfrac{3}{a-b} + \dfrac{3}{b-a} = 0$ h. $\dfrac{5}{4t} + \dfrac{3}{5t} = \dfrac{12t + 25t}{20t}$

   i. $\dfrac{5x}{9} - \dfrac{4x}{5} = \dfrac{x}{4}$ j. $\dfrac{d}{3} - \dfrac{d}{3} = 0$

   k. $\dfrac{2}{c} - \dfrac{9}{c} = -\dfrac{7}{c}$ l. $\dfrac{5}{t-3} - \dfrac{9}{t-3} = \dfrac{-4}{t-3}$

   m. $\dfrac{3f}{2} - \dfrac{4f}{5} = \dfrac{7f}{10}$ n. $\dfrac{7r}{5} - \dfrac{3r-5}{5} = \dfrac{4r-5}{5}$

   o. $\dfrac{3d+5}{8} - \dfrac{9-7d}{3}$ p. $\dfrac{5}{3a} - \dfrac{6+c}{2} = \dfrac{10 - 18a + 3ac}{6a}$

   $= \dfrac{9d + 15 - 72 + 56d}{24}$

3. Write a single rational expression that is equivalent to each of the following. State the answers in simplest form without using parentheses.

   a. $\dfrac{5}{x} + \dfrac{2}{3}$ b. $\dfrac{3}{x} + \dfrac{2}{a}$

   c. $\dfrac{5a}{7b} + \dfrac{3x}{7b}$ d. $\dfrac{3a}{5} + \dfrac{2}{7}$

7.5 Theorems about Addition and Subtraction

e. $\dfrac{2x+4}{5} + \dfrac{7x}{5}$ f. $\dfrac{5a}{9} + \dfrac{3c}{2}$

g. $\dfrac{5a}{7} + \dfrac{3x}{4m}$ h. $\dfrac{5a}{7b} + \dfrac{3x}{4m}$

i. $\dfrac{5a}{7b} + \dfrac{3x}{4a}$ j. $\dfrac{6x}{23} + \dfrac{12-2x}{23}$

k. $\dfrac{22}{5(a-7)} + \dfrac{4}{5(a-7)}$ l. $\dfrac{3-4t}{5} + \dfrac{6+3t}{2}$

m. $\dfrac{(7-k)}{4} + \dfrac{(k-8)}{4}$ n. $\dfrac{3}{a-t} + \dfrac{5}{6}$

o. $\dfrac{5}{-7} + \dfrac{-2(c-4)}{a+b}$ p. $\dfrac{k+5}{2k-4} + \dfrac{3}{-7}$

q. $\dfrac{5x}{6} + \dfrac{x}{2}$ r. $\dfrac{x-2}{2x} + \dfrac{2x+1}{x}$

s. $\dfrac{2(x+1)}{-3c} + \dfrac{3(x-1)}{-12c}$ t. $-\dfrac{5}{x} - \dfrac{2}{3}$

u. $\dfrac{3}{t} - \dfrac{5}{u}$ v. $\dfrac{f}{7} - \dfrac{f}{3}$

w. $\dfrac{5a}{2b} + \dfrac{3x}{2b}$ x. $\dfrac{9}{g} - \dfrac{5}{g}$

y. $\dfrac{x-7}{3} - \dfrac{x-5}{3}$ z. $\dfrac{2x+5}{4} - \dfrac{4+3x}{4}$

a'. $\dfrac{2t+3}{5} - \dfrac{3t+1}{6}$ b'. $\dfrac{4}{a-2} - \dfrac{3}{7}$

c'. $\dfrac{3x}{4} - \dfrac{x}{2}$ d'. $\dfrac{5}{6c} - \dfrac{3}{4}$

e'. $\dfrac{x-2}{8} - \dfrac{x}{4}$ f'. $\dfrac{2x+1}{6} - \dfrac{3x}{8}$

g'. $\dfrac{5}{6c} - \dfrac{3}{4c}$ h'. $\dfrac{x+y}{4a} - \dfrac{x-y}{10a}$

i'. $\dfrac{2(x+y)}{9a} - \dfrac{x-y}{15a}$

4. Tell which of the following are true for all permissible replacements of the variables.

a. $(6r - 12s) \div 6 = r - 2s$ b. $\tfrac{1}{3}(4x - 5y) = \tfrac{4}{3}x - \tfrac{5}{3}y$

c. $\dfrac{-(2r+5s)}{2} = -r + \tfrac{5}{2}s$ d. $\tfrac{2}{3}r \div \tfrac{4}{5}s = \dfrac{5r}{6s}$

e. $\dfrac{\frac{3}{5}x}{\frac{2}{8}t} = \dfrac{12x}{5t}$

f. $\dfrac{3(x-y)}{2(y-x)} = \dfrac{3}{2}$

g. $\dfrac{6}{5}r \div \dfrac{5}{3}s = 2\dfrac{r}{(s)}$

h. $\dfrac{r-2y}{3} = \dfrac{r}{3} - \dfrac{2}{3}y$

i. $\dfrac{3xy - 8xr}{2x} = \dfrac{3}{2}y - 4r$

j. $(x+5) \div (-5) = -\dfrac{x}{5} - 1$

k. $\dfrac{5}{6}(24x - 3y) = 20x - \dfrac{5}{2}y$

l. $\dfrac{1}{5}(x-y) - \dfrac{1}{3}(y-x) = \dfrac{8}{15}x + \dfrac{2}{15}y$

m. $\dfrac{4r-3s}{5} - \dfrac{2r+5s}{7} = \dfrac{18}{35}r - \dfrac{8}{7}s$

n. $\dfrac{2}{7}(4x-7) + \dfrac{1}{7}(3x+8) = \dfrac{11}{7}x + \dfrac{6}{7}$

o. $\dfrac{-5x-11y}{5} - \dfrac{4x+3y}{7} = -\dfrac{11}{7}x - \dfrac{92}{35}y$

5. Write a single rational expression equivalent to each of the following. State the answers in simplest form without using parentheses.

a. $\dfrac{5}{x^2-y^2} + \dfrac{3}{x+y}$
b. $\dfrac{-3}{x^2-y^2} + \dfrac{7}{x-y}$

c. $\dfrac{x}{x^2-y^2} - \dfrac{y}{x-y}$
d. $\dfrac{2}{a(b+c)} + \dfrac{1}{a}$

e. $\dfrac{a}{x+y} - \dfrac{b}{x-y}$
f. $\dfrac{x+y}{x-y} + \dfrac{x-y}{x+y}$

g. $\dfrac{-2(a+b)}{a} - \dfrac{3(a+b)}{b}$
h. $\dfrac{3(a+b+c)}{a} + \dfrac{4(a+b+c)}{b}$

6. For each expression give a simpler equivalent expression.

a. $\dfrac{1}{2}(3r-7s) - \dfrac{4}{3}(r+5s)$
b. $\dfrac{2}{3}(2x+4y-5r) - \dfrac{5}{6}r$

c. $\dfrac{2}{3}(\dfrac{5}{8}r - \dfrac{15}{7}s) + \dfrac{2}{9}(r+2s)$
d. $(5x - \dfrac{3}{7}y) \div \dfrac{4}{3}$

e. $\dfrac{\frac{5}{8}x}{\frac{3}{4}y}$
f. $\dfrac{4}{5}(3x-2y) \div 5$

g. $\dfrac{\frac{7}{3}(y-r) + \frac{1}{3}y}{5}$

7.6 Complex Rational Expressions

Below are several examples of *complex rational expressions*.

$$\frac{\frac{2}{3}}{\frac{1}{2}} \qquad \frac{\frac{a}{b}}{\frac{c}{d}} \qquad \frac{x+\frac{y}{2}}{3x-\frac{7}{3}} \qquad \frac{\frac{x}{8}}{7} \qquad \frac{\frac{a}{b+2}}{\frac{c}{4}}$$

$$\frac{\frac{2x+3}{2}}{\frac{x+y}{z}} \qquad \frac{\frac{x}{y}}{3} \qquad \frac{\frac{a}{b}+\frac{c}{d}}{\frac{x}{y}+\frac{r}{s}} \qquad \frac{\frac{x+y}{x-y}}{\frac{a}{b}}$$

We would like to find expressions which are equivalent to those above and are not as complicated. We have already had some experience in doing this. The key for this work is Theorem 7.10, which states that $\frac{x}{y} \div \frac{r}{s} = \frac{x}{y} \times \frac{s}{r}$. For example, according to this theorem,

$$\frac{\frac{x}{8}}{7} = \frac{x}{8} \div 7 = \frac{x}{8} \times \frac{1}{7} = \frac{x}{56} \qquad \frac{\frac{2}{3}}{\frac{1}{2}} = \frac{2}{3} \div \frac{1}{2} = \frac{2}{3} \times \frac{2}{1} = \frac{4}{3}.$$

There is another method for simplifying complex expressions. It is based on Theorem 7.7:

$$\frac{xr}{yr} = \frac{x}{y} \qquad [y \neq 0, r \neq 0],$$

and on the Property of One for Multiplication:

$$1 \times x = x.$$

Below are examples of a second method of simplifying complex rational expressions. Study these examples and give particular attention to the various names for 1 chosen to accomplish the purpose in the quickest way possible. Keep in mind that only permissible values of the variables are considered.

Example 1 $\quad \dfrac{\frac{x}{y}}{3} = \dfrac{\frac{x}{y} \times y}{3 \times y} = \dfrac{x}{3y}.$

Notice that the step $\dfrac{\frac{x}{y}}{3} = \dfrac{\frac{x}{y} \times y}{3 \times y}$ is justified by Theorem 7.7.

We replaced $\frac{x}{y} \times y$ by x on the basis of the following proof.

$$\frac{x}{y} \times y = \left(x \times \frac{1}{y}\right)y \qquad \text{Def. div.}$$
$$= x\left(\frac{1}{y} \times y\right) \qquad \text{APM}$$
$$= x(1) \qquad \text{Prop. mult. inv.}$$
$$= x. \qquad \text{P1M}$$

Example 2 $\quad \dfrac{\frac{a}{b}}{\frac{c}{d}} = \dfrac{\frac{a}{b} \times bd}{\frac{c}{d} \times bd} = \dfrac{ad}{bc}.$

Example 3 $\quad \dfrac{\frac{6}{7}}{\frac{7}{11}} = \dfrac{6 \times 11}{\frac{7}{11} \times 11} = \dfrac{66}{7}.$

Example 4 $\quad \dfrac{x + \frac{y}{2}}{3x - \frac{7}{3}} = \dfrac{\left(x + \frac{y}{2}\right)(2 \times 3)}{\left(3x - \frac{7}{3}\right)(2 \times 3)}$
$$= \dfrac{6x + 3y}{18x - 14}.$$

Example 5 $\quad \dfrac{\frac{a}{b+2}}{\frac{c}{4}} = \dfrac{\frac{a}{b+2} \times 4(b+2)}{\frac{c}{4} \times 4(b+2)}$
$$= \dfrac{4a}{c(b+2)}.$$

Example 6 $\quad \dfrac{\frac{2x+3}{2}}{\frac{x+y}{z}} = \dfrac{\frac{2x+3}{2} \times 2z}{\frac{x+y}{z} \times 2z}$
$$= \dfrac{(2x+3)z}{(x+y)2}$$
$$= \dfrac{z(2x+3)}{2(x+y)}.$$

There is a shorter way of simplifying expressions like those above. Theorems 7.4 and 7.10 give the following result:

7.6 Complex Rational Expressions

$$\frac{\frac{x}{y}}{\frac{r}{s}} = \frac{xs}{yr}.$$

The following two examples illustrate the direct application of this result.

State the nonpermissible replacements of the variables.

Example 1 $\quad \dfrac{\dfrac{x+y}{x-y}}{\dfrac{a}{b}} = \dfrac{(x+y)b}{(x-y)a}.$

Example 2 $\quad \dfrac{\dfrac{a}{b} + \dfrac{c}{d}}{\dfrac{x}{y} + \dfrac{r}{s}} = \dfrac{\dfrac{ad+bc}{bd}}{\dfrac{xs+ry}{ys}}$

$$= \frac{(ad+bc)ys}{bd(xs+ry)}.$$

EXERCISE 7.5

1. Using the method you prefer, simplify each of the following:

a. $\dfrac{\frac{1}{2}}{\frac{2}{3}}$
b. $\dfrac{\frac{3}{7}}{\frac{4}{5}}$
c. $\dfrac{\frac{1}{3}}{\frac{5}{7}}$

d. $\dfrac{\frac{3}{2}}{5}$
e. $\dfrac{\frac{1}{7}}{10}$
f. $\dfrac{\frac{3}{2}}{15}$

g. $\dfrac{\frac{1}{3}}{12}$
h. $\dfrac{\frac{4}{7}}{2}$
i. $\dfrac{\frac{5}{9}}{3}$

j. $\dfrac{5}{\frac{1}{3}}$
k. $\dfrac{6}{\frac{2}{7}}$
l. $\dfrac{7}{\frac{1}{4}}$

m. $\dfrac{-2}{\frac{3}{2}}$
n. $\dfrac{-4}{-\frac{5}{3}}$
o. $\dfrac{-\frac{1}{2}}{6}$

p. $\dfrac{-\frac{5}{6}}{-7}$

2. Using the method you prefer, find a simpler equivalent expression for each of the following. In each case specify the restriction on the replacements of the variables.

a. $\dfrac{\frac{a}{2}}{\frac{3}{5}}$

b. $\dfrac{\frac{x}{3}}{\frac{1}{3}}$

c. $\dfrac{\frac{2}{m}}{\frac{2}{5}}$

d. $\dfrac{\frac{4}{5}}{\frac{a}{3}}$

e. $\dfrac{\frac{1}{7}}{\frac{3}{c}}$

f. $\dfrac{\frac{a}{b}}{\frac{2}{3}}$

g. $\dfrac{\frac{1}{2}}{\frac{m}{n}}$

h. $\dfrac{\frac{a}{b}}{\frac{c}{d}}$

i. $\dfrac{13}{\frac{x}{3}}$

j. $\dfrac{12}{\frac{2}{n}}$

k. $\dfrac{9}{\frac{x}{y}}$

l. $\dfrac{m}{\frac{a}{b}}$

m. $\dfrac{\frac{3}{2}}{x}$

n. $\dfrac{\frac{a}{3}}{k}$

o. $\dfrac{\frac{a}{b}}{c}$

p. $\dfrac{\frac{a+b}{2}}{6}$

q. $\dfrac{5}{\frac{x-y}{k}}$

r. $\dfrac{\frac{a}{b+c}}{x}$

s. $\dfrac{x+y}{\frac{a}{b}}$

t. $\dfrac{\frac{m}{n}}{x-y}$

u. $\dfrac{\frac{a-b}{3}}{\frac{x}{y}}$

v. $\dfrac{6-\frac{7}{r}}{3+\frac{2}{r}}$

w. $\dfrac{6x-y}{\frac{3}{y}}$

x. $\dfrac{\frac{3}{x}-4}{x+2}$

y. $\dfrac{1}{\frac{x}{y}}$

z. $\dfrac{2-\frac{7}{s}}{s-\frac{3}{2}}$

3. Simplify. Specify restrictions on the replacements of the variables.

a. $\dfrac{x - \frac{3-y}{x}}{2x - 3}$

b. $\dfrac{\frac{-x}{y}}{\frac{r}{t}}$

c. $\dfrac{3 + \dfrac{r}{5}}{r - \dfrac{6}{7}}$

d. $\dfrac{3 + \dfrac{5}{x+y}}{7 + \dfrac{3}{x+y}}$

e. $\dfrac{\dfrac{a-b}{a+b}}{\dfrac{a+b}{a-b}}$

f. $\dfrac{\dfrac{x^2 - y^2}{x+y}}{\dfrac{x-y}{x+y}}$

g. $\dfrac{r + \dfrac{s}{t}}{r - \dfrac{s}{t}}$

h. $\dfrac{2x - y}{\dfrac{3}{2} - \dfrac{x}{3}}$

i. $\dfrac{7 - \dfrac{1}{2x}}{6 + \dfrac{1}{x^2}}$

j. $\dfrac{r - \dfrac{7}{9}}{\dfrac{2}{r} - 3}$

k. $\dfrac{\dfrac{a+b+c}{2}}{\dfrac{a+b+c}{3}}$

l. $\dfrac{\dfrac{(x+y)^3}{n}}{\dfrac{(x+y)^3}{m}}$

7.7 Equations with Rational Expressions

Some equations which we encounter in algebra are of the form $\dfrac{x}{y} = \dfrac{z}{w}$ $[y \neq 0, w \neq 0]$. We can prove a theorem which will be quite helpful in solving such equations.

THEOREM 7.22 If $\dfrac{x}{y} = \dfrac{z}{w}$, then $xw = yz$ $[y \neq 0, w \neq 0]$.

Proof.
$\dfrac{x}{y} = \dfrac{z}{w}$

$\dfrac{x}{y} \times yw = \dfrac{z}{w} \times yw$ Rt. hand mult. prop. for equations

$\dfrac{xyw}{y} = \dfrac{zyw}{w}$ Theorem 7.1

$xw = zy$ Theorem 7.7

$xw = yz.$ CPM

Thus, if $\dfrac{x}{y} = \dfrac{z}{w}$, then $xw = yz$ $\quad [y \neq 0, w \neq 0]$.

We now illustrate the use of this theorem in solving an equation.

$$\frac{2t+1}{3} = \frac{3t+8}{-2}$$
$$(2t+1)(-2) = 3(3t+8)$$
$$-4t - 2 = 9t + 24$$
$$-4t - 2 + 4t = 9t + 24 + 4t$$
$$-2 = 13t + 24$$
$$-2 - 24 = 13t + 24 - 24$$
$$-26 = 13t$$
$$-2 = t.$$

Check.

$\dfrac{2t+1}{3}$	$\dfrac{3t+8}{-2}$
$\dfrac{2(-2)+1}{3}$	$\dfrac{3(-2)+8}{-2}$
$\dfrac{-4+1}{3}$	$\dfrac{-6+8}{-2}$
$\dfrac{-3}{3}$	$\dfrac{2}{-2}$
-1	-1

Thus, $\{-2\}$ is the solution set of the given equation.

EXERCISE 7.6

Using Theorem 7.22, solve each equation.

1. $\dfrac{x}{2} = \dfrac{3}{2}$
2. $\dfrac{9}{x} = \dfrac{2}{3}$
3. $\dfrac{5}{3} = \dfrac{x}{6}$
4. $\dfrac{-2}{3} = \dfrac{-6}{x}$
5. $\dfrac{\frac{1}{3}}{x} = \dfrac{1}{12}$
6. $\dfrac{\frac{1}{2}}{x} = \dfrac{1}{-6}$
7. $\dfrac{\frac{2}{3}}{x} = \dfrac{-2}{9}$
8. $\dfrac{-\frac{3}{4}}{x} = \dfrac{3}{20}$
9. $\dfrac{x}{\frac{1}{3}} = 6$
10. $\dfrac{x}{\frac{1}{2}} = -10$

11. $\dfrac{x}{-\frac{2}{3}} = 6$

12. $\dfrac{x+1}{2} = \dfrac{2x-1}{3}$

13. $\dfrac{-2}{2(x+1)} = \dfrac{3}{x-2}$

14. $\dfrac{-(x-3)}{-2} = \dfrac{-3(x+1)}{5}$

GLOSSARY

Complex rational expression: Example: $\dfrac{\frac{x-y}{3}}{\frac{x+y}{x}}$.

Multiplicative Identity Theorem: $\dfrac{xr}{yr} = \dfrac{x}{y}$ $[y \neq 0, r \neq 0]$.

Multiplicative Inverses Theorem: The multiplicative inverse of $\dfrac{x}{y}$ is $\dfrac{y}{x}$ $[x \neq 0, y \neq 0]$.

Rational expression: An expression which can be written in the form $\dfrac{P}{Q}$, where P and Q are polynomials $[Q \neq 0]$.

CHAPTER REVIEW PROBLEMS

1. For each expression, write a single expression equivalent to it. Assume that only permissible replacements of the variables may be used.

   a. $\dfrac{x}{3} + \dfrac{y}{3}$
   b. $\dfrac{a}{2} + \dfrac{b}{6}$
   c. $\dfrac{a}{x} + \dfrac{m}{3x}$

   d. $\dfrac{x}{t} - \dfrac{r}{s}$
   e. $\dfrac{x}{3} \times \dfrac{y}{5}$
   f. $\dfrac{a}{x} \times \dfrac{b}{y}$

   g. $\dfrac{a+b}{x} \times \dfrac{m+n}{y}$
   h. $\dfrac{y}{t} \div \dfrac{3}{2}$
   i. $\dfrac{a}{m} \div \dfrac{x}{y}$

2. Label each statement with T for true and F for false.

   a. $\dfrac{(-4)(-7)}{5} = -7 \times \dfrac{-4}{5}$
   b. $\dfrac{4}{-3 \times 5} = \dfrac{4}{-3} \times \dfrac{4}{5}$

   c. $\dfrac{4}{5} \div \dfrac{3}{7} = \dfrac{7}{3} \times \dfrac{4}{5}$
   d. $\dfrac{1}{-\frac{3}{5}} = -\dfrac{3}{5}$

e. $\dfrac{-3(-5)}{-7(-5)} = \dfrac{3}{7}$
f. $\dfrac{-3}{-(-4)} = \dfrac{3}{4}$
g. $\dfrac{3}{-4} + \dfrac{7}{4} = \dfrac{3+7}{-4}$
h. $-\dfrac{-(2-6)}{5} = \dfrac{6-2}{5}$
i. $\dfrac{4}{3} + \dfrac{2}{7} = \dfrac{4 \times 3 + 2 \times 7}{3 \times 7}$
j. $\dfrac{\frac{4}{5}}{\frac{3}{7}} = \dfrac{5 \times 3}{4 \times 7}$

3. Tell which of the following are true for all permissible replacements of the variables.

a. $\dfrac{x(a+b)}{a+b} = x$
b. $\left(\dfrac{1}{y}\right)[(-x)(y)] = -x$
c. $\dfrac{a-b}{-1} = b - a$
d. $(-1)(a-b) = -a - b$
e. $\dfrac{1}{x} + \dfrac{1}{y} = \dfrac{1}{xy}$
f. $\dfrac{m}{n} - \dfrac{n}{m} = 0$
g. $-\dfrac{x+y}{x-y} = \dfrac{x+y}{y-x}$
h. $\dfrac{\frac{x-2}{x}}{\frac{x-4}{x+1}} = \dfrac{(x+1)(x-2)}{x(x-4)}$
i. $r - \dfrac{1}{r} = 0$
j. $\dfrac{1+n}{2+n} = \dfrac{1}{2}$
k. $\dfrac{2-m}{1-m} = 2$
l. $\dfrac{\frac{a}{b}}{\frac{c}{d}} = \dfrac{ac}{bd}$
m. $\dfrac{\frac{m+n}{x}}{y} = \dfrac{m+n}{yx}$
n. $\dfrac{m+n}{\frac{x}{y}} = \dfrac{(m+n)x}{y}$
o. $(-2z)(\tfrac{1}{2}yx) = -xyz$

4. For each of the following, find a simpler equivalent expression.

a. $\dfrac{\frac{3}{4}}{\frac{7}{8}}$
b. $\dfrac{\frac{2}{7}}{\frac{7}{2}}$
c. $\dfrac{5}{\frac{3}{7}}$
d. $\dfrac{\frac{4}{5}}{6}$
e. $\dfrac{\frac{a}{3}}{\frac{x}{4}}$
f. $\dfrac{\frac{m}{n}}{\frac{4}{x}}$

g. $\dfrac{\dfrac{a+b}{3}}{\dfrac{c}{d}}$

h. $\dfrac{\dfrac{2a-b}{3}}{\dfrac{a+b}{2}}$

i. $\dfrac{\dfrac{x-y}{a}}{\dfrac{x+y}{b}}$

j. $\dfrac{\dfrac{a+b}{c+d}}{\dfrac{x+y}{z+w}}$

k. $\dfrac{x-\dfrac{1}{2}}{x+\dfrac{3}{4}}$

l. $\dfrac{1-\dfrac{a}{b}}{1-\dfrac{c}{d}}$

m. $\dfrac{3-\dfrac{1}{x}}{4-\dfrac{1}{y}}$

n. $\dfrac{5a-\dfrac{1}{2}}{3a+\dfrac{1}{2}}$

o. $\dfrac{\dfrac{a}{b}+\dfrac{c}{d}}{\dfrac{x}{y}}$

p. $\dfrac{\dfrac{a}{b}+\dfrac{c}{d}}{\dfrac{m}{n}+\dfrac{p}{r}}$

q. $\dfrac{\dfrac{3(a+b)}{2}}{\dfrac{6(a-b)}{8}}$

r. $\dfrac{\dfrac{4(x-y)}{3u}}{\dfrac{10(a+b)}{6}}$

5. In each expression tell what replacements of the variables are not permissible.

a. $\dfrac{2}{x-3}$

b. $\dfrac{1}{2x+7}$

c. $\dfrac{x}{3-2x}$

d. $\dfrac{x+y}{x-y}$

e. $\dfrac{2x+3y}{4x+5y}$

f. $\dfrac{x+y}{(2x+1)(3x-7)}$

6. Solve each equation.

a. $\dfrac{3}{7}=\dfrac{x}{-2}$

b. $\dfrac{x}{\frac{2}{3}}=\dfrac{6}{15}$

c. $\dfrac{\frac{1}{2}}{x}=\dfrac{-5}{6}$

d. $\dfrac{3}{7}=\dfrac{\frac{1}{14}}{x}$

7. Prove that each of the following is true for all permissible values of the variables.

a. $\dfrac{-(-x)}{y}=\dfrac{x}{y}$

b. $\dfrac{-x-y}{x+y}=-1$

c. $\dfrac{\frac{1}{x}}{\frac{1}{y}}=\dfrac{y}{x}$

d. $\dfrac{1}{x}-\dfrac{1}{y}=\dfrac{y-x}{xy}$

8
Applications of Equations and Inequalities

8.1 Equation Solving

Much of the work in algebra consists of solving equations. In cases where equations represent certain physical phenomena, their solutions give answers to situations which are significant to the scientist, the businessman, or the engineer. In this chapter we shall examine the various uses of equations.

To refresh your skills in equation solving, study the following examples.

Example 1   Solve: $4m - 1.5m = 2m + 5$
$$2.5m = 2m + 5$$
$$2.5m - 2m = 2m + 5 - 2m$$
$$.5m = 5$$
$$m = 10.$$

Check.

$4m - 1.5m$	$2m + 5$
$4(10) - 1.5(10)$	$2(10) + 5$
$40 - 15$	$20 + 5$
25	25

8.1 Equation Solving

The solution set is $\{10\}$. This is sometimes written as

$$\{m \mid 4m - 1.5m = 2m + 5\} = \{10\}$$

and is read "the set of all real numbers m such that $4m - 1.5m = 2m + 5$ is the set consisting of the number 10."

Example 2 Solve:
$$-\tfrac{1}{2}(3t - 5) = 4t - 3(t + \tfrac{1}{4})$$
$$-\tfrac{3}{2}t + \tfrac{5}{2} = 4t - 3t - \tfrac{3}{4}$$
$$-\tfrac{3}{2}t + \tfrac{5}{2} = t - \tfrac{3}{4}$$
$$-\tfrac{3}{2}t + \tfrac{5}{2} - t = t - \tfrac{3}{4} - t$$
$$-\tfrac{5}{2}t + \tfrac{5}{2} = -\tfrac{3}{4}$$
$$-\tfrac{5}{2}t + \tfrac{5}{2} - \tfrac{5}{2} = -\tfrac{3}{4} - \tfrac{5}{2}$$
$$-\tfrac{5}{2}t = -\tfrac{13}{4}$$
$$(-\tfrac{2}{5})(-\tfrac{5}{2}t) = (-\tfrac{2}{5})(-\tfrac{13}{4})$$
$$t = \tfrac{26}{20}$$
$$t = \tfrac{13}{10}, \text{ or } t = 1.3.$$

Check.

$-\tfrac{1}{2}(3t - 5)$	$4t - 3(t + \tfrac{1}{4})$
$-\tfrac{1}{2}(3 \times 1.3 - 5)$	$4 \times 1.3 - 3(1.3 + \tfrac{1}{4})$
$-\tfrac{1}{2}(3.9 - 5)$	$5.2 - 3 \times 1.55$
$-\tfrac{1}{2} \times (-1.1)$	$5.2 - 4.65$
.55	.55

The solution set is $\{1.3\}$. As before, this can be stated

$$\{t \mid -\tfrac{1}{2}(3t - 5) = 4t - 3(t + \tfrac{1}{4})\} = \{1.3\}.$$

Example 3 Solve:
$$5(y + \tfrac{2}{5}) - 3y = -5 - (-2y - 7)$$
$$5y + 2 - 3y = -5 + 2y + 7$$
$$2y + 2 = 2y + 2$$
$$2 = 2.$$

$2 = 2$ is a true statement regardless of the replacement made for y. Hence the equation

$$5(y + \tfrac{2}{5}) - 3y = -5 - (-2y - 7)$$

has as its roots *all* real numbers. Its solution set is the set of all real numbers. This can be written as

$$\{y \mid 5(y + \tfrac{2}{5}) - 3y = -5 - (-2y - 7)\} = R,$$

where R is the set of all real numbers. This set is equivalent to the set $\{y|2 = 2\}$, meaning the set of all y for which $2 = 2$. This last sentence is true regardless of the replacement made for y.

Example 4 Solve: $5(x - 2) + 2(x + 6) = 3x + 6 + 4(x + 1)$.

Simplifying,

$$5x - 10 + 2x + 12 = 3x + 6 + 4x + 4$$
$$7x + 2 = 7x + 10$$
$$2 = 10.$$

There is no replacement for x such that $2 = 10$ is true. Hence we say that the equation

$$5(x - 2) + 2(x + 6) = 3x + 6 + 4(x + 1)$$

has no roots, or that its solution set is the empty set. We are thinking of a set of values of x for which $5(x - 2) + 2(x + 6) = 3x + 6 + 4(x + 1)$ is true. This is written

$$\{x|5(x - 2) + 2(x + 6) = 3x + 6 + 4(x + 1)\} = \phi$$

and is read "the set of all x (assumed to be real numbers) such that $5(x - 2) + 2(x + 6)$ is equal to $3x + 6 + 4(x + 1)$ is the empty set." This can also be written as

$$\{x|2 = 10\} = \phi.$$

Recall that in working with expressions, we shall assume that only permissible replacements of the variables may be used.

EXERCISE 8.1

Determine the solution set of each equation.

1. $5(m + 2) = 10$
2. $3(x - 6) = -9$
3. $-2(2t - 3) = -6$
4. $-6(7y + 5) = 12$
5. $3(2a + 5) - 7(a - 6)$ $= 2(19 - 10a)$
6. $3x + 2 + x = -(x + 2x) + 18$
7. $7 + 6x + 5 = -3(2 - 2x) + 18$
8. $y + 13 = -(y + 3)$
9. $\dfrac{x + 3}{2} = 7$
10. $5(3 - 2v) = 4(2 - v) - 6v$
11. $3(x - 3) = 4(3 - x)$
12. $17(z + 7) = 15(z + 7)$
13. $5(2d - 6) = 5(6 - 2d)$
14. $1.5(12 - f) = 3 + 1.5f$
15. $-5(x + 11) = 5x + 5$
16. $-1 + 3r = 3(r - 1)$

8.1 Equation Solving

17. $\dfrac{y-7}{6} = 2y + 8$

18. $4(3v + 2) = 5(v + \frac{8}{5}) + 7v$

19. $3(x - 5) = 3(4 - x)$

20. $\dfrac{1}{v} + \dfrac{3}{v-1} = \dfrac{7}{v}$

21. $\dfrac{y}{\frac{1}{3}} = -6$

22. $\dfrac{3}{r-3} + \dfrac{5}{r} = \dfrac{25}{2r}$

23. $3x - 7 = -(7 - 3x)$

24. $\dfrac{4x + 7}{3} = \dfrac{4}{3}x + 7$

25. $\dfrac{t+6}{4} = 2(t + 2.5)$

26. $y - 1 = -y - (y + 1) + 2y$

27. $\dfrac{v-3}{v} = \dfrac{1}{4}$

28. $\dfrac{4(x-1)}{2} = 2x - 3$

29. $\dfrac{4}{3}(s - 3) = s + 2 - \left(6 - \dfrac{s}{3}\right)$

30. $\dfrac{4}{x} + \dfrac{3}{x} + \dfrac{7}{8} = \dfrac{6}{x}$

31. $4t = \dfrac{1}{3}t - \dfrac{11}{3}$

32. $\dfrac{7}{y+2} = \dfrac{11}{y}$

33. $\dfrac{3}{2-v} = \dfrac{3}{v-2}$

34. $\dfrac{3}{4-s} = \dfrac{-5}{2s-5}$

35. $\dfrac{5}{r-3} = \dfrac{9}{3r-1}$

36. $\dfrac{c+1}{c} = \dfrac{1}{2}$

37. $3(n - 5) + 4(3n - 7) = -43$

38. $\frac{1}{2}(3x - 5) = -\frac{1}{2}(3x - 5)$

39. $2 \times (3 - x) = 2$

40. $5 - 8r = \dfrac{3(7 + 13r)}{2}$

41. $\dfrac{7}{x} - \dfrac{3}{x} - \dfrac{5}{x} = \dfrac{1}{2}$

42. $\dfrac{1 - 3x}{-7x} = \dfrac{1}{5}$

43. $y + 3 = \dfrac{3y + 14}{4}$

44. $\dfrac{2x - 1}{7} = \dfrac{1 - x}{3}$

45. $7 + \dfrac{4}{y} = \dfrac{1}{y} - \dfrac{3}{y}$

46. $\dfrac{5 - t}{2} = \dfrac{t + 4}{7}$

47. $(s - 7) \times 3 + 5(6 - s) = 8 \times (s + 9)$

48. $8 \times (4x - 7) = (6 + 7x) \times (-4)$

49. $6y + 9 = y$

50. $\dfrac{4}{s} - 3 = \dfrac{5}{s} - 7$

51. $\dfrac{1}{t} + 5 = \dfrac{2}{t} + 5\frac{1}{2}$

52. $\dfrac{5x - 3}{7} = x + 4$

53. $.3r + 1.5 = 1.5r + .5$

54. $5(x - 3) - 6x = 13$

55. $\dfrac{2x + 11}{4} = \dfrac{3x - 7}{5}$

56. $\dfrac{x - 4}{5} = \dfrac{3(x - 5)}{10}$

57. $\dfrac{4x-7}{x+3} = \dfrac{1}{2}$ 58. $\dfrac{3r-1}{-7} = r + 9$

8.2 Uses of Mathematics in Science

Mathematics has been instrumental in the development of present-day science and technology. In physics, for example, it is an indispensable tool for the study of many phenomena.

When dealing with *motion* in physics, the concept of *speed* arises. To consider the average speed s, we speak in terms of distance d and time t. The formula,

$$s = \frac{d}{t}$$

expresses the relation between the three concepts. Speed may be measured in miles per hour $\left(\dfrac{\text{mi}}{\text{hr}}\right)$, feet per second $\left(\dfrac{\text{ft}}{\text{sec}}\right)$, centimeters per second $\left(\dfrac{\text{cm}}{\text{sec}}\right)$, or meters per second $\left(\dfrac{\text{m}}{\text{sec}}\right)$. That is, speed is given as a quotient of a number of distance units and a number of time units.

For example, a car which covered the distance of 96 mi in 2 hr traveled at the speed of $\dfrac{96}{2}$, or 48 miles per hour. This is also written as 48 mph.

Question: 48 mph is how many ft per sec?

To answer this question it is essential to know the following two relations between the units involved:

$$1 \text{ mi} = 5280 \text{ ft,}$$
$$1 \text{ hr} = 3600 \text{ sec.}$$

In computing the answer to the question, the scientist sets up what are called *conversion factors*. For this problem, the following two conversion factors are needed:

$$\frac{100 \text{ ft}}{1 \text{ sec}} = \frac{100(\cancel{1 \text{ ft}})}{\cancel{1 \text{ sec}}} \times \frac{1 \text{ mi}}{5280(\cancel{1 \text{ ft}})} \times \frac{3600(\cancel{1 \text{ sec}})}{1 \text{ hr}}$$
$$= \frac{100(3600) \text{ mi}}{5280 \text{ hr}}$$
$$= 68 \frac{2}{11} \frac{\text{mi}}{\text{hr}}, \text{ or } 68 \frac{2}{11} \text{ mph.}$$

So, a car traveling $100 \dfrac{\text{ft}}{\text{sec}}$ travels $68 \dfrac{2}{11}$ mph.

8.2 Uses of Mathematics in Science

One concept which is related to motion is *acceleration*. It is the rate of change of velocity. Velocity is the speed in a given direction. We will use the following symbols:

a_{av} = average acceleration,
v_i = initial velocity at time t_i,
v_f = final velocity at time t_f.

The formula showing the relation between these is

$$a_{av} = \frac{v_f - v_i}{t_f - t_i}.$$

Frequently, it is necessary to determine any one of v_f, v_i, t_f, or t_i in terms of the remaining quantities. This is accomplished by the use of equation properties which were stated in Chapter 6. To illustrate, let us solve

$$a_{av} = \frac{v_f - v_i}{t_f - t_i}$$

for t_f. Study this solution and justify each step.

$$(t_f - t_i)a_{av} = v_f - v_i$$
$$t_f \times a_{av} - t_i \times a_{av} = v_f - v_i$$
$$t_f \times a_{av} = v_f - v_i + t_i \times a_{av}$$
$$t_f = \frac{v_f - v_i + t_i \times a_{av}}{a_{av}}.$$

Another important concept in physics as well as in chemistry is that of temperature. We know that water freezes at a termperature of 32° on the Fahrenheit scale and boils at 212° on the same scale. There is another important temperature scale called the Celsius (centigrade) scale, on which the freezing point of water is 0° and the boiling point is 100°. It is often necessary to interchange temperature measurements from one of these scales to the other.

We let F stand for the Fahrenheit reading and C for the Celsius reading. A formula which defines the relationship between these two scales is

$$F = 1.8C + 32.$$

To verify this formula for the freezing point of water, we replace C in the formula by 0 to mean 0°.

$$F = 1.8 \times 0 + 32$$
$$F = 0 + 32$$
$$F = 32.$$

Thus, the temperature of 0° Celsius is the same as 32° Fahrenheit.

For the boiling point of water, we replace C by 100 to mean 100°.

$$F = 1.8 \times 100 + 32$$
$$F = 180 + 32$$
$$F = 212.$$

Thus, the temperature of 100° Celsius is the same as 212° Fahrenheit.

What Fahrenheit reading corresponds to the 25° Celsius reading? To find it, we replace C by 25 in the formula.

$$F = 1.8(25) + 32$$
$$F = 45 + 32$$
$$F = 77; \text{ or } 25°C \text{ corresponds to } 77°F.$$

If we replace F by 68, we can determine the value of C.

$$68 = 1.8C + 32$$
$$36 = 1.8C$$
$$C = \frac{36}{1.8}$$
$$C = 20; \text{ or } 68°F \text{ is equivalent to } 20°C.$$

We can express C in terms of F.

$$F = 1.8C + 32$$
$$F - 32 = 1.8C$$
$$\frac{F - 32}{1.8} = C$$
$$\text{or } C = \frac{F - 32}{1.8}.$$

We say that the formula $F = 1.8C + 32$ was *solved* for C.

You will discover that you can successfully work with scientific formulas without fully understanding the physical phenomena with which the formulas deal. In the following exercises you will perform certain mathematical manipulations which the scientist employs when seeking answers to his problems. You may wish to refer to a science book and look up the meaning of the scientific concepts mentioned in the exercises.

EXERCISE 8.2

1. A car traveling at 70 mph travels at how many ft per sec?
2. A car traveling at 50 ft per sec travels at how many mph?
3. Solve the formula $a_{av} = \dfrac{v_f - v_i}{t_f - t_i}$ for

   a. t_i **b.** v_f **c.** v_i

8.2 Uses of Mathematics in Science

4. The formula for the specific heat S of a substance in terms of heat in calories H, mass in grams m, final temperature t_2, and initial temperature t_1 is

$$S = \frac{H}{m(t_2 - t_1)}.$$

a. Determine S to one decimal place for the following given replacements:

27 for H, 16 for m, 23 for t_2, and 19 for t_1

b. Solve the formula for H in terms of other variables.
c. Solve the formula for m.
d. Solve the formula for t_2.
e. Solve the formula for t_1.

5. The formula for the distance traveled by a freely falling object is $d = \frac{1}{2}gt^2$, where g is the acceleration due to gravity and t is time. Determine d for the following given replacements:

32 for g and 6 for t

a. Express g in terms of d and t.
b. Express t in terms of d and g.

6. The formula for the greatest height h attained by a body projected vertically upward is $h = \frac{v^2}{2g}$, where v is velocity and g is the acceleration

a. Solve the formula for v.
b. Solve the formula for g.
c. If v is doubled and g remains the same, how will h change?

7. The formula for the velocity v of a liquid flowing through an opening is $v = \sqrt{2gh}$, where g is the acceleration due to gravity and h is the height of the liquid above the opening.

a. If g is replaced by 32 and h by 2, determine v.
b. Solve the formula for h.
c. If h is quadrupled, how is v changed?

8. The formula for the heat H in calories developed from a current flowing through a wire is $H = \frac{Eit}{4.18}$, where E is the difference in potential or voltage, i is the rate of flow in amperes, and t is the time in seconds.

a. Solve the formula for E.

b. Solve the formula for t.
   c. If each of E, i, and t is doubled, how will H change?
   d. If E is doubled, i is tripled, and t is quadrupled, how will H change?

9. The formula for electric current I in a circuit of two cells connected in series in terms of the electromotive force E of the cell, external resistance R, and internal resistance r is

$$I = \frac{2E}{R + 2r}.$$

   a. Determine I to one decimal place for the following given replacements:

   2.2 for E, 14 for R, and 7 for r

   b. Solve the formula for E.
   c. Solve the formula for R.
   d. Solve the formula for r.

10. The formula involving the effect of temperature and pressure on the volume of a confined gas is $\frac{P_1 V_1}{T_1} = \frac{P_2 V_2}{T_2}$, where P_1, V_1, and T_1 are the original pressure, volume, and temperature, respectively. P_2, V_2, and T_2 are the final pressure, volume, and temperature, respectively.

   a. Solve the formula for P_1.
   b. Solve the formula for V_2.
   c. Solve the formula for T_2.

11. The formula relating the density D of an object to its weight W and volume V is $D = \frac{W}{V}$.

   a. If W is replaced by 13 and V by 6.5, determine D.
   b. Express W in terms of D and V.
   c. Express V in terms of D and W.

12. The formula for the total resistance R of two conductors in a parallel hook-up is $R = \frac{r_1 r_2}{r_1 + r_2}$, where r_1 and r_2 are the resistances of the two conductors, respectively.

   a. If r_1 is replaced by 48 and r_2 by 62, determine R to one decimal place.
   b. Express r_1 in terms of r_2 and R.

13. The formula relating the rate I of flow of an electric current to the electromotive force or voltage E and resistance R of the wire through which the current flows is $I = \dfrac{E}{R}$.
   a. If E is replaced by 110 and R by 20, determine I.
   b. Solve the formula for E.
   c. Solve the formula for R.

14. The formula for the kinetic energy E of a moving object in terms of weight W, velocity v, and the acceleration due to gravity g is $E = \dfrac{Wv^2}{2g}$.
   a. Solve the formula for W.
   b. Solve the formula for v.

15. Given the formula $m_i c_i (T_i - T_m) = m_w c_w (T_m - T_w)$, solve for
   a. T_i b. T_m

8.3 Formulas as Equations

Formulas are used in mathematics, business, science, and other areas because of their usefulness in solving significant problems. Since formulas are equations, we can draw on our knowledge of equations to deal with them. In the problems below you will practice using formulas concerned with geometry.

EXERCISE 8.3

In each problem, you are given a formula. Do the required computations. In exercises that involve π, use 3.14 as an approximation for π.

Example Volume of a rectangular solid: $V = lwh$.
  Compute the volume when $l = 12$ in., $w = 9$ in., and $h = 3$ in.
  $V = 12 \times 9 \times 3$ cu in. $= 324$ cu in.
  The volume is 324 cu in.

1. Perimeter of a square: $p = 4s$.
   a. The perimeter when $s = 7$ in.
   b. The length of the side of the square when $p = 37$ in.

s

2. Perimeter of a rectangle: $p = 2l + 2w$.
   a. The perimeter when $l = 2$ ft, $w = 13$ in.
   b. The length of the rectangle if the perimeter is 100 in. and the width is 15 in.

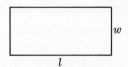

3. Volume of a cube: $V = e^3$.
   a. The volume of the cube if $e = 7$ in.
   b. The length of one edge if $V = 1000$ cu in.

4. Volume of a right circular cylinder: $V = \pi r^2 h$.
   a. The volume when $r = 3.5$ in., $h = 17$ in.
   b. The height when $V = 255$ cu in., $r = 2$ in.

5. Volume of a sphere: $V = \frac{4}{3}\pi r^3$.
Compute the volume when $r = 5$ in.

6. Surface area of a sphere: $A = 4\pi r^2$.
Compute the area when $r = 1.5$ ft.

Solve each formula for the indicated letter.

Example $V = lwh$.

    a. for l b. for w c. for h

    $l = \dfrac{V}{wh}$ $w = \dfrac{V}{lh}$ $h = \dfrac{V}{lw}$

7. $p = 4s$;   for s 8. $p = 2l + 2w$
                                     a. for l
                                     b. for w

9. $V = e^2$;   for e 10. $V = \pi r^2 h$
                                     a. for h
                                     b. for r

11. $V = \frac{4}{3}\pi r^3$;   for r

8.4 Using Equations to Solve Problems

We have seen that in order to solve a problem by using an equation, we need to write an equation that fits the problem, find the root of the equation, and then convince ourselves that the root of the equation "works" in the problem.

It is very important that we first read the problem thoughtfully to fully understand it. Then we should write the conditions of the problem in the form of a mathematical sentence.

You should always attempt to develop your own ways of solving problems. Use your ingenuity and be clever, remembering that there are often many different ways of solving the same problem. Below are three examples.

Example 1 Alice has a bank which contains $2.65 in dimes and nickels. If Alice has eleven more nickels than dimes, how many of each type of coin does she have?

One Way to Solve. Since there are only two kinds of coins, we know that the product of .05 (value of a nickel) and the number of nickels, added to the product of .10 (value of a dime) and the number of dimes will yield 2.65 ($2.65 in the bank).

We do not actually know how many dimes or nickels she has, but we do know that there are eleven more nickels than dimes. If we let x stand for the number of dimes, then $x + 11$ would stand for the number of nickels. The following equation would fit the story of the problem:

$$.05(x + 11) + .10x = 2.65.$$

Solving the equation,

$$100[.05(x + 11) + .10x] = 100 \times 2.65$$
$$5(x + 11) + 10x = 265$$
$$5x + 55 + 10x = 265$$
$$15x = 210$$
$$x = 14.$$

[*Hint:* 100 used as a multiplier clears the equation of decimals.]

Thus, there are 14 dimes and 14 + 11 or 25 nickels.

Check. Dimes $14 \times .10 = 1.40$ or $1.40
         Nickels $25 \times .05 = 1.25$ or $1.25
                                              Total $2.65

Example 2 A family vacation budget allowed three times as much for food as for camping expenses and four times as much for travel as for camping expenses. If the total allowed for food, camping expenses, and travel was $168, how much was allotted to travel?

One Way to Solve. We note the relationship between food and camping expenses, and also between camping expenses and travel.
If we let x stand for the number of dollars spent on food, then $\frac{1}{3}x$ would stand for the number of dollars spent for camping expenses. It would then follow that $\frac{4}{3}x$ would stand for the number of dollars allowed for travel. The sum of x, $\frac{1}{3}x$, and $\frac{4}{3}x$ should be 168.

$$x + \tfrac{1}{3}x + \tfrac{4}{3}x = 168$$
$$\frac{3x + x + 4x}{3} = 168$$
$$\frac{8x}{3} = 168$$
$$\frac{x}{3} = 21$$
$$x = 63.$$

Check. Food expense $63
Camping expense $21
Travel expense $84
           Total $168

Another Way to Solve. Perhaps a less complicated approach would be to let y stand for the amount spent for camping; then $3y$ would stand for food expense and $4y$ would stand for the travel expense in dollars.
  Thus, our equation would be

$$y + 3y + 4y = 168$$
$$8y = 168$$
$$y = 21.$$

The camping expense was $21. The travel expense allotment was 4×21 which is 84, or $84. The food allotment was 3×21 which is 63, or $63.

Example 3 The sum of two numbers is 97. One number is 19 greater than the other number. What are the two numbers?

One Way to Solve. One number is 19 greater than the other. If we let y stand for the smaller number, then the larger number may be represented by $y + 19$.

8.4 Using Equations to Solve Problems

Our equation is
$$y + (y + 19) = 97.$$

Solve:
$$2y + 19 = 97$$
$$2y = 78$$
$$y = 39.$$

Thus, the smaller number is 39, and the larger of the two is $39 + 19$, or 58.

Check. Smaller number 39
Larger number 58
Sum 97
The difference is $58 - 39$, or 19.

Example 4 Joe is now fifteen years younger than Jane. In six years Jane will be twice as old as Joe will be then. What are their ages now?

One Way to Solve. We know that the difference in their ages is fifteen years. If we let x stand for Jane's present age in years, then we can express Joe's age in terms of x. It is $x - 15$. We may represent Jane's age six years from now by $x + 6$, and Joe's age six years from now by $(x - 15) + 6$, or $x - 9$. Are these expressions, $x + 6$ and $x - 9$, equivalent? Rereading the problem reveals that Jane's age six years from now is twice as great as Joe's age. That is, Joe's age must be doubled (multiplied by 2) to make it equal to Jane's age. Thus, we have

$$2(x - 9) = x + 6.$$

Solve:
$$2(x - 9) = x + 6$$
$$2x - 18 = x + 6$$
$$x = 24.$$

Thus, Jane is 24 years old, and Joe is $24 - 15$ or 9 years old.

Check. Ages now   Ages six years from now
  Jane 24 years   Jane 30 years
  Joe   9 years   Joe   15 years

Difference: 15 years.

Jane will be twice as old as Joe six years from now.

Example 5 The number of tickets sold for a church supper was 215.

Adult tickets were $.75 each and children's tickets were $.35 each. The sale of both kinds of tickets amounted to $135.25. How many adult tickets were sold?

One Way to Solve. The product of .75 and the number of adults, added to the product of .35 and the number of children will yield 135.25. (We are dropping the dollar signs temporarily to simplify writing.) If we sold 100 adult tickets, we would have 215 − 100 or 115 children. If we had 150 adults, we would have 215 − 150 or 65 children. Thus, if we let r stand for the number of adult tickets, then 215 − r will stand for the number of children's tickets.

Thus, our equation is

$$.75r + .35(215 - r) = 135.25.$$

Solve:

$$100[.75r + .35(215 - r)] = 100 \times 135.25$$
$$75r + 35(215 - r) = 13,525$$
$$75r + 7525 - 35r = 13,525$$
$$40r + 7525 = 13,525$$
$$40r = 6000$$
$$r = 150.$$

Thus, 150 adult tickets and 215 − 150, or 65 children's tickets were sold.

Check. $150 \times .75 = 112.50$ or $112.50
$65 \times .35 = 22.75$ or $ 22.75
Total $135.25

EXERCISE 8.4

Solve each problem and check.

1. Miss Thrifty has $9.00 in nickels and dimes. She has twice as many dimes as she has nickels. How many coins of each kind does Miss Thrifty have?

2. Mr. Tight has a jar full of nickels. If he removes 50 nickels and replaces them with 70 dimes, he will have $1\frac{1}{2}$ times the original amount of money. How much money does Mr. Tight have in the jar?

3. A jar full of pennies and nickels contains three times as many nickels as pennies. The total amount of money in the jar is $5.60. How many coins of each kind are there in the jar?

4. Mr. Games has dice, marbles, and coins—136 of these objects in all.

8.4 Using Equations to Solve Problems

If the number of marbles is twice the number of dice, and the number of coins is five times the number of dice, how many of each kind does he have?

5. Mrs. Homemaker buys three cans of different vegetables. Can A weighs $2\frac{1}{2}$ ounces more than can B, and can C weighs $10\frac{1}{3}$ ounces more than can B. The total weight of the three cans is $48\frac{5}{6}$ ounces. How many ounces does can A weigh?

6. Multiplying a number by 3 gives the same result as adding 4 to the number. What is the number?

7. Taking one-half of a number gives the same result as adding 5 to the number. What is the number?

8. Dividing a number by 5 gives the same result as adding 16 to the number. What is the number?

9. What two numbers have 59 as their sum and 11 as their difference?

10. Joe Old is presently five years older than Suzie Young. When Joe's age doubles, Suzie's age will triple. How old is each of them?

11. John is six years older than Jerry. In nine years Jerry will be two-thirds as old as John. What is John's age now?

12. Ann has $4.75 made up of half dollars and quarters. If there is a total of thirteen coins, how many coins of each kind does she have?

13. A bank contained four fewer dimes than quarters, ten more nickels than dimes, and three times as many half dollars as quarters. The total value of the coins was $11.30. How many dimes were there?

14. College Y has 1431 students, and it has $3\frac{1}{2}$ times as many boys as girls. What is the enrollment of girls?

15. The number of women at a mixed gathering was 4 less than one-third of the entire group present. This number of women was also equal to one-fourth of the entire number of people present. How many women were present?

16. There are 36 students in a mathematics class. The girls outnumber the boys two to one. How many boys are there in the class?

17. The combined ages of Ken and Tom are 23. If Tom is five years younger than Ken, how old is each boy?

18. Sharon's age now is three times what it was eight years ago. How old is Sharon now?

19. Ken's age five years ago was twice one-third of his present age. How old is Ken now?

20. A number is multiplied by 4. The quotient of 8 and 4 is added to this product. The result is 0. What was the original number?

21. A bank contains $4.50 made up of nickels and dimes. If there are fifteen more dimes than nickels, how many coins of each kind are there?

22. An equilateral triangle and a square have the same perimeter. If each side of the equilateral triangle is 16 in. long, how long is each side of the square?

23. An equilateral triangle and a rectangle have the same perimeter. The length of the rectangle is twice its width. If each side of the triangle is 22 in. long, what are the width and the length of the rectangle?

24. The perimeters of a square and a rectangle are the same. Each side of the square is 20 in. long. The length of the rectangle is four times its width. What are the length and the width of the rectangle?

8.5 Inequalities

So far in this chapter we have explored the uses of equations in solving various problems. In Chapter 5 we learned how to solve simple inequalities using the graphing method. We now explore inequalities and their applications to a greater extent. Before using inequalities, however, we must learn to solve them algebraically. A brief deductive development will serve as a basis for our work with them.

First, recall that having a number line in standard position, as shown in Figure 8.1, enables us to tell which one of two given numbers is less than the other. It is the one whose point is to the left of the other. For example, $-4 < -1$ because point M (corresponding to -4) is to the left of point C (corresponding to -1). Similarly $-2 < 0, -6 < -5, -4 < 3, 0 < 2$, and $1 < 5$.

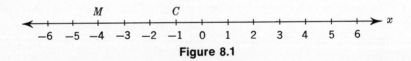

Figure 8.1

Now we pose the following two questions.

(1) Given that $x < y$, is there a unique number which we can add to x and obtain y?
(2) If so, what kind of a number is it?

To answer the first question, consider the examples we used above.

$-4 < -1$ and $-4 + 5 = -1$; so there is a unique number, 5, which when added to -4 gives -1.

$-2 < 0$ and $-2 + 2 = 0$.

8.6 Addition and Subtraction Theorems for Inequalities

$-6 < -5$; what number added to -6 gives -5?
$-4 < 3$; what number added to -4 gives 3?
$0 < 2$; what number added to 0 gives 2?
$1 < 5$; what number added to 1 gives 5?

It appears that if $x < y$, there is a unique number p such that $x + p = y$.

Now to answer the second question: what kind of a number is p? In the first example, it was 5; in the second it was 2; in the third, 1; in the fourth, 7; in the fifth, 2; and in the last example it was 4. It appears that each of these numbers is positive. This leads us to the following definitions.

DEFINITION 8.1 For any two real numbers x and y, x *is less than* y (written $x < y$) means that there is a positive real number p such that $x + p = y$.

DEFINITION 8.2 For any two real numbers x and y, x *is greater than* y (written $x > y$) means that $y < x$.

8.6 Addition and Subtraction Theorems for Inequalities

Each of the following statements is true.

If $2 < 3$, then $2 + 5 < 3 + 5$ ($7 < 8$).
If $-2 < 0$, then $-2 + 6 < 0 + 6$ ($4 < 6$).
If $-5 < -4$, then $-5 + (-3) < -4 + (-3)$ ($-8 < -7$).

The examples above suggest the following theorem.

THEOREM 8.1 If $x < y$, then $x + z < y + z$.

Proof.

$x < y$	Given
$x + p = y$ $\quad [p > 0]$	Def. of $<$
$(x + p) + z = y + z$	Rt.-hand add. prop.
$(x + z) + p = y + z$	TR
$x + z < y + z$	Def. of $<$

Thus, if $x < y$, then $x + z < y + z$.

This theorem is helpful in simplifying inequalities. For example, the inequality $x + 7 < -3$ can be simplified as follows:

$$x + 7 + (-7) < -3 + (-7)$$
$$x < -10.$$

The inequality, $x < -10$, is simpler than $x + 7 < -3$, and according to Theorem 8.1, both have the same solution set. Thus, the solution set of $x + 7 < -3$ is $\{x|x < -10\}$. Its graph is shown on the number line in Figure 8.2, assuming, of course, that the set of all real numbers is the replacement set.

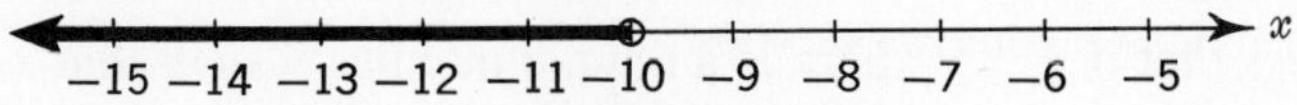

$$\xleftarrow{\quad\;|\;\;\;|\;\;\;|\;\;\;|\;\;\;|\;\;\bullet\;\;|\;\;\;|\;\;\;|\;\;\;|\;\;\;|\;}\rightarrow x$$
$$\;-15\;-14\;-13\;-12\;-11\;-10\;\;-9\;\;-8\;\;-7\;\;-6\;\;-5$$

Figure 8.2

Since addition and subtraction are related, we can prove a theorem which is analogous to Theorem 8.1, but which involves subtraction.

THEOREM 8.2 If $x < y$, then $x - z < y - z$.

Proof.
$x < y$ — Given
$x + (-z) < y + (-z)$ — Theorem 8.1
$x - z < y - z$. — Def. subt.

Thus, if $x < y$, then $x - z < y - z$.

In some inequalities, it is more convenient to use Theorem 8.2 than Theorem 8.1, as is shown in the following example:

$$x + 26 < 10$$
$$x + 26 - 26 < 10 - 26$$
$$x < -16.$$

Thus, the solution set of $x + 26 < 10$ is $\{x|x < -16\}$.

From our work with equations, we know that the relation *is equal to* (=) has three properties, which are named as follows:

REFLEXIVE PROPERTY OF EQUALITY $x = x$.

SYMMETRIC PROPERTY OF EQUALITY If $x = y$, then $y = x$.

TRANSITIVE PROPERTY OF EQUALITY If $x = y$ and $y = z$, then $x = z$.

Any relation having the reflexive, symmetric, and transitive properties is called an *equivalence relation*.

EXERCISE 8.5

1. For each number on the left of $<$, give the positive number which when added to the number will give the number on the right.

8.7 Multiplication Theorems for Inequalities

Example $-12 < 16$
             28, since $-12 + 28 = 16$.

a. $-1 < 0$ b. $-2 < 3$ c. $-10 < -9$
d. $2 < 17$ e. $-16 < 16$ f. $0 < 20$

2. Determine the solution set of each inequality and graph it.
   a. $x + 6 < 9$ b. $8 + x < 5$
   c. $7 < x + 1$ d. $-7 < 3 + x$
   e. $x - 9 > 5$ f. $12 + x > -4$
   g. $-4 > x - 7$ h. $6 > -2 + x$
   i. $-1 > -5 + x$ j. $-2 > x - 13$

3. Using Definition 8.1, state a theorem about $>$ which is analogous to
   a. Theorem 8.1   b. Theorem 8.2

4. Prove that if $x + z < y + z$, then $x < y$. [*Hint:* Use Theorem 8.1 and add $-z$ to each member of $x + z < y + z$.]

5. State the theorem about $>$ which is analogous to the theorem you proved in Problem 4.

6. a. Give an example to show that $<$ does not have the reflexive property.
    b. Give an example to show that $<$ does not have the symmetric property.
    c. Is $<$ an equivalence relation? Why or why not?

7. Prove that if $x < y$ and $y < z$, then $x < z$. That is, prove that $<$ is transitive. [*Hint:* Use the definition of $<$ and the transitive property of $=$.]

8. State the transitive property of $>$.

8.7 Multiplication Theorems for Inequalities

We shall make the following two definitions for future use.

DEFINITION 8.3 $a > 0$ means a is *positive*.

DEFINITION 8.4 $a < 0$ means a is *negative*.

Now observe the pattern displayed by the following true statements.

If $6 < 8$, then $6 \times 3 < 8 \times 3$ $(18 < 24)$.
If $-2 < 0$, then $-2 \times 5 < 0 \times 5$ $(-10 < 0)$.
If $-4 < -1$, then $-4 \times 2 < -1 \times 2$ $(-8 < -2)$.

These statements suggest the following theorem.

THEOREM 8.3 If $x < y$ and $z > 0$, then $xz < yz$.

Proof.
$x < y$ and $z > 0$	Given
$x + p = y \quad [p > 0]$	Def. of $<$
$(x + p)z = yz$	Rt.-hand mult. prop.
$xz + pz = yz$	RDPMA
$pz > 0$	Clos. of pos. nos. under mult.
$xz < yz$.	Def. of $<$

Thus, we see that each member of $x < y$ may be multiplied by a positive number z, and the inequality $xz < yz$ results.

Observe the pattern displayed by the following true statements.

If $4 < 6$, then $4(-2) > 6(-2) \quad (-8 > -12)$.
If $-5 < 0$, then $-5(-1) > 0(-1) \quad (5 > 0)$.
If $-3 < -1$, then $-3(-4) > -1(-4) \quad (12 > 4)$.

These statements suggest the following theorem.

THEOREM 8.4 If $x < y$ and $z < 0$, then $xz > yz$.

Proof.
$x < y$ and $z > 0$	Given
$x + p = y \quad [p > 0]$	Def. of $<$
$(x + p)z = yz$	Rt.-hand mult. property
$xz + pz = yz$	RDPMA
$xz = yz + [-(pz)]$	Rt.-hand add. property
$yz + [-(pz)] = xz$	Symm. property $=$
$-(pz) > 0$	$p > 0$ and $z < 0$
$yz < xz$	Def. of $<$
$xz > yz$.	Def. 8.2

Thus, if $x < y$ and $z < 0$, then $xz > yz$.

Using Theorem 8.4 and the definition of division, we obtain two more theorems.

THEOREM 8.5 If $x < y$ and $z > 0$, then $\frac{x}{z} < \frac{y}{z}$.

Proof. $x < y$ and $z > 0$
$\frac{1}{z} > 0$

8.7 Multiplication Theorems for Inequalities

$$x \times \frac{1}{z} < y \times \frac{1}{z}$$
$$\frac{x}{z} < \frac{y}{z}.$$

Thus, if $x < y$ and $z > 0$, then $\frac{x}{z} < \frac{y}{z}$.

The second step in the proof may have been somewhat of a puzzle to you. Let us prove that if $z > 0$, then $\frac{1}{z} > 0$.

We know that the square of every nonzero real number is positive, that is, $\left(\frac{1}{z}\right)^2 > 0$.

(Notice the convenient use of $\frac{1}{z}$ as any nonzero number.)

Thus, $\frac{1}{z^2} > 0$.

Now given that $z > 0$, we have

$$z \times \frac{1}{z^2} > 0 \times \frac{1}{z^2}$$
$$\frac{z}{z^2} > 0$$
$$\frac{1}{z} > 0.$$

THEOREM 8.6 If $x < y$ and $z < 0$, then $\frac{x}{z} > \frac{y}{z}$.

Proof. $x < y$ and $z < 0$.
$$\frac{1}{z} < 0$$
$$x \times \frac{1}{z} > y \times \frac{1}{z}$$
$$\frac{x}{z} > \frac{y}{z}.$$

Note that in the second step above we use the fact that the multiplicative inverse of a negative number is negative.

In reading about inequalities, we will encounter statements of the form $a < b < c$, which is an abbreviation for $a < b$ and $b < c$. This abbreviation is used in the next three theorems.

THEOREM 8.7 If $x < y < z$, then $x + m < y + m < z + m$.

Proof. $x < y < z$
$x < y$ and $y < z$
$x + m < y + m$
$y + m < z + m$
$x + m < y + m < z + m$.

THEOREM 8.8 If $x < y < z$ and $m > 0$, then $xm < ym < zm$.

Proof. $x < y < z$ and $m > 0$
$x < y$ and $y < z$
$xm < ym$
$ym < zm$
$xm < ym < zm$.

THEOREM 8.9 If $x < y < z$ and $m < 0$, then $xm > ym > zm$.

Proof. $x < y < z$ and $m < 0$
$x < y$ and $y < z$
$xm > ym$
$ym > zm$
$xm > ym > zm$.

EXERCISE 8.6

1. Using the definition of *is greater than* ($>$), state the four theorems for $>$ which are analogous to Theorems 8.3, 8.4, 8.5, and 8.6.

2. Solve each inequality.

Example 1 $5x < 12$
$$\frac{5x}{5} < \frac{12}{5}$$
$$x < \frac{12}{5}.$$

Thus, $\{x | x < \frac{12}{5}\}$ is the solution set of $5x < 12$.

Example 2 $-4x < 20$
$$\frac{-4x}{-4} > \frac{20}{-4}$$
$$x > -5.$$

Thus, $\{x | x > -5\}$ is the solution set of $-4x < 20$.

a. $2x < 6$ b. $-3x < 18$
c. $4x > 8$ d. $-5x > 15$
e. $\frac{1}{2}x < 6$ f. $-\frac{2}{3}x > 8$
g. $\frac{4}{3}x < 12$ h. $-\frac{2}{5}x > -4$

3. Restate Theorems 8.7, 8.8, and 8.9 in terms of $>$.
4. Determine the solution set of each inequality.
   a. $x + 3 < 2x - 1$ b. $5 - 3x > x - 3$
   c. $1 + x < 2x - 7$ d. $6 - 2x < -3 + 7x$
   e. $2(x - 1) < 3x + 12$ f. $3(x + 2) > 2(x - 1)$
   g. $-3(2x - 1) + 2 < 2(4x + 1)$ h. $-2(3 - 4x) + 1 > 4 - 3(1 - 2x)$
5. Determine the solution set of each inequality.
   a. $-10 < 3x - 2 < 7$ b. $-1 < \frac{x - 4}{5} < 1$
   c. $-1 \le 4x + 3 \le 11$ d. $0 \le 5x + 1 < 6$
   e. $-6 \le 3 - x \le 9$ f. $-1 < 5 - 2x < 7$
   g. $x < x + 2 < x + 5$ h. $x < x - 3 < x - 7$
   i. $x \le 2x \le 3x$ j. $x - 1 \le 2x + 2 \le 4x$

8.8 Absolute Value In Inequalities

Before examining the techniques of solving inequalities involving absolute value, let us recall the definition of absolute value:

$$|x| = x \text{ for all real numbers } x \ge 0,$$
$$|x| = -x \text{ for all real numbers } x < 0.$$

Thus, the absolute value of a real number is a nonnegative number.

We know that the equation $|x| = 3$ has two solutions, 3 and -3, since $|3| = 3$ and $|-3| = 3$. Thus,

$$|x| = 3$$

is equivalent to

$$x = 3 \quad \text{or} \quad x = -3.$$

The graph of the solution set $\{3, -3\}$ is shown in Figure 8.3.

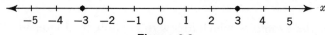

Figure 8.3

Now consider the inequality $|x| < 3$. This inequality will yield true statements for all real-number replacements of x between -3 and 3. Thus,

$$|x| < 3$$

is equivalent to

$$-3 < x < 3.$$

Its solution set is $\{x|-3 < x < 3\}$. The graph of this set is shown in Figure 8.4. It is an open segment.

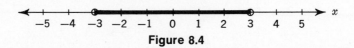

Figure 8.4

Now consider the inequality $|x| > 3$. It will yield true statements for all replacements of x which are greater than 3 or which are less than -3. Thus,

$$|x| > 3$$

is equivalent to

$$x > 3 \quad \text{or} \quad x < -3.$$

Its solution set is $\{x|x > 3 \text{ or } x < -3\}$. The graph of this solution set is shown in Figure 8.5.

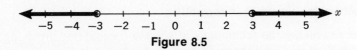

Figure 8.5

Other examples of inequalities are given below.

If $|x| < 8$, then $-8 < x < 8$.
If $|x| < 50$, then $-50 < x < 50$.
If $|x| > 8$, then $x < -8$ or $x > 8$.
If $|x| > 50$, then $x < -50$ or $x > 50$.

These examples suggest a pattern which we state as follows:

INEQUALITY PROPERTY OF $|x|$ For every $a > 0$, if $|x| < a$, then $-a < x < a$. For every $a > 0$, if $|x| > a$, then $x < -a$ or $x > a$.

Observe that a is a positive number and therefore $-a$ is a negative number.

Study the examples which follow to see how the Inequality Property of $|x|$ is used in solving inequalities which are a little more complicated than the ones we have solved thus far.

8.8 Absolute Value in Inequalities

Example 1 Solve: $|4x| < 8$.

$$-8 < 4x < 8$$
$$\frac{-8}{4} < \frac{4x}{4} < \frac{8}{4}$$
$$-2 < x < 2.$$

Thus, the solution set is $\{x | -2 < x < 2\}$. Its graph is shown in Figure 8.6. It is more difficult to check inequalities than equations, since solution sets are frequently infinite. Here is a suggested way to check the inequality above.

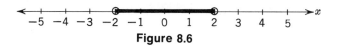

Figure 8.6

Check. Substituting first -2 then 2 in $|4x| = 8$, we have

$$|4 \times (-2)| = |-8| = 8$$
$$|4 \times 2| = |8| = 8.$$

This tells us that we have the correct "endpoints." Now try a value of x such that $-2 < x < 2$, say -1.

$$|4 \times (-1)| = |-4| = 4, \text{ and } 4 < 8.$$

This tells us that we probably have the correct "interval." Now try one number outside the "interval," say -3.

$$|4 \times (-3)| = |-12| = 12, \text{ and } 12 > 8.$$

Thus -3 is not in the solution set; this is as it should be.

Example 2 Solve: $|4x| > 8$.

$$4x < -8 \quad \text{or} \quad 4x > 8$$
$$\frac{4x}{4} < \frac{-8}{4} \quad \text{or} \quad \frac{4x}{4} > \frac{8}{4}$$
$$x < -2 \quad \text{or} \quad x > 2.$$

The solution set is $\{x | x < -2 \text{ or } x > 2\}$. Its graph is shown in Figure 8.7.

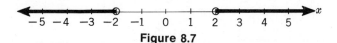

Figure 8.7

Example 3 Solve: $|2x+3| < 9$.

$$-9 < 2x + 3 < 9$$
$$-9 - 3 < 2x + 3 - 3 < 9 - 3$$
$$-12 < 2x < 6$$
$$\frac{-12}{2} < \frac{2x}{2} < \frac{6}{2}$$
$$-6 < x < 3.$$

The solution set is $\{x| -6 < x < 3\}$. Its graph is shown in Figure 8.8.

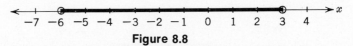

Figure 8.8

Example 4 Solve: $|2x+3| > 9$.

$$2x + 3 < -9 \quad \text{or} \quad 2x + 3 > 9$$
$$x < -6 \quad \text{or} \quad x > 3.$$

The solution set is $\{x|x < -6 \text{ or } x > 3\}$. Its graph is shown in Figure 8.9.

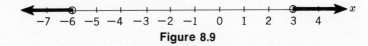

Figure 8.9

8.9 Other Patterns in Inequalities

In Problem 7, Exercise 8.5, you were asked to prove that if $x < y$ and $y < z$, then $x < z$. Since we will wish to use this, we state it as a theorem.

THEOREM 8.10 *Transitive Property of $<$* If $x < y$ and $y < z$, then $x < z$.

With this theorem available to us, we can prove the following:

THEOREM 8.11 If $x < y$ and $z < w$, then $x + z < y + w$.

Proof.
$x < y$ and $z < w$ Given
$x + z < y + z$ Theorem 8.1
$z + y < w + y$ Theorem 8.1
$y + z < y + w$ CPA
$x + z < y + w$ Theorem 8.10

Thus, if $x < y$ and $z < w$, then $x + z < y + w$.

8.9 Other Patterns in Inequalities

It is often easier to apply Theorem 8.11 if it is stated in an abbreviated vertical form:

$$\begin{array}{c} x < y \\ z < w \\ \hline x + z < y + w. \end{array}$$

You may wonder if the same pattern applies to subtraction. That is, is it true that if $x < y$ and $z < w$, then $x - z < y - w$? Before attempting to prove anything about subtraction, let us examine two examples.

Example 1 We know that $5 < 9$ and $1 < 2$ are true.
  Is $5 - 1 < 9 - 2$ true?
  Since the last statement is equivalent to $4 < 7$, we know that it is true.

Example 2 We know that $6 < 7$ and $2 < 10$ are true.
  Is $6 - 2 < 7 - 10$ true?
  Since the last statement is equivalent to $4 < -3$, we know that it is false.

Thus, we see that a pattern comparable to that stated as Theorem 8.11 does not hold for subtraction. But here is a pattern involving subtraction which can be proved.

THEOREM 8.12 If $x < y$, then $z - x > z - y$.

Proof.
$$\begin{array}{l} x < y \\ -x > -y \\ -x + z > -y + z \\ z + (-x) > z + (-y) \\ z - x > z - y. \end{array}$$

Thus, if $x < y$, then $z - x > z - y$.

From our work with real numbers, we know that the sum of two negative numbers is a negative number. We can now prove this.

THEOREM 8.13 If $x < 0$ and $y < 0$, then $x + y < 0$.

Proof. $x < 0$ and $y < 0$
  $x + y < 0 + 0$
  $x + y < 0.$

Thus, if $x < 0$ and $y < 0$, then $x + y < 0$.

It is appropriate to ask whether the pattern of Theorem 8.11 is true for multiplication. That is, is it true that if $x < y$ and $z < w$, then $xz < yw$? Again let us examine an example.

Example We know that $-10 < 1$ and $-5 < 2$ are true.
What about $-10(-5) < 1 \times 2$?
Obviously the last statement is false.

Thus, the pattern of Theorem 8.11 does not hold for multiplication.

Remember from our work with real numbers that the product of two positive numbers is a positive number. Let us prove this.

THEOREM 8.14 If $x > 0$ and $y > 0$, then $xy > 0$.

Proof. $x > 0$ and $y > 0$
$xy > 0 \times y$
$xy > 0$.

Thus, if $x > 0$ and $y > 0$, then $xy > 0$.

EXERCISE 8.7

1. Determine the solution set of each inequality and graph it.

   a. $|5x| < 15$ b. $|5x| > 15$
   c. $|x + 1| < 7$ d. $|x + 1| > 7$
   e. $|2x + 5| < 13$ f. $|2x + 5| > 13$
   g. $|5x - 10| < 0$ h. $|5x + 10| > 0$
   i. $|4 - x| < 6$ j. $|4 - x| > 6$
   k. $|2 - 3x| < 11$ l. $|2 - 3x| > 11$
   m. $|-5 - 2x| < 9$ n. $|-5 - 2x| > 9$
   o. $|-2 - 3x| < 1$ p. $|-2 - 3x| > 1$
   q. $|x + 2| < 0$ r. $|x + 2| > 0$
   s. $|3x - 1| < -7$ t. $|3x - 1| > -7$
   u. $|4x - 19| < -3$ v. $|4x - 19| > -3$

2. Label each statement with T for true and F for false.

   a. If $-2 < 0$, then $-2(-1) < 0(-1)$.
   b. If $-5 > -8$, then $5 > 8$.
   c. If $-2 < 0$ and $0 < 6$, then $-2 < 6$.
   d. If $5 > 3$ and $3 > -3$, then $5 > -3$.

8.9 Other Patterns in Inequalities

e. If $-3 < 0$ and $-17 < 0$, then $-20 < 0$.
f. If $0 > -1$ and $0 > -4$, then $0 > 4$.

3. Tell which theorem each exercise illustrates. Then tell which variable is replaced by which numerals.

   a. If $-2 < 0$ and $-17 < 0$, then $-19 < 0$.
   b. If $-6 < 2$, then $6 > -2$.
   c. If $5 > 0$ and $6 > 0$, then $30 > 0$.
   d. If $-5 < 0$ and $0 < 1$, then $-5 < 1$.

4. Using the inequality definitions, properties, and theorems, determine the solution sets of the following inequalities.

   a. $3x > -6$
   b. $2x + 2 < -4$
   c. $\frac{1}{2}y + 1 < y - 2$
   d. $\frac{m}{3} + \frac{1}{2} > m - 1$
   e. $\frac{n+1}{2} < n - 1$
   f. $\frac{3}{4}t > t - 7$
   g. $2x < 10$
   h. $3y - 1 > 17$
   i. $2z - 5 > z + 3$
   j. $\frac{5u - 3}{2} > 3u + 1$
   k. $\frac{3w - 1}{3} < \frac{2w + 3}{2}$
   l. $\frac{n}{2} + \frac{n}{3} < 1$
   m. $3(m + 1) > 5(3 - m)$
   n. $-2(x - 1) < 3(2 - x)$
   o. $|p| < 1$
   p. $|r + 1| > 15$
   q. $|s + 2| > 1$
   r. $|5h| < 0$
   s. $|x| > 0$
   t. $|y - 1| > 3$

5. Copy each statement and replace $\bigcirc$ by one of the symbols $<$ or $>$ to obtain a true statement for all real number replacements of the variables.

   a. If $x < y$, then $3x \bigcirc 3y$.
   b. If $-3a > -3b$, then $a \bigcirc b$.
   c. If $c < d$, then $c + (-a) \bigcirc d + (-a)$.
   d. If $-8x > 16$, then $x \bigcirc -2$.
   e. If $r < s$, then $-2r \bigcirc -2s$.
   f. If $-x > -y$, then $-4 - x \bigcirc -4 - y$.
   g. If $-5 + m < -5 + n$, then $m \bigcirc n$.
   h. If $6(x + y) > 6(u - w)$, then $x + y \bigcirc u - w$.

6. Tell which sentences are true for all the specified real-number replacements of the variables.

a. If $x < y$ and $y < 0$, then $xy > 0$.
b. If $x < y$ and $y < 0$, then $x + y > 0$.
c. If $x < y$ and $y < 0$, then $y - x < 0$.
d. If $x < y$ and $x > 0$, then $xy > 0$.
e. If $x < y$ and $x > 0$, then $y - x > 0$.
f. If $x < y$ and $x > 0$, then $x - y < 0$.

7. Prove each of the following.
   a. If $x < 0$, then $-x > 0$.
   b. If $x > 0$ and $y > 0$, then $x + y > 0$.
   c. If $x > 0$ and $y < 0$, then $xy < 0$.
   d. If $x < 0$ and $y < 0$, then $xy > 0$.

GLOSSARY

Equivalence relation: A relation which has the reflexive, symmetric, and transitive properties.

Reflexive property (of =): $\forall_x \ x = x$.

Symmetric property (of =): $\forall_x \forall_y$ if $x = y$, then $y = x$.

Transitive property (of =): $\forall_x \forall_y \forall_z$ if $x = y$ and $y = z$, then $x = z$.

CHAPTER REVIEW PROBLEMS

1. Find the solution set of each equation.
   a. $3(x + 2) = -3$
   b. $2(3 - 2x) = 14$
   c. $-2(x + 1) = 3(2 - x)$
   d. $\dfrac{x - 2}{3} = \dfrac{2x - 1}{4}$
   e. $\dfrac{3(x - 1)}{2} = \dfrac{-2(4 - x)}{4}$
   f. $\dfrac{5}{x - 1} = \dfrac{1}{2}$
   g. $\dfrac{3}{x + 2} = \dfrac{-2}{2x - 1}$
   h. $\dfrac{1}{x} + 2 = 1 - \dfrac{2}{x}$
   i. $\dfrac{3(x - 1)}{2(2 - 3x)} = -\dfrac{1}{2}$

2. Solve $m = \dfrac{s(t - u)}{k}$ for
   a. s
   b. k
   c. u

Chapter Review Problems

3. Solve each formula for the indicated letters.
    a. $z^2 = az + b^2$; for a; for b
    b. $x + \dfrac{x}{7} = 19$; for x
    c. $ax^2 + bx + c = 0$; for a; for b; for c
    d. $y = \dfrac{px + q}{rx + s}$; for x; for q
    e. $y = \dfrac{x(x^2 + 3D)}{3x^2 + D}$; for D
    f. $\dfrac{2E}{n} + \dfrac{2E}{r} - E = 2$; for E; for n

4. In each problem a formula is given. Do the required computations.
    a. Surface area of a cube: $A = 6e^2$, where e is the length of an edge.
        i. Solve for e.
        ii. Compute e when A is 24 sq in.
    b. Perimeter of an isosceles triangle: $p = b + 2a$, where b is the length of the base and a is the length of each of the congruent sides.
        i. Solve for a.
        ii. Compute a when p is 40 in. and b is 9 in.
    c. Surface area of a regular tetrahendron: $S = a^2\sqrt{3}$, where a is the length of each edge.
        i. Solve for a.
        ii. Compute a when S is $9\sqrt{3}$ sq ft.
    d. Interest rate, which is equivalent to a discount rate d, is $i = \dfrac{d}{1 - nd}$, where n is the number of years.
        i. Solve for d.
        ii. Compute d in percent when i is 6% and n is 25 years.
    e. Einstein's equation relating mass and energy is $e = mc^2$, where e is energy in ergs, m is mass in grams, and c is the speed of light in cm per sec.
        i. Solve for m.
        ii. Solve for c.
        iii. If the light travels at 30 billion cm per sec, compute the number of ergs produced by the conversion of 1 gram of mass to energy.

5. Which of the following is an example of the Transitive Property of <?
   a. If $2 < 4$, then $2 + 3 < 4 + 3$.
   b. If $1 < 3$ and $3 < 7$, then $1 < 7$.
   c. If $5 < 9$, then $5 \times 6 < 9 \times 6$.
6. Determine the solution set of each of the following inequalities. Then graph each solution set on a number line.
   a. $r + 5 > 12$
   b. $2y - 3 < 14$
   c. $\dfrac{m+4}{3} < 1$
   d. $3(x - 6) < 0$
   e. $\dfrac{n-1}{2} > \dfrac{n+2}{3}$
   f. $|t| < 2$
   g. $|s - 1| > 2$
   h. $|3n - \tfrac{7}{8}| < -3$
7. Prove each of the following inequalities.
   a. If $x < 0$ and $y > 0$, then $y > x$.
   b. If $x \neq 0$, then $x^2 > 0$.
8. Solve each problem.
   a. The sum of a number and 1 is equal to the product of 3 and the number. What is the number?
   b. A number divided by the sum of the number and 1 is equal to 3. What is the number?
   c. If 5 is multiplied by the sum of a number and 1, and 12 is subtracted from the product, the result is 38. What is the number?
   d. One-half of the sum of 2 and a number is equal to the number. What is the number?
   e. The sum of one-fifth of a number and two-thirds of the number is 36. What is the number?
   f. One-half of a number plus one-fourth of the number is 9. What is the number?
   g. Multiplying a number by 3 and adding 3 to the product gives the same result as multiplying this number by 2 and adding 10 to the result. What is the number?
   h. There are 49 marbles in a box. If the number of red marbles is twice the number of white marbles, and the number of green marbles is twice the number of red marbles, how many marbles of each color are there in the box?
   i. The length of a rectangle is equal to twice the width plus 1. If the perimeter of the rectangle is equal to 20 in., determine the length and the width of the rectangle.

j. The perimeter of a rectangle is equal to 72 ft. If the length is eight times the width, determine the length and the width of the rectangle.

k. The perimeter of a square is the same as the length of a rectangle. The width of the rectangle is one-fourth of the length of the side of the square. If the width of the rectangle is 2 in., what is the area of the square? of the rectangle?

l. The width of a rectangle is one-sixth of its length. Determine the length and the width of the rectangle if the perimeter is equal to 28 in.

m. Each of the two congruent sides of an isosceles triangle is one-half as long as the base of the triangle. If the perimeter of the triangle is equal to 18 in., what is the measure of each side of the triangle?

n. Mary is now five years younger than her brother Ken. Ken was twice as old as Mary was eight years ago. How old is Mary now?

o. The enrollment in a college is 6000 students with the ratio of boys to girls three to one. How many girls are there in the college?

p. Miss Coins has $4.50 in nickels and dimes. She has three times as many nickels as dimes. How many nickels and how many dimes does Miss Coins have?

9

Relations and Functions

9.1 Solution Sets of Open Sentences

We have studied open sentences such as the following:

$$2x - 9 = 13 \qquad |r| \leq 9 \qquad 3s + 5 > 17.$$

In each case above, the variables hold places for numerals. The numerals which are chosen as replacements name the elements of a set called the *replacement set*. We use the letter U in referring to a replacement set, which is also called a *universal set*.

An open sentence such as

$$2x - 9 = 13$$

partitions a chosen replacement set into two subsets. One of the subsets contains the elements for which $2x - 9 = 13$ results in false statements. The other subset of U contains elements for which $2x - 9 = 13$ yields true statements. We call this latter subset the *solution set* of the given open sentence.

If the set of positive integers is chosen as the replacement set, the following sentences have the solution sets listed on the right.

9.2 Ordered Pairs

1. $2x - 9 = 13$ $\{x \mid x = 11\}$, or $\{11\}$
2. $|r| \leq 9$ $\{r \mid 1 \leq r \leq 9\}$, or $\{1, 2, 3, 4, 5, 6, 7, 8, 9\}$
3. $3x + 5 > 17$ $\{s \mid s > 4\}$, or $\{5, 6, 7, 8, \ldots\}$

Recall that each number can be associated with a point on the number line. Thus, the sets above can be graphed as in Figure 9.1.

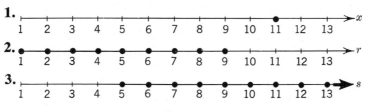

Figure 9.1

If the replacement set were chosen to be the set of nonnegative real numbers, the same open sentences would have the solution sets listed on the right below.

1. $2x - 9 = 13$ $\{x \mid x = 11\}$
2. $|r| \leq 9$ $\{r \mid 0 \leq r \leq 9\}$
3. $3s + 5 > 17$ $\{s \mid s > 4\}$

These graphs are shown in Figure 9.2.

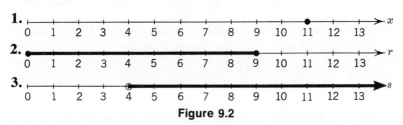

Figure 9.2

9.2 Ordered Pairs

Suppose the replacement set is the set of nonnegative integers less than 4, that is,

$$U = \{0, 1, 2, 3\}.$$

We can form what we call *ordered pairs* from this set.

DEFINITION 9.1 An *ordered pair* of elements is a pair in which one member is designated as the *first* member and one member as the *second* member. (*a, b*) denotes an ordered pair in which *a* is the first member and *b* is the second member.

The following definition tells us the condition for the equality of two ordered pairs.

DEFINITION 9.2   (*a, b*) = (*c, d*) means that *a* = *c* and *b* = *d*.

From this definition we see that, for example, (2, 4) ≠ (4, 2), because 2 ≠ 4. Thus, while 2 and 4 constitute one pair of numbers, the ordered pair (2, 4) is not the same as the ordered pair (4, 2).

The members of an ordered pair do not have to be numbers. Here is an example of an ordered pair in which not both members are numbers.

(circle *A*, 1)

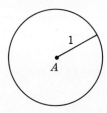

The first member in this pair is a set of points, namely circle *A*, and the second member is the number 1, which could be, for example, the length of the radius of circle *A*.

Now let us return to our replacement set $U = \{0, 1, 2, 3\}$ and form all possible ordered pairs from the members of *U*.

(0, 0)	(1, 0)	(2, 0)	(3, 0)
(0, 1)	(1, 1)	(2, 1)	(3, 1)
(0, 2)	(1, 2)	(2, 2)	(3, 2)
(0, 3)	(1, 3)	(2, 3)	(3, 3)

This set of all ordered pairs of numbers from *U* is called the *Cartesian set* of *U*, and is written $U \times U$ [read: *U* cross *U*].

9.3 The Coordinate Plane and Ordered Numbered Pairs

We shall next turn our attention to the graphs of sets of ordered pairs of numbers. To do this, we shall establish a one-to-one correspondence between *real-number pairs* and the *points in the plane*. We take two number lines that cross each other at their zero points. It is customary to take one horizontal and one vertical line. The point where the lines meet is

9.3 The Coordinate Plane and Ordered Numbered Pairs

called the *origin*. The surface on which the points are located is called a *plane*.

Now to each point we can assign a number pair, and to each number pair we can assign a point. It is the usual practice to assign number pairs as in Figure 9.3. The first number in the number pair is called the *first component*, and the second number, the *second component*. We are now ready to make an agreement by which a one-to-one correspondence between number pairs and points will be established.

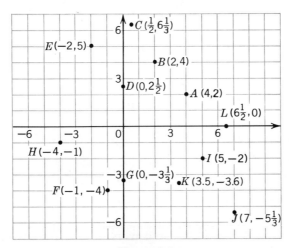

Figure 9.3

Consider the number pair (4, 2). To find the point corresponding to this pair, we shall agree that the first number tells the number of units we are to move from the origin to the right or to the left—positive number, to the right; negative number, to the left. The second number tells us how many units to move up or down—positive number, up; negative number, down.

Thus, (4, 2) means move 4 units to the right, 2 units up. The point thus obtained is the point A in Figure 9.3. It is easy to see that there is exactly one point which corresponds to the ordered pair (4, 2). And generally, there is exactly one point corresponding to each ordered number pair.

If we agree that we shall name number pairs in the form (x, y), then we can refer to the horizontal line as the *x*-axis and to the vertical line as the *y*-axis.

The numbers in the pair are the *coordinates* of that point. The first number is called the *first coordinate*, or *abscissa*, and the second number is called the *second coordinate,* or *ordinate.* For example, in the number pair (4, 2), 4 is the abscissa of the point corresponding to (4, 2), and 2 is the ordinate of that point. The abscissa is called the *x-coordinate,* and the ordinate is called the *y-coordinate.*

The plane in which we locate points corresponding to given ordered number pairs is called the *coordinate plane,* or the *Cartesian plane,* after the name of the French mathematician, Descartes, who first introduced the idea. The two axes superimposed on the plane are referred to as the *rectangular coordinate system.*

EXERCISE 9.1

1. For each open sentence:
   i. Use the set of negative real numbers as the replacement set and name the members of the solution set. Graph each set.
   ii. Use the set of real numbers as the replacement set and name the members of the solution set. Graph each set.

   a. $2x + 2 = 14$
   b. $2 < |m| \leq 4$
   c. $|y| \leq 1$
   d. $|n + 1| < 2$
   e. $1 < 2r + 3 < 9$
   f. $\dfrac{3t - 2}{t} < 10$

2. Which of the following ordered pairs are equal and which are not equal?

   a. $(0, -2)$; $(\frac{0}{3}, -1 - 1)$
   b. $(\frac{1}{2}, .5)$; $(.5, \frac{1}{2})$
   c. $(\sqrt{.25}, 1)$; $(.05, -[-1])$
   d. $(\sqrt[3]{1}, \sqrt[3]{8})$; $(1, -2)$
   e. $(\sqrt{4}, -1)$; $(2, \sqrt{1})$
   f. $(.1, \sqrt[3]{27})$; $(\frac{1}{10}, 3)$
   g. $(-2, -3^2)$; $(-\sqrt{4}, 9)$
   h. $(2^4, -[-2 - 3])$; $([-2]^4, 2 + 3)$

3. For each set U given below, give the set $U \times U$.

   a. $U = \{0, 5\}$
   b. $U = \{1\}$
   c. $U = \{0\}$
   d. $U = \{-\frac{1}{2}, \frac{1}{2}\}$

4. Give an example of an ordered pair in which not both members are numbers.

5. If $U = \{1, 2, 3\}$, how many members are there in $U \times U$?

6. If $U' = \{$all real numbers r such that $1 \leq r \leq 3\}$, how many members are there in $U' \times U'$?

7. Given the two sets $A = \{1, 2\}$ and $B = \{3, 4, 5\}$,
   a. What is $A \times B$?
   b. What is $B \times A$?
   c. Is $A \times B = B \times A$ true?

9.3 The Coordinate Plane and Ordered Numbered Pairs

8. Given $A = \{0\}$ and $B = \{1\}$,
   a. What is $A \times B$?
   b. What is $B \times A$?
   c. What must be true of the sets A and B in order for $A \times B$ to be equal to $B \times A$?

9. Make a coordinate system on a sheet of paper and locate points corresponding to the following ordered numbered pairs. Label points as shown in Figure 9.3.
   a. (4, 1) b. (1, 4) c. $(3\frac{1}{2}, 3\frac{1}{2})$
   d. (0, 7) e. (−1, 2) f. (0, 0)
   g. $(-3\frac{1}{2}, -4)$ h. $(0, -1\frac{1}{2})$ i. (2, −4.5)
   j. (6, 0) k. $(-7\frac{1}{2}, 0)$ l. (0, −3)

10. In the coordinate system in Figure 9.4 the points are named by capital letters. List each letter on your paper, and next to it write the name of the ordered number pair corresponding to the point. (The exercise is done for point A.)

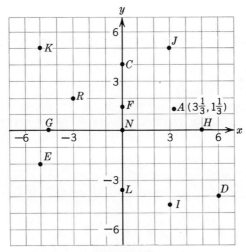

Figure 9.4

11. You have probably observed that the two axes of the rectangular coordinate system separate the plane into four subsets. It is customary to name these subsets as shown in Figure 9.5. Thus, we can say that the point corresponding to the ordered number pair (1,175), for example,

belongs to the first quadrant. The points *on* the *x*-axis or *on* the *y*-axis do *not* belong to any quadrant. Tell to which quadrant each of the points corresponding to the ordered number pairs below belongs.

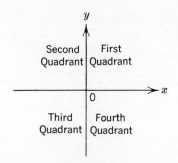

Figure 9.5

a. $(-16, -137)$ b. $(1007, -2)$
c. $(10, 100)$ d. $(-.007, 137.99)$
e. $(\frac{13}{29}, -.003)$ f. $(-111, -3)$
g. $(-37, -678)$ h. $(-\frac{1}{3}, 99)$
i. $(127, 12)$ j. $(-13, 13)$
k. $(13, -13)$ l. $(-128, 378)$
m. $(-998, 998)$ n. $(1027, -368)$
o. $(11, -11)$ p. $(2006, 2006)$
q. $(|-17|, |-1389|)$ r. $(|\frac{1}{2}|, |-\frac{1}{2}|)$

12. Examine your answers in Problem 11. What can you tell about the coordinates of points in the first quadrant? in the second quadrant? in the third quadrant? in the fourth quadrant?

13. What can you tell about the coordinates of the points on the *x*-axis? on the *y*-axis?

14. Using Figure 9.6, complete each statement.

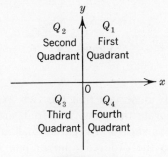

Figure 9.6

a. $Q_1 \cap Q_2 = ?$
b. $(Q_1 \cap Q_2) \cap Q_3 = ?$
c. $(Q_1 \cap Q_2) \cap (Q_3 \cap Q_4) = ?$
d. $(Q_1 \cup Q_2) \cup (Q_3 \cup Q_4) \cup (x\text{-axis} \cup y\text{-axis}) = ?$
e. x-axis $\cap$ y-axis $= ?$

9.4 Open Sentences with Two Variables

We have seen that every ordered pair of numbers has one point in the coordinate plane associated with it. We do not always, however, consider the entire coordinate plane. For example, given $U = \{0, 1, 2, 3\}$, the graph of $U \times U$ consists of 16 points, as shown in Figure 9.7.

Figure 9.7

Now let U be the set of all real numbers r for which $0 \leq r \leq 3$. The graph of $U \times U$ is the graph of all ordered number pairs (x, y) for which

$$0 \leq x \leq 3 \quad \text{and} \quad 0 \leq y \leq 3.$$

This set can be described as follows:

$$U \times U = \{(x, y) | x \in R \text{ and } y \in R, 0 \leq x \leq 3 \text{ and } 0 \leq y \leq 3\},$$

where R is the set of all real numbers. The graph of the set is shown in Figure 9.8.

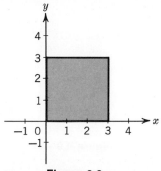

Figure 9.8

It should be obvious that the graph of $R \times R$ is the entire plane.

We shall now turn our attention to open sentences which contain two variables. Here are examples of such sentences.

$$x - 5 = y \qquad x + 5 < y$$
$$2x - 3y = 5 \qquad |x| + |y| = 7.$$

Suppose we decide to use the set $U = \{0, 1, 2, 3, 4, 5, 6, 7, 8\}$ as the replacement set. We can find all pairs (x, y), where x and y are members of U, which yield true statements upon replacement in the sentences above. As an example, the solution set of $x - 5 = y$ is

$$\{(x, y) | x \in U \text{ and } y \in U \text{ and } x - 5 = y\}.$$

The members of this set are (5, 0), (6, 1), (7, 2), (8, 3). Of course, the solution set must be a subset of $U \times U$. The open sentence $x - 5 = y$ selects a subset of $U \times U$, which is the solution set of this sentence.

The solution set of the sentence $2x - 3y = 5$ is

$$\{(x, y) | x \in U \text{ and } y \in U \text{ and } 2x - 3y = 5\}.$$

It has only two members, (4, 1) and (7, 3).

Here are the members of the solution set of $x + 5 < y$

$$(0, 6) \quad (0, 7) \quad (0, 8) \quad (1, 7) \quad (1, 8) \quad (2, 8).$$

The members of the solution set of $|x| + |y| = 7$ are

$$(0, 7) \quad (1, 6) \quad (2, 5) \quad (3, 4) \quad (4, 3) \quad (5, 2) \quad (6, 1) \quad (7, 0).$$

Examine the graphs in Figure 9.9 for each of these four sets. Keep in mind that in each case $U \times U$ is the Cartesian set of $U = \{0, 1, 2, 3, 4, 5, 6, 7, 8\}$.

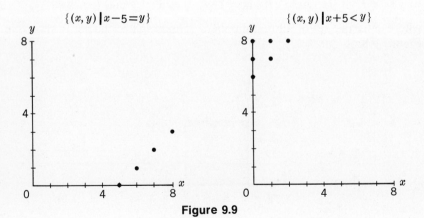

Figure 9.9

9.4 Open Sentences with Two Variables

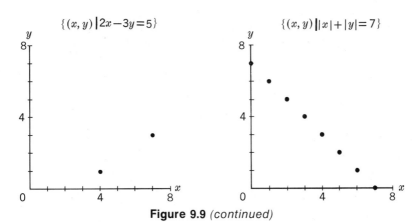

Figure 9.9 *(continued)*

EXERCISE 9.2

1. For each set U given below, how many ordered pairs are there in $U \times U$?
- **a.** $U = \{1, 2\}$
- **b.** $U = \{0, 1\}$
- **c.** $U = \{0\}$
- **d.** $U = \phi$

2. Tell whether each of the following is a sentence in one variable, in two variables, or in three variables.
- **a.** $2(x + 1) > 7$
- **b.** $\dfrac{x + 2}{x - 1} = 2$
- **c.** $2(x + 1) = 3(y - 2)$
- **d.** $2(x + y + z) = -1$
- **e.** $\dfrac{x + y}{y} = \dfrac{x + z}{z}$
- **f.** $2(x + y) < z$
- **g.** $|x + y| > |1 + z|$
- **h.** $-3(x + 1) = 4(3 - 2x)$

3. Let $U = \{0, 1, 2, 3, 4\}$, and graph the following sets.
- **a.** $U \times U$
- **b.** $\{(x, y) | x = y\}$
- **c.** $\{(x, y) | y > x\}$
- **d.** $\{(x, y) | y < x\}$
- **e.** $\{(x, y) | x + y = 4\}$
- **f.** $\{(x, y) | x + y < 4\}$
- **g.** $\{(x, y) | x + y > 4\}$
- **h.** $\{(x, y) | 2x + y = 3\}$
- **i.** $\{(x, y) | 2x + y < 3\}$
- **j.** $\{(x, y) | 2x + y > 3\}$
- **k.** $\{(x, y) | 2y = x\}$
- **l.** $\{(x, y) | y = 3x\}$

4. Make a graph of $U \times U$, where $U = \{(x, y) | x \in R \text{ and } y \in R, 0 \leq x \leq 4 \text{ and } 0 \leq y \leq 4\}$.

5. Let $U = \{-3, -2, -1, 0, 1, 2\}$.
   - *i.* Name the members of the solution set of each sentence given below.
   - *ii.* Graph each solution set.

   a. $x + y = 0$
   b. $y = x$
   c. $y > x$
   d. $y < x$
   e. $2x + y = 0$
   f. $2x + y < 0$
   g. $2x + y > 0$
   h. $y = -x$
   i. $y > -x$
   j. $y < -x$
   k. $x - y = 0$
   l. $x - y = -2$
   m. $|y| = |x|$
   n. $|x| + |y| \leq 2$
   o. $|y| = -|x|$
   p. $|y| = 2 \times |x|$

9.5 Relations and Functions

We have seen that given a set U, $U \times U$ is a set of ordered pairs. We now give a definition of this concept.

DEFINITION 9.3 A *relation* is a set of ordered pairs.

According to this definition, each of the following sets is a relation.

Example 1 $\{(5, 0), (6, 1), (7, 2), (8, 3)\}$.

Example 2 $\{(4, 1), (7, 3)\}$.

Example 3 $\{(0, 6), (0, 7), (0, 8), (1, 7), (2, 8)\}$.

Example 4 $\{(0, 7), (1, 6), (2, 5), (3, 4), (4, 3), (5, 2), (6, 1), (7, 0)\}$.

Consider, for example, the relation in Example 1. Call it A.

$$A = \{(5, 0), (6, 1), (7, 2), (8, 3)\}.$$

We can identify two sets in connection with this relation. One of these is the set of all first elements of the ordered pairs which belong to A. Call it D.

$$D = \{5, 6, 7, 8\}.$$

The other set is the set of all second elements of the ordered pairs. Call it R.

$$R = \{0, 1, 2, 3\}.$$

9.5 Relations and Functions

Since, when studying various relations we refer to these sets quite frequently, we give them special names.

DEFINITION 9.4 The *domain* of a relation is the set of all first members of all ordered pairs which belong to the relation.

DEFINITION 9.5 The *range* of a relation is the set of all second members of all ordered pairs which belong to the relation.

For Example 1 above the domain is {5, 6, 7, 8} and the range is {0, 1, 2, 3}.

Now let us consider the replacement set U = {0, 1, 2}. The graph of $U \times U$ is at the right.

$U \times U$ = {(0, 0), (0, 1), (0, 2), (1, 0), (1, 1), (1, 2), (2, 0), (2, 1), (2, 2)}.

Now consider the following four subsets of $U \times U$: K, L, M, and N.

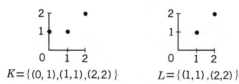

K={(0, 1),(1,1),(2,2)}   L={(1,1),(2,2)}

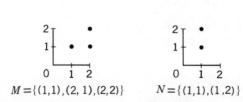

M={(1,1),(2, 1),(2,2)}   N={(1,1),(1,2)}

Note that in K each member of the domain D = {0, 1, 2} is paired with only *one* member of the range R = {1, 2}. The same is true in L.

Now examine M; 2, which is a member of the domain, is paired once with 1 and another time with 2. Thus the pairs (2, 1) and (2, 2) belong to M.

In N there are also two pairs which share the same first element.

To distinguish between two kinds of relations, we state the following definition.

DEFINITION 9.6 A *function* is a relation in which no two ordered pairs have the same first element.

According to this definition, the relations *K* and *L* above are functions and the relations *M* and *N* are not functions.

EXERCISE 9.3

1. Tell why every function is a relation.
2. Tell why not every relation is a function.
3. Give an example of a function in which the domain is the same as the range.
4. Let $U = \{0, 1, 2, 3\}$.
   a. List the pairs which belong to $A = \{(x, y) | y < x\}$.
   b. Graph $U \times U$.
   c. Graph the set *A*.
   d. List the members of the domain of *A* and of the range of *A*.
5. Tell whether each of the following relations is a function.
   a. $\{(2, 1), (3, 1), (4, 2), (6, 7)\}$
   b. $\{(3, 3), (4, 5), (6, 9), (7, 1), (5, 4)\}$
   c. $\{(3, 3), (4, 7), (8, 9), (3, 5), (1, 2)\}$
   d. $\{(2, 8), (6, 1), (7, 6), (3, 2), (4, 4), (5, 3)\}$
   e. $\{(1, 1), (2, 1), (3, 1), (4, 1), (5, 1), (6, 1)\}$
   f. $\{(1, 1), (2, 1), (1, 3), (1, 4), (1, 5), (1, 6)\}$
   g. $\{(3, 7), (7, 3), (8, 5), (5, 8), (9, 2), (3, 5)\}$
   h. $\{(1, 5), (1, 7), (3, 9), (8, 2), (6, 5)\}$
   i. $\{(2, 1), (5, 3), (4, 4), (8, 2)\}$
   j. $\{(6, 6)\}$
6. Each graph below is a graph of a relation in $U \times U$, where $U = \{0, 1, 2, 3, 4\}$. For each relation,
   i. Tell whether it is a function.
   ii. List its members.
   iii. List the members of its domain and range.

Example

   i. Not a function
   ii. $\{(1, 1), (2, 2), (3, 1), (3, 4)\}$
   iii. Domain: $\{1, 2, 3\}$
          Range: $\{1, 2, 4\}$

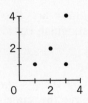

9.5 Relations and Functions

a.

e.

b.

f.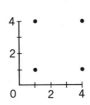

c.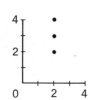

g.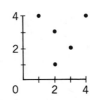

d.

h.

7. Graph the Cartesian set $U \times U$, where U is the set of integers less than 1 and greater than -4.

8. Using set A as the domain and set B as the range, list the members of $A \times B$, and graph $A \times B$. Then use B as the domain and A as the range and graph $B \times A$.

Example $A = \{1, 2\}$.
$B = \{2, 3, 4\}$.
$A \times B = \{(1, 2), (1, 3), (1, 4), (2, 2), (2, 3), (2, 4)\}$.

$B \times A = \{(2, 1), (2, 2), (3, 1), (3, 2), (4, 1), (4, 2)\}.$

a. $A = \{1, 2, 3\}$
   $B = \{3, 4, 5, 6\}$
b. $A = \{1, 2\}$
   $B = \{3, 4\}$
c. $A = \{4, 5\}$
   $B = \{5, 6\}$
d. $A = \{1, 2, 3\}$
   $B = \{1, 2, 3\}$
e. $A = \{1\}$
   $B = \{2\}$
f. $A = \{1, 2\}$
   $B = \{3, 4, 5, 6\}$

9. Given the replacement set $U = \{0, 1, 2, 3, 4, 5, 6, 7, 8\}$, graph each of the following relations. List the members of the domain and the range.

Example Relation: $\{(x, y) | 2x > y\}$.
Domain: $\{1, 2, 3, 4, 5, 6, 7, 8\}$.
Range: $\{0, 1, 2, 3, 4, 5, 6, 7, 8\}$.

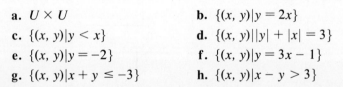

a. $U \times U$
b. $\{(x, y) | y = 2x\}$
c. $\{(x, y) | y < x\}$
d. $\{(x, y) | |y| + |x| = 3\}$
e. $\{(x, y) | y = -2\}$
f. $\{(x, y) | y = 3x - 1\}$
g. $\{(x, y) | x + y \leq -3\}$
h. $\{(x, y) | x - y > 3\}$

10. Given $U = \{-3, -2, -1, 0, 1, 2, 3\}$, graph each of the following and tell which relations are functions.

a. $\{(x, y) | y = x - 6\}$
b. $\{(x, y) | y < x\}$
c. $\{(x, y) | y \geq x\}$
d. $\{(x, y) | 2x + y = 5\}$
e. $\{(x, y) | x + y < 5\}$
f. $\{(x, y) | y = x^2\}$
g. $\{(x, y) | y = 2 \times |x|\}$
h. $\{(x, y) | y \neq x\}$
i. $\{(x, y) | 2x + y > 4\}$
j. $\{(x, y) | y = 3\}$
k. $\{(x, y) | x = 4\}$
l. $\{(x, y) | |x| = |y|\}$

9.6 Graphing Equations of the Form $Ax + By = C$

We now turn our attention to the study of a special kind of equation. We will assume that the universal set is the set of all real numbers. Let us consider the following equation:

$$2x + 5y = -3.$$

Its solution set is $\{(x, y) | 2x + 5y = -3\}$. This is read "the set of all ordered real number pairs x and y for which $2x + 5y = -3$." Here are some ordered number pairs (x, y) which satisfy this equation.

$(1, -1)$ $(-4, 1)$ $(0, -\frac{3}{5})$
$(11, -5)$ $(3.5, -2)$ $(6, -3)$

These points are shown on the graph in Figure 9.10.

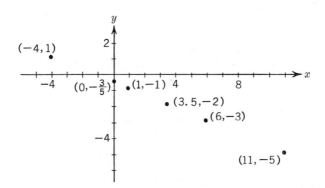

Figure 9.10

If we try to fit an edge of a ruler to these points, we notice that they line up quite well along the edge. As a matter of fact, later in our study of mathematics we will be able to prove that the graph of the equation $2x + 5y = -3$ is a straight line. Moreover, we will be able to *prove* that the graph of any equation of the form $Ax + By = C$, where the replacement set of A, B, and C is the set of real numbers (A and B not both 0), is a straight line. For now, we will *assume* that it is so. In fact, let us agree on the following two properties concerning straight lines.

LINE PROPERTY 1 In $R \times R$, where R is the set of real numbers, the graph of every equation of the form $Ax + By = C$ is a straight line [A and B are not both 0]. Equations which may be written in this form are called *linear* equations.

LINE PROPERTY 2 There is exactly one straight line which contains two given points.

According to Line Property 1, the graph of the equation $2x + 5y = -3$ in $R \times R$ is a straight line.

According to Line Property 2, we need to locate only two points whose coordinates satisfy the equation $2x + 5y = -3$ in order to locate the graph of this equation. In practice, however, it is a good idea to locate a third point as a check.

Thus, the graph of the equation $2x + 5y = -3$ in $R \times R$ is shown in Figure 9.11.

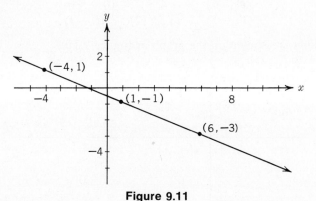

Figure 9.11

EXERCISE 9.4

1. Given the equation $5x - 4y = -3$, tell which of the following ordered pairs (x, y) satisfy it and which do not.

   a. $(1, 2)$ b. $(-1, -2)$ c. $(0, \frac{3}{4})$
   d. $(\frac{3}{5}, 0)$ e. $(2, \frac{13}{4})$ f. $(6, 8)$

2. Graph each of the following sets. Use the set of real numbers as the replacement set.

   a. $\{(x, y) | x + y = 6\}$ b. $\{(x, y) | 2x + 3y = -2\}$
   c. $\{(x, y) | 4x - 5y = 10\}$ d. $\{(x, y) | x = y + 1\}$
   e. $\{(x, y) | 5x - 6y = 2\}$ f. $\{(x, y) | 3x - 2y = -4\}$
   g. $\{(x, y) | x - 3y = 12\}$ h. $\{(x, y) | y = 2x - 4\}$
   i. $\{(x, y) | 7 = x - 7y\}$ j. $\{(x, y) | 4y + 5x = 0\}$

3. Show that each of the following is equivalent to $Ax + By = C$.

a. $Ax + By - C = 0$ **b.** $y = \dfrac{C - Ax}{B}$ **c.** $x = \dfrac{C - By}{A}$

4. Graph each of the following sets. Use the set of real numbers as the replacement set.

   a. $\{(x, y) | 3x = 4 - 7y\}$
   b. $\{(x, y) | 5(x + 3) = 2(y - 1)\}$
   c. $\{(x, y) | -2(3 - 2x) = 3(1 + 3y)\}$
   d. $\{(x, y) | -6x + 4 = -11y\}$
   e. $\left\{(x, y) \Big| \dfrac{x}{3} + y = -5\right\}$
   f. $\left\{(x, y) \Big| 5 + \dfrac{4}{3}x = -7y\right\}$
   g. $\{(x, y) | -4y + 3x = 7\}$
   h. $\{(x, y) | 2x - 5 = -13y\}$

9.7 Slope

Recall that the graph of every equation of the form $Ax + By = C$ is a straight line [A and B are not both 0]. Let us now solve the equation $Ax + By = C$ for y.

$$Ax + By = C$$
$$By = -Ax + C$$
$$y = -\dfrac{A}{B}x + \dfrac{C}{B}.$$

Since A, B, and C can be any real numbers, except $B \neq 0$, let us simplify writing by using m for $-\dfrac{A}{B}$ and b for $\dfrac{C}{B}$. We have then

$$y = mx + b.$$

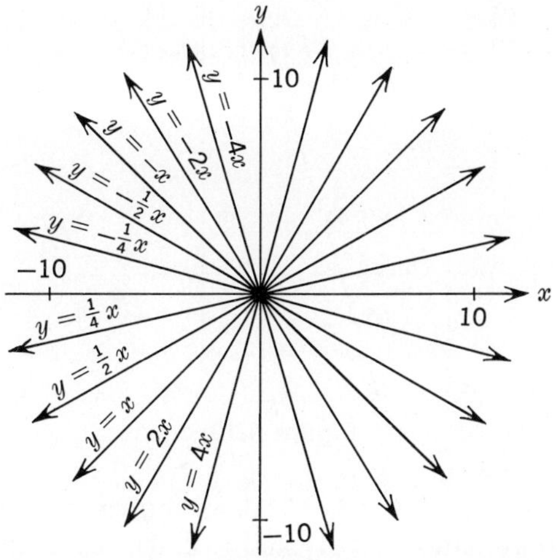

Figure 9.12

We shall now try to discover how different replacements for m in $y = mx + b$ affect the graphs of linear equations. To do this, we graph the following equations in Figure 9.12 on the same coordinate system.

1. $y = x$
2. $y = 2x$
3. $y = 4x$
4. $y = \frac{1}{2}x$
5. $y = \frac{1}{4}x$
6. $y = -x$
7. $y = -2x$
8. $y = -4x$
9. $y = -\frac{1}{2}x$
10. $y = -\frac{1}{4}x$

An examination of the graph above reveals the following:

> In the graphs of equations of the form $y = mx + b$, if $m > 0$, the line rises as we move toward the right. As $|m|$ increases, the line approaches the vertical position. As $|m|$ decreases, the line approaches the horizontal position. If $m < 0$, the line falls as we move toward the right.

In the graphs you made, observe that the replacements for m in $y = mx + b$ determine the "steepness" of the graphs. This "steepness" is determined by a number called the *slope* of the graph. This number is represented by m in the equation $y = mx + b$. Thus, the slope of the graph of the equation $y = 3x + 10$ is 3. The slope of the graph of $y = -2x - 1$ is -2.

DEFINITION 9.7 The *slope* of the line given by the equation $y = mx + b$ is the number m.

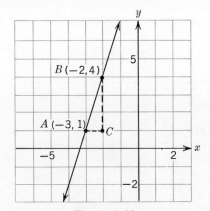

Figure 9.13

Consider the graph of $y = 3x + 10$, as shown in Figure 9.13. Points A and B with coordinates $(-3, 1)$ and $(-2, 4)$, respectively, are on the graph of $y = 3x + 10$. We will now compare the slope of the line contain-

9.7 Slope

ing A and B with the idea of "moving" from point B to point A, first along the vertical, and then along the horizontal. In moving from B to C, we cover -3 units (negative direction); then from C to A we cover -1 unit (negative direction). Note that the quotient $\frac{-3}{-1}$ is equal to 3. And 3 is the value of m in $y = 3x + 10$.

Now let us consider moving from point A to point B instead of from point B to point A. Let us trace this move, first along the vertical, and then along the horizontal. Examine Figure 9.14 as you read on. We move from point A to point H, covering 3 units (positive direction). Next we move from point H to point B along the horizontal, covering 1 unit (positive direction). Again note that the quotient 3 is equal to $\frac{3}{1}$, which is the slope of the line.

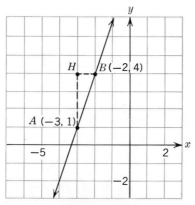

Figure 9.14

Let us now follow a similar procedure with the graph of $y = -2x - 1$, as shown in Figure 9.15. Moving from point D to point F, we cover -6 units (negative direction). Moving from F to E we cover 3 units (positive direction). The quotient $\frac{-6}{3}$ is equal to -2. And -2 is the value of m in $y = -2x - 1$.

Actually, it is possible to prove that the slope of a line may be found by considering *any* two points of the line, and taking the quotient of the vertical move to the horizontal move in going from one point to the other. We will use this concept of slope in graphing the solution sets of linear equations.

Thus, we can choose two points at random on the graph of a linear equation. Knowing the coordinates of these points, we can compute the

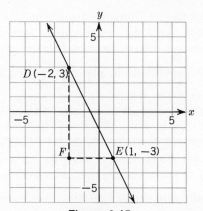

Figure 9.15

slope of the graph of the equation without even knowing what the equation is.

Next we shall attempt to discover how different values of b in $y = mx + b$ affect the graphs. To do this, we graph the following equations on the same coordinate system as shown in Figure 9.16.

1. $y = 2x$
2. $y = 2x + 1$
3. $y = 2x + 2$
4. $y = 2x + 3$
5. $y = 2x - 1$
6. $y = 2x - 2$
7. $y = 2x - 3$
8. $y = 2x - 4$

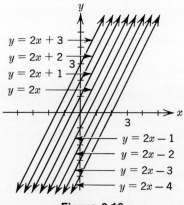

Figure 9.16

Note that each of the lines above has the slope 2. The lines are parallel. The replacement for b in $y = mx + b$ determines the point at which the line intersects the y-axis.

9.7 Slope

EXERCISE 9.5

1. Tell the slope of the line given by each of the following equations.
   - a. $y = 3x - 6$
   - b. $y = x + 7$
   - c. $y = -\frac{1}{3}x + 3$
   - d. $y = .6x - .5$

2. Tell the slope of the line given by each of the following equations by first determining an equivalent equation of the form $y = mx + b$.
   - a. $4x + y = 1$
   - b. $3x + 2y = 7$
   - c. $x - y = 5$
   - d. $2x + 3y - 1 = 0$
   - e. $3(x + 2) + y = 2$
   - f. $2(x + y) + 3(2x - y) = 0$

3. a. Graph $y = 2x + 3$.
   b. Choose two points on the graph, A and B. Moving from A to B, obtain the slope as the quotient of two moves, first along the vertical and then along the horizontal.
   c. Repeat part b, but let the moves take you from B to A.

4. Graph each of the following lines.
   - a. $y = x + 3$
   - b. $y = 2x + 5$
   - c. $y = 5x + 4$
   - d. $y = -x + 1$
   - e. $y = -2x - 3$
   - f. $y = -3x + 4$

5. For each of the following equations,
   - i. draw its graph,
   - ii. choose any two points on the line, (x_1, y_1) and (x_2, y_2), and compute the quotient $\frac{y_2 - y_1}{x_2 - x_1}$,
   - iii. decide if the quotient $\frac{y_2 - y_1}{x_2 - x_1}$ is equal to the slope of the line.
   - a. $y = 2x - 1$
   - b. $y = \frac{1}{2}x + 3$
   - c. $y = -x - 2$

6. Graph each equation by doing the following:
   - i. Write an equivalent equation of the form $y = mx + b$.
   - ii. Locate one point on the line and use the slope m to obtain the graph.
   - a. $y - 3x = -7$
   - b. $y + 2x = 3$
   - c. $6x = 3y + 9$
   - d. $x = y - \frac{1}{2}$
   - e. $2x + 4y = 1$
   - f. $3 - \frac{1}{2}y = -x$
   - g. $-\frac{1}{3}x - \frac{1}{6}y = 2$
   - h. $3x + 2y - 4 = 0$

7. Prove that if (x_1, y_1) and (x_2, y_2) belong to the solution set of the equation $y = mx + b$, then $m = \dfrac{y_2 - y_1}{x_2 - x_1}$ $[x_1 \neq x_2]$.

9.8 Intercepts

Let us continue our investigation of equations of the form

$$y = mx + b.$$

We know that each such equation has a line for its graph. We also know that if the first coordinate x of any point is 0, then this point lies on the y-axis. Thus we can conclude the following:

> If x is replaced by 0 in $y = mx + b$, the resulting value of y will tell us the point at which the graph of $y = mx + b$ intersects the y-axis.

For example, given $y = 3x + 5$, we replace x by 0:

$$y = 3 \times 0 + 5$$
$$y = 0 + 5$$
$$y = 5.$$

Similarly, we know that if we replace y by 0 in $y = mx + b$, we obtain the x-coordinate of the point at which the line intersects the x-axis. For example, given $y = 2x - 4$, if we replace y by 0, we find that $x = 2$.

These two lines are shown on the graph in Figure 9.17. Note that the graph of $y = 3x + 5$ intersects the y-axis at the point with coordinates $(0, 5)$. The graph of $y = 2x - 4$ intersects the x-axis at the point $(2, 0)$.

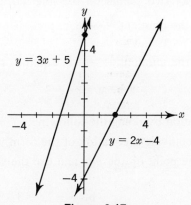

Figure 9.17

We now generalize these two ideas. First, we replace x by 0 in $y = mx + b$:

9.8 Intercepts

$$y = mx + b$$
$$y = m \times 0 + b$$
$$y = 0 + b$$
$$y = b.$$

We see then that the graph of $y = mx + b$ intersects the y-axis at the point $(0, b)$. We define the number b as follows:

DEFINITION 9.8 The second coordinate of the point at which any nonvertical line intersects the y-axis is called the *y-intercept* of the line.

In the definition we stated that the line is nonvertical, since a vertical line is parallel to the y-axis and thus does not intersect the y-axis.

Now we replace y by 0 in $y = mx + b$:

$$0 = mx + b$$
$$0 - b = mx + b - b$$
$$-b = mx$$
$$-\frac{b}{m} = x$$
$$x = -\frac{b}{m}.$$

We see then that the graph of $y = mx + b$ intersects the x-axis at the point $\left(-\frac{b}{m}, 0\right)$. We define the number $-\frac{b}{m}$ as follows:

DEFINITION 9.9 The first coordinate of the point at which any nonhorizontal line intersects the x-axis is called the *x-intercept* of the line.

In the last definition we stated that the line is nonhorizontal, since a horizontal line is parallel to the x-axis and thus does not intersect the x-axis.

The two intercepts can be used to good advantage in graphing some lines. Recall that to determine a line, we need to locate only two points of the line. In the following example we will use the x- and y-intercepts to locate the two points. We call this method of graphing the *intercept method*.

Example Graph $2x - 3y = 12$ using the intercept method.

First we replace y by 0:

$$2x - 3 \times 0 = 12$$
$$2x - 0 = 12$$
$$2x = 12$$
$$x = 6.$$

The x-intercept is 6. Thus, the line intersects the x-axis at the point (6, 0). Now we replace x by 0:

$$2 \times 0 - 3y = 12$$
$$0 - 3y = 12$$
$$-3y = 12$$
$$y = -4.$$

The y-intercept is -4. Thus, the line intersects the y-axis at the point $(0, -4)$. We locate these two points and draw a picture of the line passing through them. This line is the graph of $2x - 3y = 12$ shown in Figure 9.18.

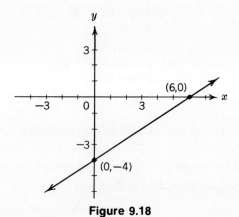

Figure 9.18

9.9 Slope-intercept Form

When a linear equation is written in the form $y = mx + b$, we say that it is in *slope-intercept* form. This form is convenient for graphing some lines, as is shown in the following example.

Example Graph the equation $y = 4x + 3$.

The equation is in the form $y = mx + b$. Since $b = 3$, the graph intersects the y-axis at the point (0, 3). The slope is 4. These two facts are used in obtaining the graph of the line which is shown in Figure 9.19.

We first located the point (0, 3). Since the slope can be written as $\frac{4}{1}$, we moved 4 units along the vertical in the positive direction and 1 unit along the horizontal in the positive direction. We would have obtained the same graph by starting at the point (0, 3) and moving 4 units

9.10 Horizontal and Vertical Lines

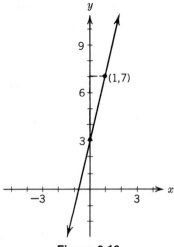

Figure 9.19

along the vertical in the negative direction and then 1 unit along the horizontal in the negative direction.

The method of graphing shown in the preceding example is called the *slope-intercept* method.

9.10 Horizontal and Vertical Lines

Consider the equation

$$y = 3.$$

This equation can also be written as

$$y = 0 \times x + 3,$$

and its solution set is

$$\{(x, y) | y = 0 \times x + 3\}.$$

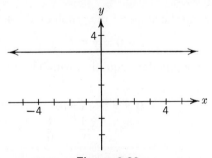

Figure 9.20

Note that x can be replaced by any real-number name, and the corresponding value of y is always 3. The domain of the solution set of this equation is the set of all real numbers and the range is $\{3\}$. Figure 9.20 is the graph of $y = 3$. By writing $y = 0 \times x + 3$, we see that the slope of the line is 0 and the y-intercept is 3. It is a horizontal line. It is easy to conclude that every line given by an equation of the form $y = b$ is horizontal and intersects the y-axis at the point $(0, b)$.

Now consider the equation
$$x = -4.$$
This equation can be written as
$$x + 0 \times y = -4,$$
and its solution set is
$$\{(x, y) | x + 0 \times y = -4\}.$$
Observe that y can be replaced by any real-number name, and the corresponding value of x is always -4. Thus, the domain of the solution set is -4 and the range is the set of all real numbers. The graph of the equation $x = -4$ is shown in Figure 9.21.

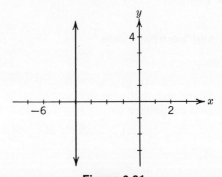

Figure 9.21

Let us try to determine the slope of the graph of $x = 4$. Consider the form
$$x + 0 \times y = 4.$$
We will try to find an equivalent equation which is of the form $y = mx + b$.
$$x + 0 \times y = 4$$
$$0 \times y = -x + 4$$
$$y = \frac{-1}{0}x + \frac{4}{0}.$$
But the last step violates a previous agreement: we may not divide by

9.10 Horizontal and Vertical Lines

zero! The symbols $\frac{-1}{0}$ and $\frac{4}{0}$ are meaningless. Thus, we will not consider a slope for the graph of $x = -4$. We shall say that it has no slope. Observe also that a vertical line has no y-intercept.

It is easy to conclude that every equation of the form $x = a$ has a vertical line for its graph.

EXERCISE 9.6

1. Tell the y-intercept for each of the following lines.
   a. $y = 2x + 3$
   b. $y = -2x + 7$
   c. $y = -2x - 3$
   d. $y = 5x$

2. Tell the x-intercept for each of the following lines.
   a. $3 + 2y = x$
   b. $x = 7y$
   c. $x = 5y - 1$
   d. $y = 5x - 1$

3. For each line described below,
      i. tell its y-intercept and its x-intercept,
      ii. draw its graph using the intercept method.
   a. $y = 3x + 3$
   b. $y = 2x - 8$
   c. $y = -3x - 9$
   d. $y = 5x + 1$
   e. $y = 4x$
   f. $y = -8x$
   g. $2y = 3x + 4$
   h. $-3y = 6x + 1$
   i. $2x + 3y = 10$
   j. $-3x + 5y = 1$
   k. $6x - 2y = 5$
   l. $-2x - 5y = -7$

4. Given the equation $y = kx + p$, tell the x-intercept in terms of k and p and the y-intercept in terms of p.

5. Given the equation $ay = bx + c$, tell the x-intercept in terms of b and c and the y-intercept in terms of a and c.

6. Given the equation $ax + by + c = 0$, tell the x-intercept in terms of a and c and the y-intercept in terms of b and c.

7. Tell the slope and the y-intercept of each of the following lines.
   a. $y = 3x + 7$
   b. $y = 2x - 3$
   c. $y = -3x + 4$
   d. $y = -x + 1$
   e. $y = x - 7$
   f. $y = x$
   g. $y = -x$
   h. $-y = x + 2$
   i. $-y = \frac{2x + 1}{2}$
   j. $y + 1 = 2x$

k. $2y = 4x + 2$ l. $y = 2(x + 1)$

m. $x = y - 1$ n. $y = \frac{1}{3}(3x - 3)$

8. For each equation, write an equivalent equation of the form $y = mx + b$.

   a. $x + y = 3$ b. $2x + y = -2$
   c. $4x - y = 1$ d. $2x - y = 0$
   e. $5x + 2y = 8$ f. $8x - 2y = 9$
   g. $-2x - 3y = 4$ h. $3(x + y) = 5$
   i. $-2(2x - y) = -1$ j. $-3(3x + 4y) + 7 = 0$

9. Using the slope-intercept method, graph each of the following lines.

   a. $y = x + 1$ b. $y = 3x + 2$
   c. $y = -2x + 3$ d. $y = -3x - 1$
   e. $\frac{1}{2}y = 5x + 4$ f. $\frac{1}{3}y = 2x - 2$
   g. $x + y = 3$ h. $2x + y = -7$
   i. $4x + 2y = 2$ j. $2y - 3x = 1$

10. Give the equation which is equivalent to $Ax + By + C = 0$ and which is in the slope-intercept form.

11. Give the equation which is equivalent to $A(x + y) + B = 0$ and which is in the slope-intercept form.

12. Give the equation which is equivalent to $A(x + y) + B(x + y) + C = 0$ and which is in the slope-intercept form.

13. For each of the following equations, tell whether its graph is a horizontal or a vertical line.

   a. $x = 2$ b. $y = -5$
   c. $x + 2 = 0$ d. $y - 6 = 0$
   e. $y - 3 = 2$ f. $2(x + 1) = 3$
   g. $4(1 - y) = 0$ h. $-2(x + 1) = 3(x + 4)$
   i. $-4(y - 3) = 5(1 - y)$ j. $-4(x + y) = -2(2x + 2)$

14. Graph each of the following lines.

   a. $x = -1$ b. $x = 4$
   c. $x + 2 = 0$ d. $y + (-7) = 0$
   e. $y = 4$ f. $y = -2$
   g. $y - 5 = 0$ h. $x + (-3) = 2$

15. Graph each of the following lines.

   a. $2(x + 1) = 3$ b. $-4(2 - x) = 10$
   c. $3(y + 1) = 7$ d. $-2(2y - 1) = 10$

9.11 Linear Functions

16. A horizontal line has only one intercept. Which intercept is it?
17. A vertical line has only one intercept. Which intercept is it?
18. What is the slope of any horizontal line?
19. Explain why no vertical line has a slope.

9.11 Linear Functions

Recall that the solution set of a linear equation such as $y = 2x - 1$ is a *relation;* that is, the solution set is a set of ordered pairs.

$$\{(x, y) | y = 2x - 1\}.$$

Now we are concerned with deciding whether the solution set of $y = 2x - 1$ is a *function*. Recall that a function is a relation such that no two ordered pairs have the same first component. In other words, a particular x value may not be paired with more than one y value.

It is easy to decide whether a relation is a function by examining its graph. Observe, as in Figure 9.22, that if at least one vertical line intersects the graph of the relation in more than one point, then the relation is not a function. This is the case because the relation contains two or more ordered pairs with the same first component.

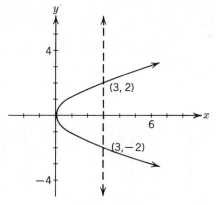

Figure 9.22

Note that in Figure 9.22 the vertical line we chose intersects the graph of the relation in two points, $(3, 2)$ and $(3, -2)$. The element 3 of the domain is paired with two different elements of the range, 2 and -2. Thus the relation whose graph is shown is *not* a function.

We can generalize this idea as follows:

VERTICAL LINE TEST FOR A FUNCTION A relation is a function if no vertical line intersects or touches the graph of the relation in more than one point.

Now consider in Figure 9.23 the graph of
$$y = 2x - 1.$$

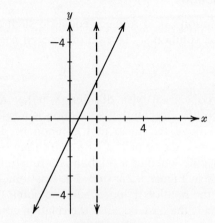

Figure 9.23

Could any vertical line intersect the graph in more than one point? Since the answer to the question is *no*, we can conclude that the relation
$$\{(x, y) | y = 2x - 1\}$$
is a function.

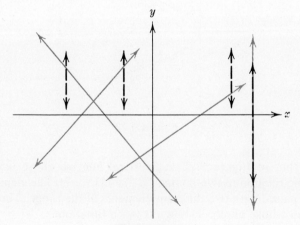

Figure 9.24

9.12 Inverses

Now we ask ourselves whether the solution set of every linear equation is a function. Since the graph of a linear equation is a line, let us consider the set of all lines in a coordinate plane. If we apply the vertical line test, we can see that all the lines in the plane are graphs of functions, except vertical lines.

Note that the vertical line graphed on the right in Figure 9.24 fails the vertical-line test, since a vertical line intersects it in an infinite number of points.

What kind of equations have graphs that are vertical lines? In the previous section, we discovered that these are the equations of the form

$$x = a$$

where a is a real number. Thus, the solution set of every linear equation, except those of the form $x = a$, is a function.

DEFINITION 9.10 A *linear function* is a relation which may be written in the form

$$\{(x, y) | y = mx + b)\}$$

where m and b are real numbers, and $m \neq 0$.

Note that we stated in the definition that $m \neq 0$. Therefore, any function specified by $y = b$ is not a *linear* function, although it is a function. In the next section we shall tell why we do not wish a function specified by an equation of the form $y = b$ to be considered a linear function.

9.12 Inverses

If $U = \{0, 1, 2, 3, 4\}$, the relation $G = \{(2, 1), (3, 2), (4, 3)\}$ is a subset of $U \times U$. The graph of G is at the right. The domain of relation G is

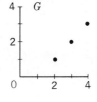

$$\{2, 3, 4\}$$

and the range is

$$\{1, 2, 3\}.$$

Now consider the relation

$$H = \{(1, 2), (2, 3), (3, 4)\}.$$

The graph of this relation is at the right. How is relation H related to G? The coordinates of each pair in H are obtained by interchanging the coordinates of the pairs in G. Thus, if $(2, 1) \in G$, then $(1, 2) \in H$.

The following display shows how H is obtained from G.

$$G = \{(2, 1), (3, 2), (4, 3)\}$$
$$H = \{(1, 2), (2, 3), (3, 4)\}.$$

H is called the *inverse* of G. Observe that if H is the inverse of G, then G is the inverse of H.

In general, we define the inverse of any relation as follows:

DEFINITION 9.11 For each relation A, the *inverse* of A is a relation B obtained by interchanging the components of each of the ordered pairs in A.

If we graph both G and H on the same set of axes, we can observe an interesting relationship between their graphs. The line segments which connect the corresponding points of the graphs of a relation and its inverse are perpendicular to the line described by the equation $y = x$, and they are bisected by the line. The graph of a relation is a *reflection* of its inverse with respect to the line $y = x$.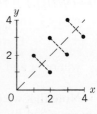

If we interchange the first and second coordinates in each ordered pair of a function, we obtain the *inverse of the function*.

In the examples which follow, examine the graph of each relation and its inverse. We will use the symbol A' for the inverse of A.

Example 1 $A = \{(1, 1), (2, 3), (4, 2)\}$. $A' = \{(1, 1), (3, 2), (2, 4)\}$.

Domain $= \{1, 2, 4\}$ Domain $= \{1, 2, 3\}$
Range $= \{1, 2, 3\}$.   Range $= \{1, 2, 4\}$.

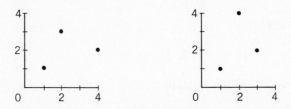

Observe that the range of the inverse of A is equal to the domain of A. The domain of the inverse of A is equal to the range of A.

In this example, A is a function and its inverse, A', is also a function.

9.12 Inverses

Example 2 $B = \{(1, 2), (3, 2), (4, 4)\}$. $B' = \{(2, 1), (2, 3), (4, 4)\}$.

Domain = $\{1, 3, 4\}$
Range = $\{2, 4\}$.

Domain = $\{2, 4\}$
Range = $\{1, 3, 4\}$.

In this case again the range of the inverse of B is equal to the domain of B, and the domain of the inverse of B is equal to the range of B.

However, in this example B is a function but its inverse is not a function.

Example 3 $C = \{(1, 2), (1, 3), (2, 2)\}$. $C' = \{(2, 1), (3, 1), (2, 2)\}$.

Domain = $\{1, 2\}$
Range = $\{2, 3\}$.

Domain = $\{2, 3\}$
Range = $\{1, 2\}$.

In this case neither C nor its inverse are functions.

Example 4 $D = \{(1, 2), (1, 3), (2, 4)\}$. $D' = \{(2, 1), (3, 1), (4, 2)\}$.

Domain = $\{1, 2\}$
Range = $\{2, 3, 4\}$.

Domain = $\{2, 3, 4\}$
Range = $\{1, 2\}$.

In this case, D is not a function, but its inverse is a function.

On the basis of Examples 1-4 above, the following conclusions can be drawn:

(1) The inverse of every relation is a relation.
(2) The inverse of a function may be a function or it may not be a function.
(3) The inverse of a relation which is not a function may be a function.

Now let us take the set of all real numbers as the universal set. In the universal set, consider the function
$$E = \{(x, y) | y = x + 3\}.$$
To obtain the inverse of E, we interchange x and y in the equation.
$$E' = \{(x, y) | x = y + 3\}$$
$$= \{(x, y) | y = x - 3\}.$$

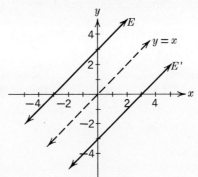

Figure 9.25

The graphs of E and E' are shown on the same axes. The graph of the line $y = x$ is also shown. From the graph in Figure 9.25 we can see that the graph of the inverse of E is a reflection of the graph of E in the graph of the line $y = x$.

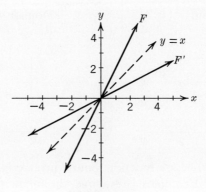

Figure 9.26

9.12 Inverses

Here is another pair of graphs of a function and its inverse. The universal set is the set of real numbers.

$$F = \{(x, y) | y = 2x\}$$
$$F' = \{(x, y) | x = 2y\}$$
$$= \{(x, y) | y = \tfrac{1}{2}x\}.$$

The last two examples show that the graph of the inverse of a relation (Figure 9.26) is a reflection of the graph of the relation with respect to the line $y = x$.

Now consider a function specified by an equation of the form $y = b$. The graph of one such function, $y = 2$, is shown in Figure 9.27.

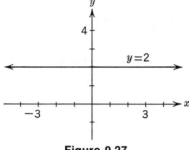

Figure 9.27

The inverse of $\{(x, y) | y = 2\}$ is $\{(x, y) | x = 2\}$. And we know that $x = 2$ defines a relation which is not a function. The graphs of $y = 2$ and $x = 2$ are shown in Figure 9.28.

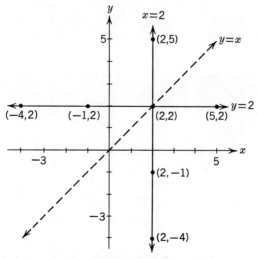

Figure 9.28

We wish for the inverse of every *linear* function to be a linear *function*. This is the reason why, in defining a *linear function*, we specified that $m \neq 0$ in $y = mx + b$. Thus, equations of the form $y = b$ do not define linear functions, although they define functions.

EXERCISE 9.7

1. Which of the following relations are also functions?
   a. $\{(1, 3), (1, 2)\}$
   b. $\{(3, 1), (2, 1)\}$
   c. $\{(-1, 0), (0, -1), (2, 0)\}$
   d. $\{(-2, 6), (3, 6), (-2, 5)\}$
   e. $\{(0, 0), (1, 1), (2, 2)\}$
   f. $\{(1, 2), (2, 4), (3, 9)\}$
   g. $\{(-1, 1), (-2, 4), (1, 1), (2, 4)\}$
   h. $\{(1, -1), (4, -2), (1, 1), (4, 2)\}$

2. Tell whether each graph below is a graph of a function. Use the vertical line test.

   a.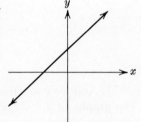
   b.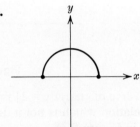
   c.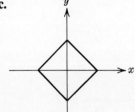
   d.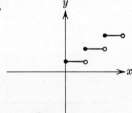
   e.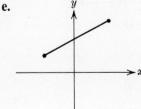
   f.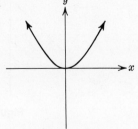

9.12 Inverses

g.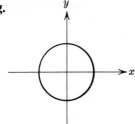

h.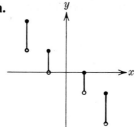

i.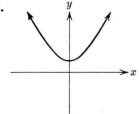

j.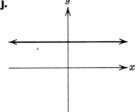

k.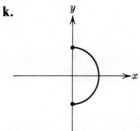

l.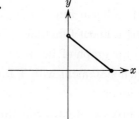

m.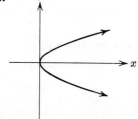

n.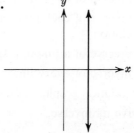

3. Tell why each of the following is a graph of a relation which is not a function.

a.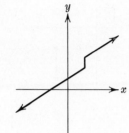

b.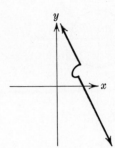

c.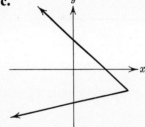

4. Is $\{(x, 3x + 2)\}$ a linear function?
5. Is $\{(x, 5)\}$ a linear function?
6. Is $\{(3, y)\}$ a linear function?
7. Tell why the graph of $\{(x, mx + b)\}$ for $b = 0$ passes through the origin.
8. Label each statement with T for true and F for false.
   a. The inverse of each function is a function.
   b. The inverse of each function is a relation.
   c. The inverse of each relation is a function.
   d. The inverse of each relation is a relation.
   e. If $Y = \{(1, 3), (2, 3)\}$, then $\acute{Y}$ is not a function.
   f. The inverse of $C = \{(1, 3), (3, 1)\}$ is equal to C.
   g. The inverse of each relation has the same number of elements as the relation.
   h. The inverse of a function specified by $y = -1$ is $x = -1$.
   i. The inverse of every linear function is a linear function.
9. For each relation graphed below,
   i. give its inverse,

9.12 Inverses

  ii. graph the inverse,
  iii. List the domain and the range of each relation and of its inverse.

a.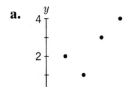

b.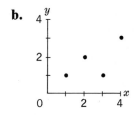

c.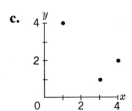

d.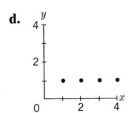

e.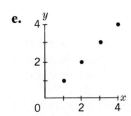

f.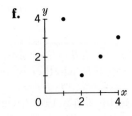

g.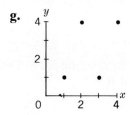

h.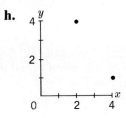

i.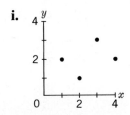

10. Using the set of all real numbers as the replacement set,
   i. graph each of the following functions,
   ii. determine the inverse of each function and graph it.

   a. $\{(x, y) | y = x + 1\}$ **b.** $\{(x, y) | y = x - 2\}$

   c. $\{(x, y) | y = 5x\}$ **d.** $\{(x, y) | y = 2x + 2\}$

   e. $\{(x, y) | y = \frac{x}{3} + 1\}$ **f.** $\{(x, y) | y = \frac{x}{2} - 5\}$

   g. $\{(x, y) | 2x = y - 2\}$ **h.** $\{(x, y) | 3x = 2y + 4\}$

   i. $\{(x, y) | 4x - 3 = 3y + 5\}$

11. Give an argument showing that the inverse of any function specified by $y = b$ is not a linear function.

9.13 Naming Functions

We are already familiar with the symbol (x, y) used in connection with ordered number pairs. Here, x denotes the first member of the pair and y the second member.

Frequently a function is named by the use of letters f, g, and others. For example, f may be the set of ordered pairs (x, y) for which $y = 2x$:

$$f = \{(x, y) | y = 2x\}.$$

The sentence $y = 2x$ *defines,* or *describes,* the function. Knowing this sentence, we can decide whether or not any ordered pair belongs to the function.

When considering a function f, we use the symbol $f(x)$ [read: f at x] to denote the element in the range which is paired with x in the domain. For example, if

$$f = \{(3, 2), (-1, 7), (0, 5)\},$$

then $f(3) = 2$, $f(-1) = 7$, and $f(0) = 5$.

The symbol $f(x)$ is often called the *value* of the function f at x. Thus, the value of function f at 3 is 2, at -1 is 7, and at 0 is 5.

Again consider the function

$$f = \{(x, y) | y = 2x\}.$$

Notice that for any replacement of x, $y = f(x)$. Thus, another way of specifying f is

$$f = \{(x, f(x)) | f(x) = 2x\}.$$

We see that the value of f at 1 is

$$f(1) = 2 \times 1 = 2.$$

9.14 Quadratic Functions

The value of f at -1 is
$$f(-1) = 2 \times (-1) = -2.$$
Also
$$f(0) = 2 \times 0 = 0$$
$$f(3) = 2 \times 3 = 6$$
$$f(\tfrac{1}{2}) = 2 \times \tfrac{1}{2} = 1.$$

Example Let h be the function defined by $h(x) = x^2 + 1$. Then
$$h(0) = 1$$
$$h(1) = 2$$
$$h(-1) = 2$$
$$h(-4) = 17$$
$$h(\tfrac{1}{2}) = 1\tfrac{1}{4}$$
$$h(\sqrt{2}) = 3.$$

9.14 Quadratic Functions

We have seen that a linear function is a set of ordered pairs $(x, f(x))$ in which $f(x) = mx + b$. Now we shall consider *quadratic functions*.

Each of the five following equations defines a quadratic function.

$$f(x) = x^2$$
$$g(u) = 3u^2 + 7$$
$$u(v) = -2v^2 + 5$$
$$t(w) = \frac{w^2}{2} + 2w$$
$$s(y) = -\tfrac{1}{3}y^2 + y - 3.$$

DEFINITION 9.12 A *quadratic function* is a relation which may be written in the form
$$\{(x, f(x)) \mid f(x) = ax^2 + bx + c\},$$
where a, b, and c are real numbers, and $a \neq 0$.

If $f(x) = ax^2 + bx + c$ and $a = 0$, then $f(x) = bx + c$ and it is a linear function [$b \neq 0$].

If $f(x) = 4x^2 + 3x + 1$, then $f(2) = 4(2)^2 + 3(2) + 1 = 23$. Thus $f(2) = 23$, and the ordered pair $(2, 23)$ belongs to the function.

Suppose we want to graph the function defined by $y = 2x^2$ when the universal set is the set of integers. Let us choose the domain D to be $\{x \mid -2 \leq x \leq 2\}$; that is, $D = \{-2, -1, 0, 1, 2\}$. To find the range, we replace x by each element of the domain.

$$f(x) = 2x^2 \qquad f(-1) = 2 \qquad f(1) = 2$$
$$f(-2) = 8 \qquad f(0) = 0 \qquad f(2) = 8.$$

Thus, the following ordered pairs make up the function: $(-2, 8)$, $(-1, 2)$, $(0, 0)$, $(1, 2)$, $(2, 8)$. Its graph is given in Figure 9.29.

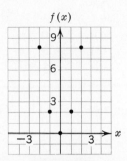

Figure 9.29

In the universe of the real numbers, if the domain is

$$D = \{x | -2 \leq x \leq 2\},$$

then there will be an infinite number of ordered pairs $(x, f(x))$ such that $f(x) = 2x^2$. In addition to those listed above, a few of these are as follows:

$$\left(-\frac{3}{2}, \frac{9}{2}\right) \quad \left(-\frac{5}{4}, \frac{25}{8}\right) \quad \left(-\frac{1}{2}, \frac{1}{2}\right) \quad \left(-\frac{1}{4}, \frac{1}{8}\right)$$
$$\left(\frac{3}{2}, \frac{9}{2}\right) \quad \left(\frac{5}{4}, \frac{25}{8}\right) \quad \left(\frac{1}{2}, \frac{1}{2}\right) \quad \left(\frac{1}{4}, \frac{1}{8}\right)$$

Plotting these and drawing a smooth curve between successive points, we get the graph shown in Figure 9.30. Curves of this type are called *parabolas*. The graph here is a part of a parabola.

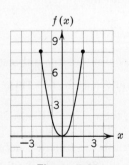

Figure 9.30

EXERCISE 9.8

1. Following are some examples of equations that describe functions. Compute the required function values.

9.14 Quadratic Functions

$$f(r) = 2r$$
$$k(s) = 5s + 3$$
$$t(w) = w^2$$
$$v(t) = 5t^2 + 3t$$
$$r(v) = \frac{1}{v^2}$$
$$h(y) = 3y - 7$$
$$g(z) = |z|$$

a. $f(3)$ **b.** $f(-\frac{1}{2})$ **c.** $f(.5)$
d. $k(1)$ **e.** $k(-2)$ **f.** $k(0)$
g. $t(0)$ **h.** $t(-2)$ **i.** $t(.1)$
j. $v(0)$ **k.** $v(-1)$ **l.** $v(1)$
m. $r(.1)$ **n.** $r(\frac{1}{2})$ **o.** $r(-1)$
p. $h(0)$ **q.** $h(-1)$ **r.** $h(3)$
s. $g(2)$ **t.** $g(-2)$ **u.** $g(0)$

2. For what value of v is the fifth function above, $r(v)$, not defined?

3. For $g(z)$ in the last example above, is it true that $g(a) = g(-a)$ for each real number a?

4. For $t(w)$ in the third example above, is it true that $t(a) = t(-a)$ for each real number a?

5. For each function defined below, compute its value at $0, -1, -2, 3, \frac{1}{2}, \frac{2}{3}$, and $\sqrt{2}$.

Example $m(u) = 2u^2 + 5$ $m(3) = 23$
  $m(0) = 5$ $m(\frac{1}{2}) = 5\frac{1}{2}$
  $m(-1) = 7$ $m(\frac{2}{3}) = 5\frac{8}{9}$
  $m(-2) = 13$ $m(\sqrt{2}) = 9$

a. $h(y) = 3y - 7$ **b.** $f(v) = 2v^2 + v$
c. $m(z) = -2 + z^2 + 2z$ **d.** $g(x) = 3x^2 + |x|$
e. $s(x) = x - |x| + x^2$ **f.** $t(w) = w^3 + w^2 + 6$

6. Classify each function as a linear or a quadratic function.
a. $f(x) = 2x + 1$ **b.** $g(y) = 3y^2 + y - 3$
c. $h(m) = -2 + 2m$ **d.** $k(s) = -1 + s^2 - 3s$
e. $f(u) = -\sqrt{2} - 6u$ **f.** $g(t) = -1 - t - t^2$

7. Let g be a function defined by $g(x) = 3x^2 - 5x + 7$. Compute each of the following.
a. $g(1)$ **b.** $g(3)$ **c.** $g(-5)$
d. $g(-\frac{1}{2})$ **e.** $g(1.5)$ **f.** $g(-2.3)$

g. $g(\sqrt{2})$ h. $g(\sqrt{3})$ i. $g\left(\dfrac{\sqrt{2}}{2}\right)$

j. $g(-\sqrt{5})$

8. Graph the functions defined below in the universe of real numbers when the domain $D = \{x \mid -3 \leq x \leq 3\}$.

   a. $y = x^2$
   b. $y = 2x^2$
   c. $y = 3x^2$
   d. $y = 4x^2$
   e. $y = -x^2$
   f. $y = -2x^2$
   g. $y = -3x^2$
   h. $y = -4x^2$
   i. $y = \dfrac{x^2}{2}$
   j. $y = \dfrac{x^2}{4}$
   k. $y = \dfrac{x^2}{8}$
   l. $y = -\dfrac{x^2}{2}$
   m. $y = -\dfrac{x^2}{4}$
   n. $y = -\dfrac{x^2}{8}$

9. The functions in Problem 8 are described by equations of the form $y = ax^2$. Describe the nature of the parabolas when

   a. a is positive
   b. a is negative
   c. $|a|$ increases.

10. Graph the quadratic functions defined below in the universe of real numbers. Plot them all on the same coordinate system and use the domain $D = \{x \mid -3 \leq x \leq 3\}$.

    a. $y = x^2$
    b. $y = x^2 + 1$
    c. $y = x^2 + 2$
    d. $y = x^2 + 3$
    e. $y = x^2 - 1$
    f. $y = x^2 - 2$
    g. $y = x^2 - 3$

11. In Problem 10, the equations are of the form $y = x^2 + c$. Tell how the parabolas are affected by different replacements for c.

12. Graph the quadratic functions defined by the following equations in the universe of real numbers. Use the domain $D = \{x \mid -3 \leq x \leq 3\}$ and graph on the same coordinate system.

    a. $y = x^2$
    b. $y = x^2 + 1$
    c. $y = x^2 + 2$
    d. $y = -x^2$
    e. $y = -x^2 + 1$
    f. $y = -x^2 + 2$
    g. $y = -x^2 - 1$
    h. $y = -x^2 - 2$

13. Examine the graphs in Problem 12 in relation to their equations. Describe the relation between the numbers named by replacements of a and c in $y = ax^2 + c$ and the graphs.

14. Find thirteen pairs belonging to each function defined below by replacing x by 0, 1, −1, 2, −2, 3, −3, 4, −4, 5, −5, 6, and −6. Plot the points. Draw a parabola for each set of points.

    a. $f(x) = x^2 + x + 1$ b. $g(x) = 2x^2 + x - 3$
    c. $m(x) = x^2 - 4x + 7$ d. $n(x) = -3x^2 + 5x - 7$
    e. $s(x) = 4x^2 - 3x - 9$ f. $t(x) = 6x^2 + 5x + 3$

GLOSSARY

Abscissa: The first coordinate of a point.

Cartesian plane: A plane with a coordinate system established in it.

Coordinate plane: The same as *Cartesian plane*.

Domain: The set of all first elements of the ordered pairs in a relation.

Function: A relation in which no two ordered pairs have the same first element.

Linear equation: An equation which can be given in the form $y = mx + b$.

Linear function: A function which can be described by $\{(x, y) | y = mx + b\}$, where $b \neq 0$.

Ordinate: The second coordinate of a point.

Quadratic function: A relation which can be written in the form $\{(x, f(x)) | f(x) = ax^2 + bx + c\}$, where $a \neq 0$.

Range: The set of all second elements of the ordered pairs in a relation.

Relation: A set of ordered pairs.

Slope (of a line): The number m in the equation of a line, $y = mx + b$.

Value (of a function): Example: The value of $f(x)$ at 2, when $f(x) = 3x + 1$ is equal to $3 \times 2 + 1$ or 7.

x-intercept: The first coordinate of the point at which a graph intersects the x-axis.

y-intercept: The second coordinate of the point at which a graph intersects the y-axis.

CHAPTER REVIEW PROBLEMS

1. On a number line, graph the solution set of each of the following. The universal set is the set of real numbers.

    a. $5x + 3 > 4$ b. $|y| \leq 3$
    c. $|m| > 5$ d. $|2p| < 1$
    e. $|3r + 2| \geq 3$ f. $|2t + 1| \leq 7$

2. For each graph below, tell whether it is a graph of a function. List all of the ordered number pairs which belong to the relation.

a.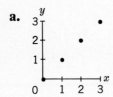

b.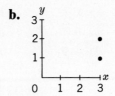

c.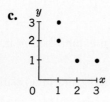

d.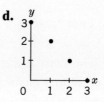

3. Using the vertical line test, tell whether each graph below is a graph of a function.

a.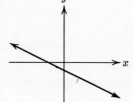

b.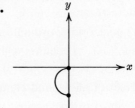

c.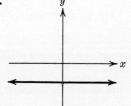

d.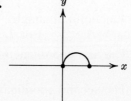

e.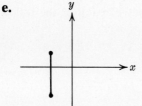

f.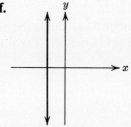

g.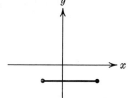
h.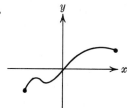

i.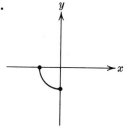

4. For each relation graphed below, draw the graph of its inverse.

a.
b.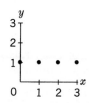

c.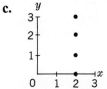
d.

e.
f.

5. For each function defined below, compute its value at 0, $-\frac{1}{2}$, and $\sqrt{3}$.

   a. $f(x) = x^3$
   b. $g(x) = 2x^2 - x$
   c. $h(x) = \dfrac{x^2 - 1}{x + 1}$
   d. $j(x) = 4x + |x|$

6. Given the universal set $U = \{-1, 0, 1, 2\}$, list the ordered pairs which are members of the solution set of each of the following.
   a. $x = y$
   b. $|x| = |y|$
   c. $x > y$
   d. $2x > y$
   e. $x < 2y$
   f. $|x| > |y|$

7. Using the set of real numbers as the universal set, graph each of the following. Tell the slope of each graph.
   a. $y = 3x - 1$
   b. $3y = 6x - 5$
   c. $2x + y = -\frac{1}{2}$
   d. $2y - 4x = 2x + y - 3$

8. For each equation, draw its graph using the given domain.
   i. $D = \{-2, -1, 0, 1, 2\}$
   ii. $D = \{x | -2 \leq x \leq 2\}$
   a. $\{(x, y) | y = x^2\}$
   b. $\{(x, y) | y = x^2 - 1\}$
   c. $\{(x, y) | y = -2x^2\}$

9. Graph each function and its inverse. Tell the slope of each graph. The universal set is the set of real numbers.
   a. $\{(x, y) | y = 2x + 1\}$
   b. $\{(x, y) | y = \frac{1}{3}x - 4\}$
   c. $\{(x, y) | y = -2x + 3\}$
   d. $\{(x, y) | y = -\frac{1}{2}x - 2\}$

10. Let the domain of each relation below be the set of real numbers. Graph each relation and tell if it is a function.
    a. $\{(x, y) | y = x\}$
    b. $\{(x, y) | y = x^2\}$
    c. $\{(x, y) | y = x^3\}$
    d. $\{(x, y) | y = x^4\}$
    e. $\{(x, y) | 2x + 3 = y\}$
    f. $\{(x, y) | \frac{1}{2}y + 1 = x\}$
    g. $\{(x, y) | |y| = x\}$

10
Systems of Equations and Inequalities

10.1 Equations in Two Variables

An example of an equation in two variables is

$$x + 2y = 5.$$

The variables in this equation are x and y. In order to obtain true statements, we shall replace the variables x and y by numerals. We agree that the two variables may be replaced by names for the same number or for two different numbers.

We pick any number whatsoever for one variable, for example, x. Replacing x by, say 1, we have

$$1 + 2y = 5.$$

Then solving for y we get

$$1 + 2y = 5$$
$$2y = 4$$
$$y = 2.$$

Thus, 1 for x and 2 for y, gives us

$$1 + 2 \times 2 = 5$$

which is a true statement.

Now let us use $\frac{1}{2}$ for x:
$$\frac{1}{2} + 2y = 5$$
$$2y = 4\frac{1}{2}$$
$$y = 2\frac{1}{4}.$$

$\frac{1}{2}$ for x and $2\frac{1}{4}$ for y gives us

$$\frac{1}{2} + 2 \times 2\frac{1}{4} = 5$$
$$\frac{1}{2} + 4\frac{1}{2} = 5,$$

which also is a true statement.

Thus, the pairs

$(1, 2)$, that is, 1 for x, 2 for y

and

$(\frac{1}{2}, 2\frac{1}{4})$, that is, $\frac{1}{2}$ for x, $2\frac{1}{4}$ for y

satisfy the equation $x + 2y = 5$.

There are many more number pairs which satisfy the equation above. Some of these are

$(-1, 3)$ $(4, \frac{1}{2})$ $(-2, 3\frac{1}{2})$
$(1.5, 1.75)$ $(-4, 4\frac{1}{2})$.

We can observe that there is no end to the number of ordered pairs which satisfy the equation $x + 2y = 5$.

More generally, an equation of the form $Ax + By = C$ has an infinite number of ordered real-number pairs which satisfy it.

The equation $3x + 2y = 7$ is of such form. This equation is obtained from $Ax + By = C$ by replacing A by 3, B by 2, and C by 7. Replacing A, B, and C by various numerals we obtain many different equations.

If either A or B is replaced by 0, an equation is obtained which has one variable in it. For example,

0 for A, -5 for B,   6 for C: $-5y = 6$
1 for A,   0 for B,   0 for C: $x = 0$.

The set of replacements which does not result in an equation with at least one variable is

0 for A and 0 for B.

An equation with two variables is not always in the form $Ax + By = C$. For example, the equation $2y = 3(x - 5) + 12 - 5y$ is *not* in this form. We can find, however, an equation which is equivalent to the above and which is of the form $Ax + By = C$. Recall that two equations are said to be equiv-

10.1 Equations in Two Variables

alent if and only if they have the same solution set. An example below shows how to obtain an equation of the form $Ax + By = C$ which is equivalent to an equation given in a different form. Each step is written out so that you will be able to identify the property which justifies it.

Example
$$2y = 3x - 15 + 12 - 5y$$
$$2y + 5y = 3x - 3 - 5y + 5y$$
$$7y = 3x - 3$$
$$7y - 3x = 3x - 3 - 3x$$
$$-3x + 7y = -3.$$

In the preceding example we have shown many details. With some practice, you will be able to "skip" some steps. Do not be too hasty in skipping steps, however, especially when you are not sure of yourself.

The last equation, $-3x + 7y = -3$, is of the form $Ax + By = C$, where A is replaced by -3, B by 7, and C by -3. This form is called the *standard form* of a linear equation in two variables.

EXERCISE 10.1

1. Label the following statements with T for true and F for false.
   a. $(3, -2)$ is a pair (x, y) that satisfies $4x + 3y = 6$.
   b. $(5, 2)$ is a pair (x, y) that satisfies $3x - 4y = 27$.
   c. $(4, -2)$ is a pair (x, y) that satisfies $3x - 4y = 20$.
   d. $(5, 3)$ is a pair (x, y) that satisfies $2y - x = 1$.
   e. $(\frac{1}{2}, \frac{1}{3})$ is a pair (x, y) that satisfies $4x + 6y = 4$.

2. Tell which of the following ordered pairs (x, y) satisfy $3x - 5y = 10$.
   a. $(9, \frac{1}{5})$
   b. $(0, 2)$
   c. $(3\frac{1}{3}, 0)$
   d. $(-2, -3.2)$
   e. $(-\frac{10}{3}, 0)$
   f. $(5, 11)$
   g. $(0, -2)$
   h. $(5, -1)$
   i. $(-5, 5)$
   j. $(-5, -1)$
   k. $(-5, 1)$
   l. $(-5, -5)$

3. For each value of x given below, determine the value of y so that (x, y) satisfies $2x + 6y = 7$.
   a. 5
   b. $2\frac{1}{2}$
   c. -1
   d. $-2\frac{1}{2}$
   e. $\frac{1}{2}$
   f. 0
   g. -3
   h. $3\frac{1}{2}$
   i. $-3\frac{1}{2}$
   j. 10
   k. -10
   l. $-5\frac{1}{2}$

4. For each equation, determine three ordered pairs (x, y) that satisfy the following equations.

a. $4x + 5y = 9$
b. $6x - 2y = 10$
c. $x + y = 6$
d. $2x + 2y = 12$
e. $x - y = 5$
f. $3x - 2y = -4$

5. Tell which of the following are true for all real-number replacements of the variables.

a. $|x - y| = |y - x|$
b. $|x + y| = |x| + |y|$
c. $|x - y| = |x| - |y|$
d. $|xy| = |x| \times |y|$
e. $\left|\dfrac{x}{y}\right| = \dfrac{|x|}{|y|}$
f. $(|x|)^2 = x^2$

6. For each equation below, find an equivalent equation in standard form. At each point you should be prepared to justify what you do by the appropriate property. In the example not every step is shown. Fill in the intermediate steps if you are unsure of them.

Example $3 - 7y + 3x = 6(x - y) + 3y$
$3 - 7y + 3x = 6x - 6y + 3y$
$3 - 7y + 3x = 6x - 3y$
$3 - 7y - 3x = -3y$
$3 - 4y - 3x = 0$
$-4y - 3x = -3$
$-3x + (-4)y = -3.$

a. $x = y + 7$
b. $2x = -3 - 8y$
c. $-5(x + y) = 17$
d. $3x = 2(y + 6)$
e. $-5(2x + y) = -3(x - 2)$
f. $-10(x - 3y) = 2y - 3x + 6$
g. $\dfrac{x + 3}{5} = \dfrac{y - 1}{2}$
h. $\dfrac{1}{2x + 1} = \dfrac{2}{3y + 4}$
i. $\dfrac{2y - 3}{-2} = \dfrac{-3x - 1}{-3}$
j. $6(y - 6x) + 17 = 13 - 4(x + 10y)$
k. $\dfrac{y - x}{y + x} = 3$
l. $\dfrac{2x - 3}{3y + 7} = -9$
m. $\dfrac{3}{5} = \dfrac{6x - y + 15}{3x + 7y - 1}$
n. $\dfrac{-5}{-y + 7} = \dfrac{2}{-4x - 3}$
o. $\dfrac{3(x - 1)}{5} = \dfrac{2(2y + 7)}{9}$
p. $\dfrac{-4}{3 - y} = \dfrac{-3}{5x - 2}$

10.2 Solving Systems of Equations by Graphing

In Chapter 9 we learned to graph a linear equation in two variables. Since the graph of such an equation is a straight line and since two points determine a line, we needed to determine only two ordered number pairs

10.2 Solving Systems of Equations by Graphing

which satisfy the equation. The two points corresponding to these number pairs determine the line.

Let us consider the equation $3x - 4y = 12$. In order to find ordered real-number pairs satisfying the equation, let us use the intercept method and first replace x by 0.

$$3 \times 0 - 4y = 12$$
$$-4y = 12$$
$$y = -3.$$

Thus, the y-intercept is -3, and $(0, -3)$ should be one pair which satisfies the equation. To check,

$$3 \times 0 - 4(-3) = 0 - (-12), \text{ or } 12,$$

so we know that $(0, -3)$ does, in fact, satisfy the equation.

Now replace y by 0.

$$3x - 4 \times 0 = 12$$
$$3x = 12$$
$$x = 4.$$

Thus, the x-intercept is 4 and $(4, 0)$ should be another pair which satisfies the equation. To check,

$$3 \times 4 - 4 \times 0 = 12 - 0, \text{ or } 12.$$

Let us locate points corresponding to the ordered number pairs $(0, -3)$ and $(4, 0)$ and draw a picture of the line through them, as in Figure 10.1. This line is the graph of the equation

$$3x - 4y = 12.$$

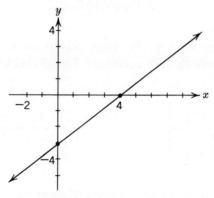

Figure 10.1

Now we are ready to consider pairs of linear equations in two variables. Each of these equations has a straight line for its graph. Thus, the graph of the two equations consists of two straight lines; and two straight lines in the same plane either intersect in one point or do not intersect at all, or else the lines are not distinct. In the last case we have two equivalent equations, that is, two equations for the same line.

Consider the following *system of two linear equations* in two variables:

$$-x + 2y = 5$$
$$3x - 2y = -13.$$

We shall be concerned with *solving* the system; that is, we wish to find all ordered pairs (x, y) which satisfy *both* equations. The set of all such ordered pairs is the *solution set* of the system.

Below are graphs of the two equations. Examine the graphs in Figure 10.2 to see how they were obtained.

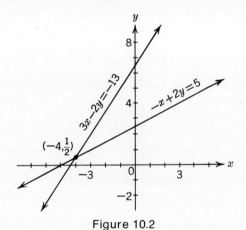

Figure 10.2

From the graph we can see that the two lines intersect at the point whose coordinates are $(-4, \frac{1}{2})$. Since this point is on both lines, its coordinates must satisfy both equations. Let us check the pair $(-4, \frac{1}{2})$ in each equation.

$-x+2y$	5	$3x - 2y$	-13
$-(-4) + 2 \times \frac{1}{2}$	5	$3(-4) - 2 \times \frac{1}{2}$	-13
$4 + 1$		$-12 - 1$	
5		-13	

Thus, $(-4, \frac{1}{2})$ satisfies each equation. We can now state that the intersection of the two solution sets is equal to $\{(-4, \frac{1}{2})\}$.

$$\{(x, y)|-x + 2y = 5\} \cap \{(x, y)|3x - 2y = -13\} = \{(-4, \tfrac{1}{2})\}.$$

10.2 Solving Systems of Equations by Graphing

We have used the *graphic method* to locate the solution set of the system of two linear equations in two variables.

Example For the system of equations

$$2x + y = -4$$
$$3y = 5x - 1,$$

(a) graph the solution set of each equation,
(b) list the coordinates of the point of intersection of the two graphs as an ordered pair of real numbers,
(c) check to see whether the coordinates of the point of intersection satisfy both equations,
(d) state the solution set.

Answers.

(a)

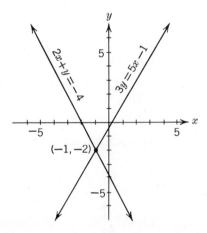

(b) $(-1, -2)$.
(c) We must remember to check both equations.

Check. -1 for x; -2 for y

$2x + y$	-4
$2(-1) + (-2)$	-4
$-2 + (-2)$	
-4	

$3y$	$5x - 1$
$3(-2)$	$5(-1) - 1$
-6	$-5 - 1$
	-6

(d) $\{(x, y) | 2x + y = -4\} \cap \{(x, y) | 3y = 5x - 1\} = \{(-1, -2)\}$.

10.3 Independent, Dependent, Inconsistent Systems

In the preceding section, we learned to find the coordinates of a point common to two straight lines which are graphs of two linear equations, each in two variables. We have found that the coordinates of the common point satisfy both equations. Thus, the solution set of the system of the two equations consists of one ordered pair. This set is the intersection of two sets, each consisting of an infinite number of ordered number pairs. The graph of each of these sets is a line.

Now consider the following system of two equations in two variables:

$$2x + y = -1$$
$$2y = -2 - 4x.$$

Let us find two ordered number pairs which satisfy the first equation and two pairs which satisfy the second equation.

First equation: $(-1, 1)$; $(4, -9)$
Second equation: $(1, -3)$; $(-3, 5)$.

Examine the graph in Figure 10.3. Observe that each of the pairs of points determines the same line; that is, both equations are equations for the same line. Let us examine them more closely.

$$2x + y = -1$$
$$2y = -2 - 4x.$$

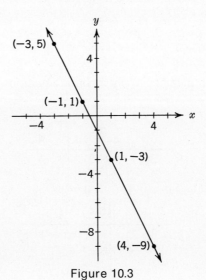

Figure 10.3

10.3 Independent, Dependent, Inconsistent Systems

We shall use equation properties to attempt to derive the second equation from the first. In doing this, we shall not fill in all of the steps. If you are not certain about any of the steps, you should fill in the details for yourself.

$$2x + y = -1$$
$$2x + y - 2x = -1 - 2x$$
$$y = -1 - 2x$$
$$2y = 2(-1 - 2x)$$
$$2y = -2 - 4x.$$

Thus we see that the two original equations are equivalent, and so their graphs are the same line. Such a pair of equations is called a *dependent* system of two linear equations in two variables. It is easy to see that the solution set of such a system is infinite. Each ordered pair of numbers which satisfies one equation also satisfies the other. Thus,

$$\{(x, y) | 2x + y = -1\} = \{(x, y) | 2y = -2 - 4x\}.$$

Now consider the following pair of equations:

$$x + 2y = 3$$
$$2x + 4y = -2.$$

The first equation is satisfied by, say,

$$(-1, 2) \text{ and } (5, -1).$$

The second equation is satisfied by, say,

$$(1, -1) \text{ and } (-1, 0).$$

Now, study the graph in Figure 10.4.

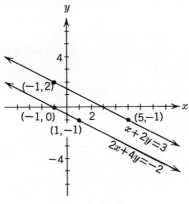

Figure 10.4

From the graph it appears that the two lines do not intersect. They are, therefore, parallel lines. Parallel lines are lines in the same plane which do not intersect no matter how far extended. In such a case there is no intersection point. If it is true that the two lines have no common point, then it is also true that there is no number pair which satisfies both equations.

It can easily be seen that the equation

$$x + 2y = -1$$

is equivalent to the equation

$$2x + 4y = -2,$$

which is our second equation. Therefore, we can use the pair of equations

$$x + 2y = 3$$
$$x + 2y = -1$$

in place of the original pair. But we see that in this pair of equations we want $x + 2y$ to yield 3 and -1 for the same replacements of x and y. This is not possible.

We can conclude, then, that the equations

$$x + 2y = 3$$
$$2x + 4y = -2$$

have no common solutions. Such a pair of equations is called an *inconsistent* system of two linear equations in two variables. We can further conclude that the solution set of such a system is the empty set. Thus,

$$\{(x, y) | x + 2y = 3\} \cap \{(x, y) | 2x + 4y = -2\} = \phi.$$

A pair of linear equations in two variables having exactly one number pair satisfying both equations is called an *independent* system of two linear equations in two variables.

Below are given three examples of systems of two equations for your study. In each example you will find the following:

(a) the graph of the solution set of each equation in the system,
(b) a statement as to whether the system is independent, dependent, or inconsistent,
(c) the solution set of an independent system, or a proof for the case of a dependent or an inconsistent system,
(d) a set description of the solution.

Example 1 $x + 2y = -1$
$2x + 4y = 4.$

10.3 Independent, Dependent, Inconsistent Systems

(a)

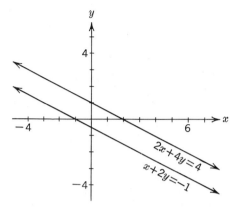

(b) The lines as graphed above appear parallel. In that case, the system is inconsistent.

(c) Let us take the second equation and obtain an equation equivalent to it.

$$2x + 4y = 4$$
$$\tfrac{1}{2}(2x + 4y) = \tfrac{1}{2}(4)$$
$$x + 2y = 2.$$

Thus, we now have the pair of equations

$$x + 2y = -1$$
$$x + 2y = 2.$$

There are no ordered pairs of numbers which will satisfy both of these equations, since they lead to the conclusion that $-1 = 2$. Therefore, this system of equations is an inconsistent system.

Since the equation

$$x + 2y = 2$$

is equivalent to the equation

$$2x + 4y = 4,$$

we can conclude that the original system

$$x + 2y = -1$$
$$2x + 4y = 4$$

is an inconsistent system.

(d) $\{(x, y) | x + 2y = -1\} \cap \{(x, y) | 2x + 4y = 4\} = \phi$

Example 2 $-m + 3n = 7$
$\phantom{\textbf{Example 2}\ \ }2m + 4n = 16.$

(a)

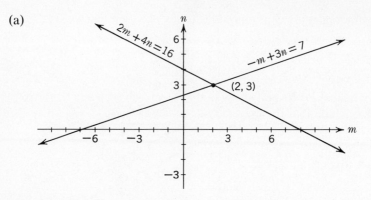

(b) The system is independent.
(c) $\{(2, 3)\}$.

Check. 2 for m; 3 for n. (We must check both equations.)

$-m + 3n$	7	$2m + 4n$	16
$-2 + 3 \times 3$	7	$2 \times 2 + 4 \times 3$	16
$-2 + 9$		$4 + 12$	
7		16	

(d) $\{(m, n)|-m + 3n = 7\} \cap \{(m, n)|2m + 4n = 16\} = \{(2, 3)\}$.

Example 3 $z\ \ \ \ = 4 - w$
$\ \ \ \ \ \ \ \ \ \ \ \ 2w + 2z = 8$.

(a)

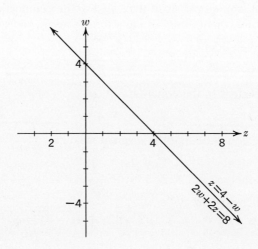

(b) The system is dependent.

10.3 Independent, Dependent, Inconsistent Systems

(c) Finding equivalent equations to $z = 4 - w$ gives us
$$w + z = 4$$
$$2(w + z) = 2 \times 4$$
$$2w + 2z = 8.$$

Thus, the first equation is equivalent to the second, and we can conclude that the system is dependent.

(d) $\{(z, w) | z = 4 - w\} = \{(z, w) | 2w + 2z = 8\}.$

EXERCISE 10.2

1. For each system of equations,
    - *i.* graph the solution set of each equation,
    - *ii.* list the coordinates of the point of intersection of the two graphs as an ordered pair of real numbers,
    - *iii.* check to see whether the coordinates of the point of intersection satisfy both equations,
    - *iv.* state the solution set.

   a. $x + y = 4$
      $2x - y = 2$

   b. $3x + 2y = 10$
      $-x + 6y = 0$

   c. $3x - y = -1$
      $x - 2y = 8$

   d. $2x = 2y + 1$
      $x + 2y = 8$

   e. $7y = x - 4$
      $x = y + 4$

   f. $-2 - 3x = 4y$
      $y = -x$

   g. $y = x - 1$
      $2y = 2x + 2$

   h. $6x = 2y - 7$
      $2y = 13x$

   i. $x = y$
      $y = -x$

   j. $x = 0$
      $x = y$

   k. $y = 0$
      $x = -y$

   l. $-x = -y$
      $x = -2$

   m. $x = 4$
      $y = -2$

2. For each system of two linear equations,
    - *i.* graph the solution set of each equation,
    - *ii.* tell whether the system is independent, dependent, or inconsistent,
    - *iii.* in the case of an independent system, determine the solution set and check; in the case of a dependent or inconsistent system, use equation properties to prove your decision,
    - *iv.* write a set description of the solution.

a. $u + v = 2$
$2u + 5v = 7$

b. $2x = 3 - 5y$
$10y = 6 - 4x$

c. $-s = 2t - 4$
$4t + 2s = 1$

d. $\frac{1}{2}k + \frac{1}{3}m = -6$
$k + m = -15$

e. $2n + 5 = r$
$-3r = -15 - 6n$

f. $7 - 3p = 4t$
$2t + 1.5p = 17$

g. $x = -y$
$x = y$

h. $2t - 3u = 1$
$3t - 2u = 1$

i. $2(a - 2b) = 3$
$2b - a = 1$

j. $5w - 4v = 6$
$3 + 2v = 2.5w$

3. Consider two lines in space. State two different conditions under which a pair of lines has no point of intersection.

10.4 Graphing Inequalities

We have agreed to use the set of real numbers as the universal set from which to obtain solutions of equations and inequalities. We consider the equation

$$x + y = 4$$

to be a *selector* of ordered pairs of real numbers which are solutions of this equation. The set of all such ordered pairs is given as

$$A = \{(x, y) | x + y = 4\}.$$

By plotting points which correspond to these ordered pairs, we obtain a *graph* of $x + y = 4$. It is shown in Figure 10.5.

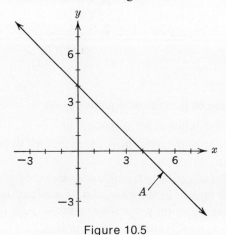

Figure 10.5

10.4 Graphing Inequalities

Now let us consider the set

$$B = \{(x, y) | x + y > 4\}.$$

To graph the inequality $x + y > 4$, we first graph the equation $x + y = 4$. We use dashed marks as is done in the graph in Figure 10.6. The boundary line, $x + y = 4$, separates the coordinate plane into two half-planes. One of these half-planes is a graph of $x + y > 4$. To decide which one of the two half-planes it is, we need to test the coordinates of just one point. If they satisfy $x + y > 4$, then this point belongs to the half-plane which is a graph of $x + y > 4$. We test $(5, 3)$.

$$5 + 3 > 4$$

is true; thus, $(5, 3) \in \{(x, y) | x + y > 4\}$.

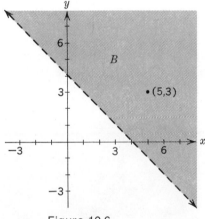

Figure 10.6

It is easy to conclude now that the graph of

$$C = \{(x, y) | x + y < 4\}$$

is the other half-plane which is shown in the graph in Figure 10.7. The point $(-1, 3)$ is plotted, and its coordinates satisfy $x + y < 4$, because the following is true:

$$-1 + 3 < 4.$$

Thus, $(-1, 3) \in \{(x, y) | x + y < 4\}$.

From the two groups above you should be able to conclude that the intersection of B and C is the empty set, that is $B \cap C = \phi$.

Now examine the last three graphs. You should be able to see that the union of A, B, and C is the entire plane. That is, $A \cup B \cup C$ is equal to

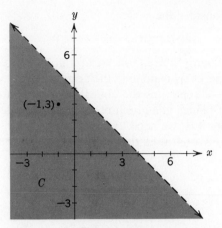

Figure 10.7

$$\{(x, y)|x + y = 4\} \cup \{(x, y)|x + y > 4\} \cup \{(x, y)|x + y < 4\} = p,$$

where p is the entire plane. We can show the graphs of all three sets on one coordinate system as shown in Figure 10.8.

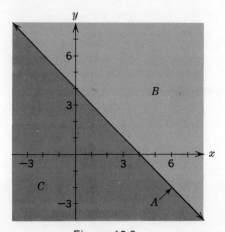

Figure 10.8

In order to graph the set $D = \{(x, y)|x + y \geq 4\}$, we would graph the union of A and B. That is,

$$A \cup B = \{(x, y)|x + y = 4\} \cup \{(x, y)|x + y > 4\}$$
$$= \{(x, y)|x + y \geq 4\}.$$

To show that the line $x + y = 4$ belongs to the graph, we use a solid rather than a dashed line, as is shown in Figure 10.9.

10.4 Graphing Inequalities

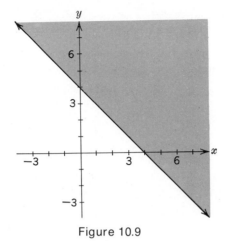

Figure 10.9

The method of graphing inequalities by first graphing the corresponding equation is an effective one. Here is a second example of the use of this method in graphing the set $\{(x, y)|x + 5 < y\}$. Notice that we first graph $\{(x, y)|x + 5 = y\}$. It is shown as a dashed line, because the line does not belong to the graph of $\{(x, y)|x + 5 < y\}$.

Before deciding which half-plane is the desired graph, we tested the point $(-4, 6)$ whose coordinates satisfy $x + 5 < y$, because the following is true:

$$-4 + 5 < 6.$$

Thus, the shaded region of Figure 10.10 is the graph of $\{(x, y)|x + 5 < y\}$.

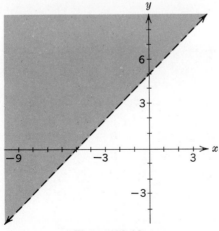

Figure 10.10

EXERCISE 10.3

1. By merely looking at the description of each of the following sets, tell whether the boundary does or does not belong to the graph of the set.
   a. $\{(x, y) | x + y < 1\}$
   b. $\{(x, y) | 2x + y > -2\}$
   c. $\{(x, y) | y \geq 2x + 7\}$
   d. $\{(x, y) | \frac{x + y}{2} \leq 0\}$
   e. $\{(x, y) | 2(3y - x) < 0\}$
   f. $\{(x, y) | -3(y - x) > 2\}$

2. In each problem, you are given a set and the coordinates of a point. Verify whether or not the coordinates satisfy the given inequality.
   a. $\{(x, y) | 2x + y > 3\}$;   (1, 2)
   b. $\{(x, y) | -x + 2y < -1\}$;   (2, 0)
   c. $\{(x, y) | 6x > y - 3\}$;   (0, 4)
   d. $\{(x, y) | y > 2x - 5\}$;   (2, −2)
   e. $\{(x, y) | 3x + 1 < y\}$;   (−2, −6)

3. Graph each of the following sets.
   a. $\{(x, y) | x + y = 2\}$
   b. $\{(x, y) | x + y > 2\}$
   c. $\{(x, y) | x + y < 2\}$
   d. $\{(x, y) | x + y \geq 2\}$
   e. $\{(x, y) | 2x - y = -1\}$
   f. $\{(x, y) | 2x - y > -1\}$
   g. $\{(x, y) | 2x - y < -1\}$
   h. $\{(x, y) | 2x - y \geq -1\}$
   i. $\{(x, y) | x = -y\}$
   j. $\{(x, y) | x > -y\}$
   k. $\{(x, y) | x < -y\}$
   l. $\{(x, y) | x \geq -y\}$
   m. $\{(x, y) | 2x + 3y = 5\}$
   n. $\{(x, y) | 2x + 3y > 5\}$
   o. $\{(x, y) | 2x + 3y < 5\}$
   p. $\{(x, y) | 2x + 3y \leq 5\}$
   q. $\{(x, y) | x - 4y = 0\}$
   r. $\{(x, y) | x - 4y > 0\}$
   s. $\{(x, y) | x - 4y < 0\}$
   t. $\{(x, y) | x - 4y \geq 0\}$

4. For each pair of sets A and B given below, give the simplest name for $A \cap B$.
   a. $A = \{(x, y) | x + y > 2\}$; $B = \{(x, y) | x + y < 2\}$
   b. $A = \{(x, y) | 2x - y < 1\}$; $B = \{(x, y) | 2x - y \leq 1\}$
   c. $A = \{(x, y) | 4x \leq y + 3\}$; $B = \{(x, y) | 4x = y + 3\}$
   d. $A = \{(x, y) | x < y - 1\}$; $B = \{(x, y) | x > y - 1\}$
   e. $A = \{(x, y) | 2y \geq x - 3\}$; $B = \{(x, y) | 2y \leq x - 3\}$
   f. $A = \{(x, y) | x - y > -1\}$; $B = \{(x, y) | x - y < -1\}$

10.5 Graphing Unions and Intersections of Sets

Let us consider the set

$$A = \{(x, y) | -1 < x < 3\}.$$

Recall that $-1 < x < 3$ is an abbreviation for the conjunction $-1 < x$ and $x < 3$. We have graphed each of the following sets in Figure 10.11:

$$\{(x, y) | x > -1\} \qquad \{(x, y) | x < 3\}.$$

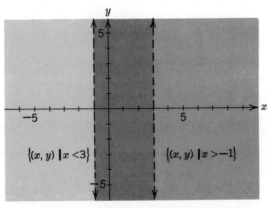

Figure 10.11

We are interested in the intersection of the two sets

$$A = \{(x, y) | x > -1\} \cap \{(x, y) | x < 3\} = \{(x, y) | -1 < x < 3\}.$$

Its graph is shown in Figure 10.12.

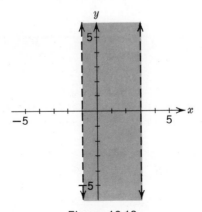

Figure 10.12

Now suppose we wish to obtain the graph of $C = \{(x, y)|x = y$ or $-1 < x < 3\}$. This set is the *union* of the following two sets:

$$A = \{(x, y)|-1 < x < 3\}$$
$$B = \{(x, y)|x = y\}.$$

The graph of C consists of the two parts shown in the graph in Figure 10.13. These two parts represent the graph of $A \cup B$.

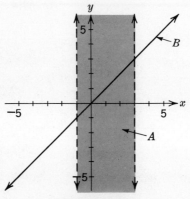

Figure 10.13

What would be the intersection of sets A and B above? It consists of all those elements which are in both sets. Its graph, which is an open segment, is shown in Figure 10.14.

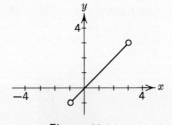

Figure 10.14

For another example, consider the set $D = \{(x, y)|x > 2$ and $y < -3\}$. This set is the intersection of the following two sets:

$$E = \{(x, y)|x > 2\}$$
$$F = \{(x, y)|y < -3\}.$$

The graphs of these sets are given in Figure 10.15.

10.5 Graphing Unions and Intersections of Sets

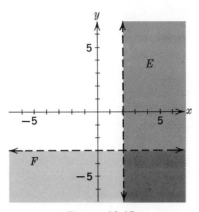

Figure 10.15

The graph in Figure 10.16(a) shows the intersection of the two sets

$$E \cap F = \{(x, y) | x > 2 \text{ and } y < -3\}.$$

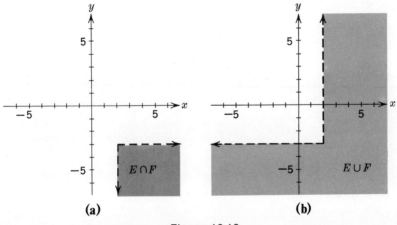

Figure 10.16

Observe that each point belonging to this graph is such that its first coordinate is greater than 2 $[x > 2]$, and its second coordinate is less than -3 $[y < -3]$.

The graph in Figure 10.16(b) shows the union of E and F. It extends indefinitely to the right of the line $x = 2$ and below the line $y = -3$.

EXERCISE 10.4

1. Write each of the following as a conjunction.

Example $0 < x < 3$.

The conjunction is $0 < x$ and $x < 3$.

a. $1 < y < 5$
b. $-3 < x < 0$
c. $-1 > x > -5$
d. $10 > y \geq 0$
e. $-5 \leq x \leq 5$

2. Graph each of the following sets.
    a. $\{(x, y) | 2 < x < 5\}$
    b. $\{(x, y) | x + y = 3\}$
    c. $\{(x, y) | 2 < x < 5 \text{ or } x + y = 3\}$
    d. $\{(x, y) | 2 < x < 5 \text{ and } x + y = 3\}$
    e. $\{(x, y) | x > -1\}$
    f. $\{(x, y) | y < 2\}$
    g. $\{(x, y) | x > -1 \text{ or } y < 2\}$
    h. $\{(x, y) | x > -1 \text{ and } y < 2\}$
    i. $\{(x, y) | -3 < x < 2\}$
    j. $\{(x, y) | -3 < y < -1\}$
    k. $\{(x, y) | -3 < x < 2 \text{ or } -3 < y < -1\}$
    l. $\{(x, y) | -3 < x < 2 \text{ and } -3 < y < -1\}$
    m. For Problems 2c, g, and k, describe the same sets as the union of two sets.
    n. For Problems 2d, h, and l, describe the same sets as the union of two sets.
    o. Explain why $\{(x, y) | x > 3 \text{ and } x < 1\}$ is the empty set.
    p. Explain why $\{(x, y) | y < -1 \text{ and } y > 0\}$ is the empty set.

10.6 Linear Programming

Recall that in solving systems of equations we looked for the intersection of the solution sets of the given equations. The situation in solving *systems of inequalities* is the same; that is, we look for the intersection of the solution sets of the given inequalities. We illustrate this in the following examples.

Example 1 Graph the solution set of the following system of inequalities:

$$x \geq 0$$
$$y \geq 0$$

10.6 Linear Programming

$$x \leq 4$$
$$x + y \leq 8.$$

It is not necessary to show the entire graph of each inequality. Examine the graph (shaded portion) in Figure 10.17 to see that the coordinates of each point satisfy *each* of the inequalities. The shaded portion and its boundaries make up a set of points which is called a *polygonal region*. Since the sentences which are graphed have the symbol $\geq$ or $\leq$ in them, the boundaries of the polygonal region in the graph belong to the graph of the solution set.

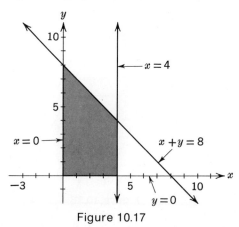

Figure 10.17

Now, in Example 2 we will add other conditions to the polygonal region graphed in Example 1.

Example 2 Graph several lines whose equations are of the form $2x + y = k$ on the graph of the system of inequalities of Example 1. Determine the maximum (largest) and the minimum value of k in $2x + y = k$ for which the graphs of $2x + y = k$ intersect the polygonal region of Example 1.

If we express the equation $2x + y = k$ in slope-intercept form, we have

$$y = -2x + k.$$

The following lines of the form $y = -2x + k$ are shown on the graph in Figure 10.18.

$$y = -2x + 14$$
$$y = -2x + 12$$
$$y = -2x + 8$$
$$y = -2x + 4$$

$$y = -2x + 0$$
$$y = -2x + (-4).$$

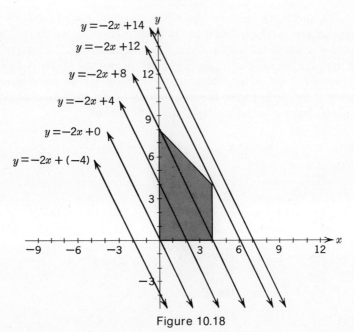

Figure 10.18

Note that -2 is the slope of each line, and the different values of k give the y-intercepts for the lines.

An examination of the graph reveals the following information:

(1) The lines $y = -2x + 14$ and $y = -2x + (-4)$ do not intersect the polygonal region.
(2) The line $2x + y = 12.01$ would not intersect the region.
(3) The line $2x + y = -.01$ would not intersect the region.
(4) More generally, no line whose equation is of the form $2x + y = k$, where $k > 12$ or $k < 0$ would intersect the region.

Now we shall answer the two questions we asked at the outset.

What is the maximum value of k in $2x + y = k$ for which the graph of $2x + y = k$ intersects the region?

It is 12, since the graph of $2x + y = 12$ intersects the region, and the graph of $2x + y = k$ for every $k > 12$ will not intersect the region.

What is the minimum value of k in $2x + y = k$ for which the graph of $2x + y = k$ intersects the region?

10.6 Linear Programming

It is 0, since the graph of $2x + y = 0$ intersects the region, and the graph of $2x + y = k$ for every $k < 0$ will not intersect the region.

Graphs of equations and inequalities, like those considered in Examples 1 and 2, are useful in solving a variety of problems encountered in business, industry, and government. This technique of problem solving is called *linear programming*. Many of the problems are concerned with decision-making situations, where management is faced with several choices, one of which would lead, say, to the maximum profit. The following example illustrates one such problem.

Example 3 The Econo-Company manufactures two sizes of TV screens, size A and size B. The screens are made by two machines. The old machine makes one size A screen in 4 minutes and one size B screen in 1 minute. The new machine makes one size A screen in 2 minutes and one size B screen in 6 minutes. Each machine can operate at most 2 hours a day. If the size A screen sells at a profit of $2 per screen and the size B at a profit of $3 per screen, how many of each size screen should be manufactured per day to bring in the maximum profit?

Let us summarize in a table the given information.

	Old Machine Time	New Machine Time	Profit
Size A	4 min	2 min	$2
Size B	1 min	6 min	$3
Maximum Time Per Day	2 hr	2 hr	

We shall now write the system of equations and inequalities which represent the essential information. We let the following variables represent the pertinent items involved in the problem:

x = the number of size A screens manufactured per day,
y = the number of size B screens manufactured per day,
p = the profit in dollars per day's production.

$$p = 2x + 3y$$
$$4x + 1y \leq 120$$
$$2x + 6y \leq 120.$$

To produce size A and size B screens, the old machine can use at most 2 hours (120 min) per day.

We also should keep in mind that the replacement set for x and y is the set of whole numbers.

We seek the values of x and y which satisfy all of the conditions above and yield the maximum profit, p. Let us first graph the system of inequalities.

$$\begin{aligned} x &\geq 0 \\ x &\geq 0 \end{aligned}\Big\}$$
$$4x + y \leq 120$$
$$2x + 6y \leq 120.$$

The solution set of the system is shown in the graph in Figure 10.19.

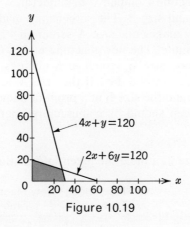

Figure 10.19

Now let us consider the equation

$$p = 2x + 3y.$$

The equivalent equation in slope-intercept form is

$$y = -\frac{2}{3}x + \frac{p}{3}.$$

Replacing p by various numerals will give us graphs which are lines with slope $-\frac{2}{3}$. If p is replaced by 3, the y-intercept will be 1, since the equation of this line will be

$$y = -\frac{2}{3}x + 1.$$

The problem calls for selecting that line which intersects the shaded region and has the greatest possible y-intercept. Thus, we are seeking all ordered pairs (x, y) in the intersection of the shaded region and the line $y = -\frac{2}{3}x + \frac{p}{3}$ which will produce the maximum value of p.

10.6 Linear Programming

The graph in Figure 10.20 shows several lines whose equations are of the form $y = -\frac{2}{3}x + \frac{p}{3}$.

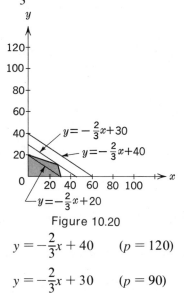

Figure 10.20

$$y = -\frac{2}{3}x + 40 \qquad (p = 120)$$

$$y = -\frac{2}{3}x + 30 \qquad (p = 90)$$

$$y = -\frac{2}{3}x + 20 \qquad (p = 60).$$

From Figure 10.20 it appears that the line which intersects the shaded region with the greatest y-intercept is $y = -\frac{2}{3}x + 30$. The value of p for this line is 90. The coordinates of the vertex where the region and the line intersect cannot be read with complete accuracy, but they appear to be approximately equal to

$$(x, y) \doteq (28, 11).$$

We can conclude then that the Econo-Company should manufacture about 28 of size A screens and 11 of size B screens per day, realizing the maximum profit of \$90. Substitute 28 for x and 11 for y in $p = 2x + 3y$ to see that p will be approximately 90.

EXERCISE 10.5

1. Graph the solution sets of the following systems of inequalities.

   a. $x \geq 0$
      $y \geq 0$
      $x \leq 2$
      $y \leq 5$

   b. $x \geq 0$
      $y \geq 0$
      $x \leq 3$
      $5x + y \leq 30$

c. $x \geq 0$
   $y \geq 0$
   $x \leq 2$
   $y + 3x \leq 12$

d. $x \geq -4$
   $y \geq -4$
   $x \leq 0$
   $y \leq 0$
   $x + y \geq -4$

2. On the graph for Problem 1a, graph several lines whose equations are of the form $x + y = p$. Determine the maximum and the minimum value of p in $x + y = p$ for which the graphs of $x + y = p$ intersect the polygonal region.

3. On the graph for Problem 1b, graph several lines whose equations are of the form $2x + y = p$. Determine the maximum and the minimum value of p in $2x + y = p$ for which the graphs of $2x + y = p$ intersect the polygonal region.

4. On the graph for Problem 1c, graph several lines whose equations are of the form $x + y = p$. Determine the maximum and the minimum values of p in $x + y = p$ for which the graphs of $x + y = p$ intersect the polygonal region.

5. On the graph for Problem 1d, graph several lines whose equations are of the form $2x + y = p$. Determine the maximum and the minimum value of p in $2x + y = p$ for which the graphs of $2x + y = p$ intersect the polygonal region.

6. Using x for the number of size 1 buttons and y for the size 2 buttons, write a mathematical sentence which fits the following conditions: The total cost of x size 1 buttons and y size 2 buttons is at most $8.00, if the price of size 1 buttons is 10¢ and of size 2 buttons 8¢.

7. For Problem 6, let c be the total cost of x size 1 buttons and y size 2 buttons. Write an equation for the total cost of these buttons in terms of x and y.

8. The Bolt Company manufactures two kinds of handmade, special purpose bolts, zero-bolts and one-bolts. Bolt-maker A makes zero-bolts at the rate of 1 per 2 min and one-bolts at 1 per 4 min. Bolt-maker B makes zero-bolts at the rate of 1 per 3 min and one-bolts at 1 per 1 min. The profit on each zero-bolt is $3 and on each one-bolt is $4. The maximum time devoted to the production of bolts by each maker is 3 hr per day.

	Maker A Time	Maker B Time	Profit
Zero-Bolt	2 min	3 min	$3
One-Bolt	4 min	1 min	$4
Maximum Time	3 hr	3 hr	

10.7 Solving Number-theory Problems

a. Using the following variables,

$x =$ for the number of zero-bolts manufactured per day,
$y =$ for the number of one-bolts manufactured per day,
$p =$ for the profit in cents per day's production,

write an equation showing the relation between p, and x and y.

b. Write a sentence showing that it takes bolt-maker A at most 3 hr (180 min) to produce x zero-bolts and y one-bolts.

c. Write a sentence showing that it takes bolt-maker B at most 3 hr to produce x zero-bolts and y one-bolts.

d. Graph the sentences in parts b and c. What is the replacement set for x and y?

e. Graph several lines which have the slope indicated by the equation in part a.

f. Find the point of intersection which gives the maximum value of p.

g. State the approximate maximum profit.

9. A manufacturer stores bicycles in his warehouses in Detroit and in Chicago. The Detroit warehouse has 600 bicycles and the Chicago warehouse 800 bicycles. The manufacturer received an order from Atlanta for 500 bicycles and from St. Louis for 400 bicycles. The profit on each bicycle shipped from Chicago to Atlanta is $30 and from Detroit to Atlanta is $20. The profit on each bicycle shipped from Chicago to St. Louis is $25 and from Detroit to St. Louis is $15. What is the maximum profit the manufacturer can make?

10.7 Solving Number-theory Problems

Number theory deals with problems which involve relationships between numbers. Here are some examples of such problems.

Example 1 Find three consecutive positive integers such that seven times the third integer diminished by two times the second integer equals 47 more than twice the third integer.

This problem deals with *consecutive* integers. Two integers are said to be *consecutive* if their difference is 1.

One Way to Solve. We use variables as follows:

$x =$ the first integer,
$x + 1 =$ the second integer,
$x + 2 =$ the third integer.

Then,

$7(x + 2)$	Seven times the third integer
$2(x + 1)$	Two times the second integer
$7(x + 2) - 2(x + 1)$	Seven times the third integer diminished by two times the second integer
$2(x + 2) + 47$	Forty-seven more than twice the third integer.

Therefore an equation for the problem is

$$7(x + 2) - 2(x + 1) = 2(x + 2) + 47.$$

Solve.
$$7x + 14 - 2x - 2 = 2x + 4 + 47$$
$$5x + 12 = 2x + 51$$
$$3x = 39$$
$$x = 13$$
$$x + 1 = 14$$
$$x + 2 = 15.$$

Thus, the three consecutive integers are 13, 14, and 15.

Example 2 The sum of three consecutive even integers is 384. What are the integers?

If the least of the three even integers is denoted by x, then the next two even integers are $x + 2$ and $x + 4$. We have the following equation:

$$x + (x + 2) + (x + 4) = 384.$$

Solve.
$$3x + 6 = 384$$
$$3x = 378$$
$$x = 126.$$

Thus, the three integers should be 126, 128, and 130.

Example 3 In a two-digit numeral, the sum of the numbers represented by the tens digit and ones digit is 7. If the number represented by the ones digit is 1 greater than the number represented by the tens digit, what is the number?

Let x be the ones digit.

Then $7 - x$ is the tens digit.

Thus, an equation for the problem is

$$x = (7 - x) + 1.$$

10.7 Solving Number-theory Problems

Solve. $x = 8 - x$
$2x = 8$
$x = 4.$

We have found that 4 is the ones digit. Therefore, $7 - 4$, or 3, is the tens digit.

Thus, the number is 34.

EXERCISE 10.6

Solve each problem.

1. The sum of three consecutive integers is 45. What are the integers?
2. The sum of three consecutive integers is 51. What are the integers?
3. The sum of four consecutive even integers is 276. What are the integers?
4. The sum of three consecutive odd integers is 273. What are the integers?
5. Why is it impossible to have three even integers whose sum is 67?
6. Why is it impossible to have three odd integers whose sum is 98?
7. In a two-digit numeral the sum of the numbers represented by the tens digit and ones digit is 11. If the number represented by the tens digit is 3 greater than the number represented by the ones digit, what is the number?
8. The sum of two numbers is 177. Their differences is 41. What are the numbers?
9. The difference of two numbers is 65. Doubling one number and adding it to the other number results in the sum of 158. What are the two numbers?
10. The difference of two numbers is 145. Three times one number added to the other number results in the sum of 213. What are the two numbers?
11. The sum of four consecutive odd integers is 104. What are the integers?
12. Find three consecutive integers such that ten times the middle integer plus three times the largest integer is 72 more than six times the smallest integer.
13. Find three consecutive even integers such that the sum of five times the smallest integer and six times the second integer equals 128 less than three times the third integer.
14. Three consecutive integers are such that three times the smallest

is equal to 47 more than the sum of the other two integers. What are the three integers?

10.8 Problems Involving Distance and Rate

We now consider problems which involve distances and rates of travel. Study the three examples below for suggested ways of solving such problems.

Example 1 A car traveled for 3 hr at a certain speed. Then it traveled for 5 more hours at a speed which was 9 mph less than twice the speed at which it traveled the first 3 hr. The entire distance covered was 423 mi. Find the speed of the car during the 3-hr period and during the 5-hr period.

Let r be the speed in miles per hour of the car during the first 3 hr of travel. Then,

$3r$ Distance covered in the 3-hr period
$2r - 9$ Speed during the 5-hr period
$5(2r - 9)$ Distance covered during the 5-hr period.

Recall the formula

$$d = rt,$$

where d is the distance, r is the speed (rate), and t is the time.

Now we write this equation:

$$3r + 5(2r - 9) = 423.$$

Solve. $3r + 10r - 45 = 423$
$13r - 45 = 423$
$13r = 468$
$r = 36.$

The car traveled at 36 mph for 3 hr; then it traveled at 63 mph for 5 hr.

Check.   The distance covered in 3 hr at 36 mph is 108 mi.
    The distance covered in 5 hr at 63 mph is 315 mi.
    Total: $108 + 315 = 423$.

Example 2 Mary and Elaine start out bicycling from the same spot in the same direction. Mary averages 12 mph and Elaine 16 mph. After how many hours will they be 10 mi apart?

10.8 Problems Involving Distance and Rate

One Way to Solve. Let t be the time each girl travels until they are 10 mi apart.
Distance covered by Elaine $= 16t$.
Distance covered by Mary $= 12t$.

$$16t - 12t = 10$$
$$4t = 10$$
$$t = 2\tfrac{1}{2}.$$

Thus, they are 10 mi apart after $2\tfrac{1}{2}$ hr.

Check. In $2\tfrac{1}{2}$ hr Elaine travels $16 \times 2\tfrac{1}{2}$, or 40 mi.
Mary travels in $2\tfrac{1}{2}$ hr $12 \times 2\tfrac{1}{2}$, or 30 mi.
It follows that they are 10 mi apart after $2\tfrac{1}{2}$ hr.

Example 3 Jim and Joe ride their motorbikes in opposite directions from Joe's house on the highway. They start at the same time. We find them 19 mi apart 15 min later. The average speed of Joe's bike is 8 mph less than the average speed of Jim's bike. Determine the average speed of Joe's bike.

We know that the two bikes end up 19 mi apart after 15 min. We observe that both bikes have been traveling for the same period of time, but they have covered different distances because of traveling at unequal rates. However, the sum of their distances will be equal to the distance they are apart. If we let z stand for the speed of Jim's bike in mph, then $z - 8$ is the speed of Joe's bike.

To find the distance traveled by Joe, we find the product of Joe's speed, $(z - 8)$ mph, and the number of hours he traveled, $\tfrac{15}{60}$, or $\tfrac{1}{4}$ hr. Thus, $\tfrac{1}{4}(z - 8)$ represents the miles traveled by Joe.

Similarly, we determine that $\tfrac{1}{4}z$ represents the distance traveled by Jim in the same length of time. We may now write the following equation:

$$\tfrac{1}{4}(z - 8) + \tfrac{1}{4}z = 19.$$

Solve.
$$4[\tfrac{1}{4}(z - 8) + \tfrac{1}{4}z] = 4 \times 19$$
$$(z - 8) + z = 76$$
$$2z - 8 = 76$$
$$2z = 84$$
$$z = 42.$$

Thus, Jim's average speed was 42 mph, and Joe's average speed was $42 - 8$, or 34 mph.

Check. Joe = ($\frac{1}{4}$ hr) × (34 mph) = $\frac{34}{4}$ mi
Jim = ($\frac{1}{4}$ hr) × (42 mph) = $\frac{42}{4}$ mi
Total = $\frac{34}{4}$ + $\frac{42}{4}$ = $\frac{76}{4}$, or 19 mi

EXERCISE 10.7

Solve each problem.

1. Mr. Fleet can run three times as fast as Mr. Sluggish. They started from the same spot at the same time and ran in opposite directions. The distance between them after 5 min was $1\frac{1}{4}$ mi. What distance did Mr. Fleet run?

2. A freight train left New York City at 7 A.M. traveling at 40 mph. At 9 A.M. a passenger train left the same station traveling at 50 mph. After how many hours will the passenger train overtake the freight train?

3. A new car leaves Detroit traveling at an average speed of 45 mph. Another car leaves Detroit on the same route $1\frac{1}{2}$ hr later. If the second car catches up with the first car in $4\frac{1}{2}$ hr, find the speed of the second car.

4. On a trip to the city, Mr. Go traveled at an average speed of 45 mph. His average speed on the return trip was 55 mph. If the time of going and returning over the same route amounted to 6 hr, how far was the city from Mr. Go's house?

5. A plane makes a trip of 900 mi. If the average speed had been increased by 75 mph, the plane could have covered 1350 mi in the same time. Determine the average speed of the plane.

6. Two small planes leave cities 970 mi apart and travel toward each other. One plane's average speed is 40 mph greater than that of the other plane. If they meet in $2\frac{1}{2}$ hr, determine the speed of each plane.

7. John vacationed at a river resort featuring boat trips. His motorboat ordinarily travels at an average speed of 12 mph. He found that he could travel 18 mi upstream in the same time that he could travel 24 mi downstream. Compute the speed of the current. [*Hint:* In traveling upstream, the actual speed of the boat is the difference of its speed in still water and the speed of the current; in traveling downstream, the actual speed of the boat is the sum of the two speeds.]

GLOSSARY

Dependent system of equations: A system whose graph consists of one line. The solution set has an infinite number of ordered number pairs.

Inconsistent system of equations: A system whose graph consists of two parallel lines. The solution set is the empty set.

Independent system of equations: A system whose graph consists of two intersecting lines. The solution set consists of one ordered number pair.

CHAPTER REVIEW PROBLEMS

1. For each value of y given below, determine the value of x so that (x, y) satisfies $3x - 7y = -1$.

    a. 1   **b.** 0   **c.** 2

    d. -1 **e.** $-\frac{1}{7}$ **f.** $\frac{1}{7}$

    g. -5 **h.** -2

2. For each equation, find an equivalent equation in standard form, that is, in the form $Ax + By = C$. Then solve for y to obtain an equation of the form $y = mx + b$.

    a. $4x + 3y = 3x + 6$
    b. $7y - 13 = 5x - 3y$
    c. $17 + 14y = 6x - 11 - 5y$
    d. $3(x - y) + 17 = 7 + 2x$
    e. $8(x + y) - 5y = 6x + 8$
    f. $14 - 6(2x + 3) = 7x - 3(y + 1)$
    g. $13(x - 3y) + 5(2y - x) + 4 = 7(7 + y) - 3x$
    h. $5(y + 2x) - 3x + 14 = 5(x - 3y) + 17$
    i. $\dfrac{3(x + 4)}{y - 5} = 5$

3. Graph the solution set of each of the following systems of equations. In which quadrant does each solution set lie?

    a. $-2x + y = -6$ **b.** $x - 3y = 4$
         $x - 3 = -5y$ $x - 2y = -4$

4. For each pair of equations,

        i. graph the solution set,
        ii. tell whether the system of equations is independent, dependent, or inconsistent.

    a. $-3y + 5 = -7x$ **b.** $-6x + 2 = 4y$
         $2x + 5 = -3y$ $-2y = 2(3 + 1.5x)$

c. $\dfrac{x+2y}{3} = 2x+1$
   $2y = 5x+3$

d. $x+3y=7$
   $3x+4=-9y$

e. $5x+9=-3y$
   $5y+4=-2x$

f. $5x+3=-2y$
   $4y+6=-10x$

5. Graph each of the following sets.
   a. $A = \{(x,y) | x \geq 5 \text{ and } y \leq 2\}$
   b. $B = \{(x,y) | x \geq y \text{ or } x \geq 1\}$
   c. $C = \{(x,y) | x+y \geq -1 \text{ and } y \geq 3\}$
   d. $D = \{(x,y) | 2x+y \geq 5 \text{ or } x \geq 2\}$
   e. $E = \{(x,y) | -1 \leq x \leq 3 \text{ and } 2 \leq y \leq 4\}$
   f. $F = \{(x,y) | 3 \leq x \leq 4 \text{ and } -4 \leq y \leq -1\}$

6. Graph the solution set of each of the following systems of inequalities.
   a. $x \geq 0$
      $y \geq 0$
      $x \leq 3$
      $y \leq 4$

   b. $x \geq 0$
      $y \geq 0$
      $x \leq 4$
      $y \leq 6$

   c. $x \geq 0$
      $y \geq 0$
      $x \leq 3$
      $x+y \leq 6$

   d. $x \geq 0$
      $y \geq 0$
      $x \leq 3$
      $y \leq 2x+1$

7. Solve each problem.
   a. If x is an even integer, then how do we represent the next two even integers?
   b. Find three consecutive integers such that the sum of the smallest integer, twice the second integer, and three times the third integer is 70 less than nine times the smallest integer.
   c. Find three consecutive odd integers such that the sum of seven times the largest, three times the next integer, and two times the smallest equals 91 more than nine times the smallest integer.
   d. The sum of three consecutive integers is 237. What are the integers?
   e. The sum of an integer and twice that integer is 111. What is the integer?
   f. The sum of an integer, three times the integer, and five more than the integer is 160. What is the integer?
   g. Multiplying a number by 5 and adding 20 to the product gives 5 as the result. What is the number?

h. Mr. Good and Mr. Bye start out from the same point and drive their cars in opposite directions. Mr. Good averages 45 mph. After 7 hr they are 749 mi apart. What was the average speed at which Mr. Bye was traveling?

i. Mr. Leisure starts out from Oil City at 8 A.M. driving west at 40 mph. Mr. Rush starts out from the same city 3 hr later and catches up with Mr. Leisure at 5 P.M. What was Mr. Rush's average speed?

j. Car A travels 10 mph faster than car B. A travels 320 mi in the same time that B travels 240 mi. What is the average speed of each car?

11
Algebraic Methods of Solving Systems

11.1 Weakness of Graphic Method

We have learned that an independent system of two linear equations in two variables has one ordered pair of numbers for its solution. Let us graph the two lines given by the following equations:

$$4x - 3y = 2$$
$$-3x + 7y = -10.$$

In making the graph, shown in Figure 11.1, we shall use the following pairs for the respective equations:

$$(2, 2) \text{ and } (-1, -2) \text{ for } 4x - 3y = 2$$
$$(-6, -4) \text{ and } (1, -1) \text{ for } -3x + 7y = -10.$$

Examination of the graph reveals that the point of intersection of the two lines has coordinates somewhere in the neighborhood of $(-1, -2)$. It is difficult, however, to tell from the graph the *exact* coordinates of the point of intersection. In other words, it is difficult to tell the solution set of the system of equations from the graph, but it is important to realize that graphing will always yield an approximation to the solution set and often provides an easy way to do a rough check.

11.2 Comparison Method

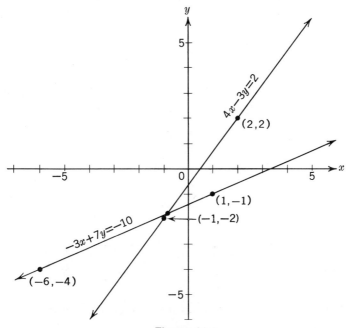

Figure 11.1

We must develop a way to determine the solution sets of systems of equations without having to make graphs. We will learn some methods which will enable us to find the exact solutions of any independent system of linear equations in two variables.

11.2 Comparison Method

We shall now illustrate one method of determining the exact solutions of a system of equations without making graphs. Consider the following system:

$$2x - 5y = 9$$
$$2x + 3y = 1.$$

We solve each equation for $2x$.

$$2x = 5y + 9$$
$$2x = -3y + 1.$$

Now, if each of the equations is to be satisfied by the same pair of numbers, then the following must be true:

$$5y + 9 = -3y + 1.$$

We solve this equation for y.

$$8y + 9 = 1$$
$$8y = -8$$
$$y = -1.$$

Now, we replace y by -1 in either the first or the second equation. We shall use the second equation.

$$2x + 3(-1) = 1$$
$$2x - 3 = 1$$
$$2x = 4$$
$$x = 2.$$

We check the pair $(2, -1)$ in each equation.

Check.	$2x - 5y$	9	$2x + 3y$	1
	$2 \times 2 - 5(-1)$	9	$2 \times 2 + 3(-1)$	1
	$4 + 5$		$4 - 3$	
	9		1	

Thus, $\{(2, -1)\}$ is the solution set of the given system; that is,

$$\{(x, y) | 2x - 5y = 9\} \cap \{(x, y) | 2x + 3y = 1\} = \{(2, -1)\}.$$

The method which we have just used in determining the solution set of the system is called the *comparison method*. This name is suggestive of the nature of the method.

Observe that, due to the nature of the comparison method, it is a good one to use in systems of equations where both equations have the same *coefficient* of one of the two variables. Observe that in $2x + 3y = 1$, 2 is the coefficient of x and 3 is the coefficient of y. In $2x - 5y = 9$, 2 is the coefficient of x and -5 is the coefficient of y, because $2x - 5y = 9$ is equivalent to $2x + (-5)y = 9$. Since the coefficient of x in both equations is the same, namely 2, the comparison method led us quickly to the solution set.

It is possible to use the comparison method even if the equations do not have the same coefficient of one variable. We can multiply by convenient numbers to obtain coefficients which will meet the required condition. This is illustrated in the following example:

$$2x + 3y = 5$$
$$7x + 6y = 4.$$

We obtain an equation equivalent to the first equation by multiplying by 2.

$$2(2x + 3y) = 2 \times 5$$
$$4x + 6y = 10.$$

11.2 Comparison Method

Now we have the new system,

$$4x + 6y = 10$$
$$7x + 6y = 4,$$

which is equivalent to the given system. We solve each equation for $6y$.

$$6y = 10 - 4x$$
$$6y = 4 - 7x.$$

Thus,

$$10 - 4x = 4 - 7x$$
$$3x = -6$$
$$x = -2.$$

Substituting -2 for x in the first equation,

$$2(-2) + 3y = 5$$
$$-4 + 3y = 5$$
$$3y = 9$$
$$y = 3.$$

Thus, $\{(-2, 3)\}$ is the solution set of the original system; that is,

$$\{(x, y)|2x + 3y = 5\} \cap \{(x, y)|7x + 6y = 4\} = \{(-2, 3)\}.$$

EXERCISE 11.1

1. Using the comparison method, solve each of the following systems of equations and check.

   a. $x = -y + 4$
      $x = y - 2$

   b. $y = x - 1$
      $y = 2x$

   c. $2m = n + 3$
      $3m = n + 3$

   d. $s + t = -6$
      $s - t = 2$

   e. $x + y = 3$
      $x - y = -2$

   f. $2h + 4k = 8$
      $2h - 3k = 1$

   g. $k + \frac{3}{2}m = 0$
      $k + m = 2$

   h. $4m - 6t = 6$
      $4m - t = 1$

   i. $2k + s = -2$
      $s + k = -1$

   j. $2a + b = 1$
      $b = 3 - a$

2. Using x for the first number and y for the second number, write an expression for each of the following.

   a. the sum of twice the first number and three times the second number

b. the difference between one-half of the second number and one-third of the first number

   c. the product of the sum of the first number and the second number and the difference of the first number and the second number

   d. one-half of the first number subtracted from four times the second number

   e. one-third of the sum of the first and second numbers

   f. twice the difference of the second and the first numbers

   g. the product of the quotient of the first number and the second number and the sum of the two numbers

   h. the square of the sum of the first and second numbers

3. Solve each system of equations by first multiplying by an appropriate number to obtain a system in which the coefficient of one of the variables is the same. Then use the comparison method.

   a. $3m = \frac{1}{2}n + 3$
   $n = 2 - m$

   b. $2p - 2r = 1$
   $p = 7r - 3$

   c. $\frac{u + w}{3} = 2u$
   $2w = 7u + \frac{5}{2}$

   d. $-4s = -3t + 1$
   $3 = s - t$

   e. $7.5c - 3d = 5$
   $4d = 3c - 6$

   f. $v + t = 3v - t + 6$
   $7 - t = 8 + v$

4. Solve each problem.

   a. A number is doubled and 1 is added to the result. The same result is obtained when multiplying this number by 4 and adding 7 to the result. What is the number?

   b. Diminishing the product of a number and 2 by 4 gives the same result as dividing the product of the number and 6 by 7. What is this number?

   c. There are three consecutive integers such that the product of the second and the largest integer exceeds the product of the smallest and the second integer by 26. What are the integers?

   d. There are four consecutive integers such that the product of the two largest integers exceeds the product of the two smallest integers by 70. What are the integers?

11.3 Substitution Method

We shall now illustrate another method of solving systems of two linear equations in two variables, in which graphing is not required. Consider the system

11.3 Substitution Method

$$2x - y = 1$$
$$x + 4y = -3.$$

We solve the second equation for x.

$$x + 4y = -3$$
$$x = -4y - 3.$$

Now we *substitute* $-4y - 3$ for x in the first equation.

$$2x - y = 1$$
$$2(-4y - 3) - y = 1.$$

The resulting equation is in one variable, y. We solve it for y.

$$2(-4y - 3) - y = 1$$
$$-8y - 6 - y = 1$$
$$-9y = 7$$
$$y = -\tfrac{7}{9}.$$

From the second equation above, we found that $x = -4y - 3$. Substituting $-\tfrac{7}{9}$ for y in this equation, we obtain

$$x = -4(-\tfrac{7}{9}) - 3$$
$$x = \tfrac{28}{9} - 3$$
$$x = \tfrac{28 - 27}{9}$$
$$x = \tfrac{1}{9}.$$

Thus, $\{(\tfrac{1}{9}, -\tfrac{7}{9})\}$ is the solution set of the system.

Note that we first solved the *second* equation for x. This was the easiest thing to do because the coefficient here is 1.

The method we used in solving this system is different from the comparison method. It is called the *substitution method*. This name is also suggestive of the nature of the method.

EXERCISE 11.2

1. Solve each system of equations using the substitution method.

   a. $x - 4y = 17$
      $2x + y = -2$

   b. $3x - 5y = -9$
      $4x + y = -12$

   c. $y + 6x = 0$
      $x - 5y = -11$

   d. $5x + 4y = 0$
      $x - y = 9$

   e. $3x - 4y = 4$
      $y + x = -1$

   f. $2x + 3y = 5$
      $3x - 4y = -1$

g. $10x - 3y = 2$
$5x - 2y = 3$

h. $5y = 2x - 12$
$3x = 4y + 18$

i. $3x - 2y = 12$
$4x - 5y = 16$

j. $3y - 2 = x$
$4x - 3 = y$

2. Solve each problem.

Example In a two-digit numeral, the number represented by the ones digit is three greater than the number represented by the tens digit. The number named by reversing the digits is 20 less than twice the original number. What is the number?

Let t be the tens digit and u be the ones digit of the original numeral. Thus, we have the equation

$$u = t + 3.$$

$10t + u$ is the value of the original numeral. $10u + t$ is the value of the numeral after reversing the digits. Thus, we have a second equation,

$$10u + t = 2(10t + u) - 20$$
$$10u + t = 20t + 2u - 20$$
$$8u - 19t = -20.$$

Now we solve the system.

$$u = t + 3$$
$$8u - 19t = -20.$$

We shall substitute $t + 3$ for u in the second equation.

$$8(t + 3) - 19t = -20$$
$$8t + 24 - 19t = -20$$
$$-11t = -44$$
$$t = 4.$$

Now we substitute 4 for t in the original equation.

$$u = 4 + 3$$
$$u = 7.$$

Thus, the original number is 47.

Check. By reversing the digits we obtain 74. It is true that the new number (74) is 20 less than twice the original number (47).

a. In a two-digit numeral the number represented by the ones digit is 4 greater than the number represented by the tens digit. If the

digits are interchanged, the sum of the resulting number and the original number is 154. Determine the original number.
   b. In a two-digit numeral the number represented by the ones digit is 2 greater than the number represented by the tens digit. A new number is formed by adding 3 to the number represented by the tens digit and subtracting 3 from the number represented by the ones digit. Sixty-five less than three times the original number is equal to the new number. What was the original number?
   c. The ones digit in a two-digit numeral represents a number 3 greater than the number represented by the tens digit. The number is 39 more than twice the sum of the numbers represented by the two digits. What is the number?
   d. The number named by the ones digit of a certain number is 1 more than twice the number named by the tens digit. When the digits are interchanged, the resulting number is 2 more than twice the original number. What is the number?
   e. In a three-digit numeral, the ones and the hundreds digit are the same. Interchanging the ones and the tens digit results in a numeral whose value is 45 less than that of the original number. What was the original number if the sum of the values of the ones digit and the tens digit is 9?

11.4 Addition Method

Examine the following arithmetic examples.

$$\begin{array}{r} 4 = 3 + 1 \\ 7 = 5 + 2 \\ \hline 4 + 7 = (3 + 1) + (5 + 2) \\ 11 = 11. \end{array}$$

$$\begin{array}{r} 17 = 25 + (-8) \\ 15\tfrac{1}{2} = 16 + (-\tfrac{1}{2}) \\ \hline 17 + 15\tfrac{1}{2} = [25 + (-8)] + [16 + (-\tfrac{1}{2})] \\ 32\tfrac{1}{2} = 32\tfrac{1}{2}. \end{array}$$

The property illustrated by these two examples is a result of the basic properties of a field. It may be stated using variables as follows: For all real numbers w, x, y, and z,

if $w = x$ and $y = z$, then $w + y = x + z$.

This property is somewhat easier to use when it is written in vertical form.

$$w = x$$
$$\frac{y = z}{w + y = x + z}$$

This addition property can be used in solving systems of equations. Consider, for example, the system

$$2x - y = 3$$
$$x + 5y = 7.$$

We know that every equation has many equivalent equations. We shall take the second equation,

$$x + 5y = 7,$$

and find an equation which is equivalent to it, but we will choose an equivalent equation which turns out to be particularly well suited for our purpose. In this case it will be the equation in which the coefficient of x is the additive inverse of the coefficient of x in the first equation, $2x - y = 3$. This equation is

$$-2(x + 5y) = -2 \times 7,$$

which is equivalent to

$$-2x - 10y = -14.$$

The last equation is equivalent to

$$x + 5y = 7.$$

We know that two equivalent equations have exactly the same solution set; that is, the solution set of $x + 5y = 7$ is also the solution set of $-2x - 10y = -14$.

Let us take the first equation in our original system,

$$2x - y = 3,$$

and the newly obtained equation,

$$-2x - 10y = -14,$$

which is equivalent to the second equation in the original system. We know that whatever pair of numbers satisfies the system,

$$2x - y = 3$$
$$-2x - 10y = -14,$$

also satisfies the original system.

Now we use the addition property stated previously.

11.4 Addition Method

$$2x - y = 3$$
$$-2x - 10y = -14$$
$$-11y = -11$$
$$y = 1.$$

Multiplication of each member of the second equation by -2 resulted in an equivalent equation so that the sum of the two equations resulted in an equation in one variable.

Now we replace y by 1 in one of the original equations.

$$2x - 1 = 3$$
$$2x = 4$$
$$x = 2.$$

Thus, the pair $(2, 1)$ should satisfy the original system of two equations.

Check.

$2x - y$	3
$2 \times 2 - 1$	3
$4 - 1$	
3	

$x + 5y$	7
$2 + 5 \times 1$	7
$2 + 5$	
7	

Thus, the solution set of this system of equations is $\{(2, 1)\}$.

This method for solving a system is called the *addition method*.

EXERCISE 11.3

1. Solve each system using the addition method.

   a. $x + y = 1$
   $x - y = -3$

   b. $2x + 3y = 7$
   $5x - 3y = 7$

   c. $x + 2y = 10$
   $3x + 2y = -8$

   d. $5x - y = 13$
   $y + x = -1$

   e. $y = x$
   $3x - y = -4$

   f. $x = y - 8$
   $-3x = -y + 16$

   g. $2(x + y) = 3x$
   $x + 6y = 0$

   h. $3(x + 5y) = 0$
   $15y - 4x = 35$

   i. $2(x + y) = 3y - 2x$
   $4x = y + 2$

   j. $3(x + y) + 6 = -2(x - 2y) - 3$
   $5x - 9y = -1$

2. Solve each system using one of the three algebraic methods which you consider the most appropriate for that system.

   a. $x + y = 0$
   $x - y = 2$

   b. $2x - y = 0$
   $2x + y = -4$

c. $3y = 2 - x$
$-2x = 4y + 2$

d. $7y - \frac{1}{2}x = 2$
$-2y = x + 4$

e. $2(x + 4y) = x + 3$
$3(5y + x) = 0$

f. $\frac{x+y}{3} = 2$
$\frac{x-y}{2} = -1$

g. $\frac{2x+y}{5} = x + y$
$\frac{4y+3x}{7} = y + 3$

h. $\frac{2(x+6y)}{3} = 2$
$x + 10y = -1$

i. $2x = 4y + 1$
$x = y$

j. $10x - 6y = 2x + 1$
$2x = 12y + 9$

3. Solve each of the following problems.

   a. The sum of two numbers is 193 and their difference is 31. What are the two numbers?

   b. The sum of twice one number and three times another number is 249. The difference of the two numbers is 22. What are the two numbers?

   c. A number and twice another number are added. Twice this sum is 346. The sum of the two original numbers is 118. What are the two numbers?

   d. One-half of the sum of two numbers is equal to 91. Twice the larger number subtracted from three times the smaller number is 51. What are the two numbers?

11.5 Problems Leading to Systems of Equations

We have already had some practice in writing systems of equations which fit a given problem, and by solving the system, obtaining the solution to the given problem. Below is one more example worked out, followed by more problems of this kind.

Example The sum of two numbers is 121. The sum of twice one of the numbers and three times the other number is 17. What are the two numbers?

If one number is x and the other y, then we have the following two equations:

$$x + y = 121$$
$$2x + 3y = 17.$$

11.6 Solving Problems about Angle Measure

We can use any one of the methods we know to solve this system of equations. Let us use the substitution method. From the first equation,

$$x + y = 121$$

we obtain $x = 121 - y.$

Replacing x by $121 - y$ in the second equation, we have

$$2(121 - y) + 3y = 17$$
$$242 - 2y + 3y = 17$$
$$242 + y = 17$$
$$y = -225.$$

Now replacing y by -225 in the first equation (if you wish, you may replace y by -225 in the second equation instead),

$$x + (-225) = 121$$
$$x = 346.$$

Thus, the numbers should be 346 and -225.

Check. The sum is $346 + (-225)$, or 121.
Twice the first number is 2×346, or 692.
Three times the second number is $3 \times (-225)$, or -675.
The sum of twice the first and three times the second is $692 + (-675)$, or 17.

Thus, we know that the two numbers are 346 and -225.

11.6 Solving Problems about Angle Measure

In this section we will consider problems dealing with angle measures in geometric figures. First we will review some simple geometric concepts which we will need to recall in order to understand the problems we will be solving.

Two angles are said to be *complementary* if the sum of their measures is equal to 90°. For example, angles whose measures are 25° and 65° are complementary, since $25° + 65° = 90°$.

Two angles are *supplementary* if the sum of their measures is equal to 180°. For example, angles whose measures are 150° and 30° are supplementary, since $150° + 30° = 180°$.

A *right triangle* is a triangle which has one angle measuring 90°. This angle is called a *right* angle.

An *isosceles triangle* is a triangle with at least two congruent sides, that is, sides having the same length. The congruent sides are called *legs*. The third side is called the *base*. The angles opposite the congruent sides have the same measure and are called *base* angles. The third angle is called the *vertex* angle.

If an isosceles triangle has all three sides of the same length, then it is called an *equilateral triangle*. Each angle of an equilateral triangle has the same measure; therefore, such a triangle is also called an *equiangular triangle*.

Remember that the sum of the measures of the three angles of a triangle is 180°. We shall write $m \angle A$ to mean *the degree-measure of angle A*. Thus, in any triangle ABC,

$$m \angle A + m \angle B + m \angle C = 180°.$$

It follows that the sum of the measures of the two angles which are not right angles in a right triangle is 90°.

Study the following examples.

Example 1   In $\triangle ABC$, $m \angle A$ is 3 times $m \angle B$, and $m \angle C$ is 5 times $m \angle B$. What is the measure of each angle?

Let $m \angle B = x$. Then

$$m \angle A = 3x$$
$$m \angle C = 5x.$$

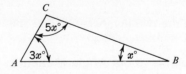

Thus, we have the equation

$$x + 3x + 5x = 180.$$

Solve.

$$9x = 180$$
$$x = 20.$$

Thus,

$$m \angle B = 20°$$
$$m \angle A = 60°$$
$$m \angle C = 100°.$$

Example 2   The measure of the vertex angle in an isosceles triangle is

11.6 Solving Problems about Angle Measure

three times the measure of each base angle. What is the measure of each angle?

Let the measure of each base angle be x. Then

$$x + x + 3x = 180$$
$$5x = 180$$
$$x = 36.$$

The measures of the three angles are 36°, 36°, and 108°.

EXERCISE 11.4

1. Write an equation or a system of equations to fit each problem and then solve. Check the solution in the original problem.
   a. The sum of two numbers is 12. Their difference is 14. What are the two numbers?
   b. The sum of twice a number and another number is 34. The difference of the first and the second numbers is 12. What are the two numbers?
   c. Twice the sum of two numbers is 134. The difference of the two numbers is 35. What are the two numbers?
   d. Three times the sum of two numbers is 285. The difference of the two numbers is 89. What are the two numbers?
   e. The product of 4 and the sum of two numbers is equal to 140. The difference of the two numbers is 13. What are the two numbers?
   f. A shelf contains 220 books. If the number of science books is $1\frac{1}{2}$ times the number of mathematics books, and the number of fiction books is twice the number of science books, how many books of each of the three kinds are there on the shelf?
   g. The total number of businesses which failed in 1963 and 1964 was 29,460. If there were 460 more failures in 1964 than in 1963, how many businesses failed in each of the two years?
   h. The sum of the lengths of two line segments is 7 in. If the length of the shorter line segment is tripled, the resulting segment is 1 in. shorter than the longer segment. Find the measure of each segment.
   i. Twelve oranges and nine grapefruit cost $1.02. Twenty oranges and four grapefruit cost $1.04. What is the price of one orange and one grapefruit?
   j. Eight cans of sardines and five cans of pears cost $2.61. Three

cans of sardines and seven cans of pears cost $2.67. What is the cost of each can of sardines and each can of pears?

k. A college received a shipment of algebra and geometry books costing $558.75. There were 25 more algebra books than geometry books. How many algebra and how many geometry books did the college purchase, if an algebra book costs $2.55 and a geometry book $2.95?

l. The combined ages of Tim and Jack are 26 years. Three times Tim's age is 10 more than Jack's age. How old is each of them?

m. Jim is 5 years older than Tom. Six years ago Jim was twice as old as Tom. How old is Jim?

n. Sally's father is three times as old as Sally is. Five years ago her father was five times as old as she was then. How old is Sally and how old is her father?

o. Miss Dollars has $24.00 in dimes and half-dollars. If she has three times as many fifty-cent coins as ten-cent coins, how many of each does she have?

2. a. Complete the following table. The first line is already filled in.

Measure of angle	(1) Complement of angle	(2) Supplement of angle	Difference (2)—(1)
5°	85°	175°	175 − 85 = 90
38°			
	89°		
		126°	
	30°25′		
		179°59′	

b. Examine your answers in the column headed Difference above. What can you conclude on the basis of the answers in this column?

3. Solve each problem.
   a. In $\triangle ABC$, $m \angle A = 2(m \angle B)$ and $m \angle B = 3(m \angle C)$. What is the measure of each angle?
   b. In $\triangle ABC$, $m \angle A = 5(m \angle B)$ and $m \angle C = 9(m \angle B)$. What is the measure of each angle?
   c. In a right triangle, one acute angle is twice as large as the other acute angle. What is the measure of each acute angle?
   d. What is the measure of each of two supplementary angles if the measure of one angle is five times the measure of the other?

11.7 Solving Problems about Perimeters

e. What is the measure of each of two complementary angles if the measure of one angle is nine times the measure of the other?

f. What is the measure of each of two complementary angles if the measure of one angle is 20° more than the measure of the other?

g. In an isosceles triangle the measure of the vertex angle is 17°. What is the measure of each base angle?

h. In an isosceles triangle the measure of one base angle is 35°. What is the measure of the vertex angle?

i. In $\triangle ABC$, $m \angle A = 2(m \angle B)$ and $m \angle B = m \angle C + 20$. What is the measure of each angle?

j. The measure of one of two complementary angles is 15° more than twice the measure of the other angle. What is the measure of each angle?

k. The measure of one of two supplementary angles is 30° more than four times the measure of the other angle. What is the measure of each angle?

l. $\angle A$ and $\angle B$ are supplementary angles. Twice the measure of $\angle A$ is equal to two-thirds of the measure of $\angle B$. Compute the measures of $\angle A$ and $\angle B$.

m. The measure of one acute angle in a right triangle is 30° more than three times the measure of the other angle. What are the measures of the two acute angles?

n. In an isosceles triangle the measure of the vertex angle is three times the measure of each base angle. What is the measure of each angle?

o. One angle of a triangle measures twice as much as another angle. The sum of the measures of the two angles is 115°. What is the measure of each angle?

11.7 Solving Problems about Perimeters

The perimeter of any polygon may be determined by adding the measures of the sides. Thus, the perimeter of a rectangle is given by the formula

$$p = l + w + l + w$$

or

$$p = 2(l + w),$$

where l and w are the length and the width of the rectangle, respectively.
The perimeter of a square is

$$p = s + s + s + s$$

or

$$p = 4s,$$

where s is the measure of one of the sides.

The perimeter of any triangle ABC, in which the measures of the three sides are a, b, and c, is

$$p = a + b + c.$$

Study the following problem.

Example The length of a rectangle is five times its width. If the perimeter is 120 in., what are the width and the length of the rectangle?

Let x represent the number of inches in the width of the rectangle. Then $5x$ is the number of inches in the length. The perimeter is 120 in. We shall use the formula $p = 2(l + w)$.

$p = 2(l + w)$
$120 = 2(5x + x)$
$120 = 2(6x)$
$120 = 12x$
$10 = x$

Thus, the width is 10 in., and the length is 5(10), or 50 in.

EXERCISE 11.5

1. If the perimeter of $\triangle ABC$ is 180 in., what is the length of each side?

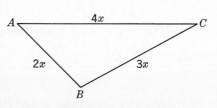

2. The length of a rectangle is three times its width. If the perimeter of the rectangle is 320 in., determine the length and the width.

3. The length of a rectangle is four times its width. The perimeter is 300 in. Determine the length and the width.

4. The perimeter of an isosceles triangle is 34 in. If the length of each leg is four times the length of the base minus 1 in., what is the length of each side?

5. The length of each leg of an isosceles triangle is $2\frac{1}{2}$ times the length of the base. What is the length of each side of the triangle if the perimeter is 38 in.?

6. The sum of the perimeters of two squares is 100 in. If the perimeter

11.8 Solving Mixture Problems

of one square is three times the perimeter of the other, find the measure of the side of each square.

7. The sum of the perimeters of two equilateral triangles is 30 in. Determine the length of the sides of each triangle if a side of the larger triangle is nine times as long as a side of the smaller triangle.

8. The length of a rectangle is 5 in. longer than its width. Its perimeter is 26 in. How long is the side of the square whose area is equal to the area of the rectangle?

9. Each leg of an isosceles triangle has the same measure as a side of a square. The base of the triangle is 5 in. long. What is the measure of each side of the square, if the triangle and the square have the same perimeter?

10. The perimeter of a rectangle is equal to 34 in. If the width is doubled and the length increased by 3 in., a new rectangle is created with a perimeter of 50 in. What are the width and the length of the first rectangle?

11.8 Solving Mixture Problems

We now turn our attention to problems which involve the concept of mixture. To learn techniques for solving such problems, study the example below.

Example A seed dealer mixed grass seed selling at $1.20 per lb with clover seed selling at $1.60 per lb. If he wanted a mixture of 90 lb to sell at $1.24 per lb, how many pounds of each did he use in the mixture?

(1) What is the cost of the mixture?

$$1.24 \times 90 = 111.60.$$

The cost of mixture is $111.60.

(2) Let x be the number of pounds of grass seed. Then $90 - x$ is the number of pounds of clover seed.

(3) What is the cost of the grass seed in dollars?

$$1.20x$$

(4) What is the cost of the clover seed in dollars?

$$1.60(90 - x)$$

(5) What is the equation for this problem?

$$1.20x + 1.60(90 - x) = 111.60.$$

Solve. $100[1.20x + 1.60(90 - x)] = 100 \times 111.60$
$120x + 160(90 - x) = 11{,}160$
$120x + 14{,}400 - 160x = 11{,}160$
$-40x = -3240$
$x = 81.$

Thus, he used 81 lb of grass seed and $90 - 81$, or 9 lb, of clover seed.

Check. Grass 81 lb at $1.20 per lb $97.20
Clover 9 lb at $1.60 per lb $14.40
Total cost $111.60

Now consider the following example which involves a solution of antifreeze and water. By a 20% antifreeze solution, we mean a solution which consists of 20% antifreeze and 80% water. That is, the solution contains one volume of pure antifreeze to every four volumes of water. Or, the solution contains one volume of pure antifreeze in every five volumes of solution.

Example If we have 10 qt of 20% antifreeze solution, how much water should we add to the solution to obtain a 15% antifreeze solution?

(1) How much antifreeze is there in the solution?

$$.20 \times 10 = 2.$$

There are 2 qt of antifreeze.

(2) Let x be the number of quarts of water we add to the solution. Then what is the number of quarts in the new solution?

$$10 + x.$$

(3) Of $10 + x$ qt, what percent is antifreeze?

$$15\%$$

(4) How many quarts of antifreeze are there?

$$2$$

(5) What is the equation?

$$.15(10 + x) = 2.$$

Solve. $15(10 + x) = 200$
$150 + 15x = 200$
$15x = 50$
$x = 3\frac{1}{3}.$

11.8 Solving Mixture Problems

Thus, if $3\frac{1}{3}$ qt of water are added to our original 10 qt of 20% antifreeze solution, a 15% solution is obtained.

Check. We check the first answer, 2 qt of antifreeze, by seeing if it is the correct volume in 10 qt of solution which is already a 20% solution.

$$\frac{2 \text{ qt of antifreeze}}{10 \text{ qt of solution}} = 20\% \text{ solution}.$$

We check the second answer, $3\frac{1}{3}$ qt of water, by seeing if it is the correct volume to *add* to the original 10 qt of solution to make a 15% solution. We see that we obtain $13\frac{1}{3}$ qt of weaker solution.

$$\frac{2 \text{ qt of antifreeze}}{13\frac{1}{3} \text{ qt of weaker solution}} = \frac{2}{\frac{40}{3}}$$
$$= \frac{6}{40}$$
$$= .15$$
$$= 15\%.$$

Thus, the new solution contains 15% of antifreeze.

EXERCISE 11.6

Solve each problem.

1. Low-quality grass seed selling at $1.80 per lb was mixed with high quality grass seed selling at $2.40 per lb. If a mixture of 150 lb was sold at $2.20 per lb, how many pounds of each quality grass seed were used in the mixture?

2. A candy store proprietor mixed 50 lb of hard candy selling at $.30 per lb with 100 lb of chocolate candy. The mixture sold at $.50 per lb. What was the price of the chocolate candy?

3. A coffee importer had coffee worth $.86 per lb and coffee worth $1.08 per lb. He made a blend of 220 lb to sell at $.98 per lb. How many pounds of each kind of coffee did he use to make the blend?

4. A candy store proprietor wishes to mix caramels selling at $.95 per lb with 32 lb of creams selling at $1.10 per lb. If he wishes to sell the resulting mixture for $1.00 per lb, how many pounds of caramels should he use in the mixture?

5. A mixture of 30 lb of candy sells for $1.10 per lb. The mixture consists of chocolates worth $.90 per lb and chocolates worth $1.50 per lb. How many pounds of each kind were used to make the mixture?

6. A farmer has some large potatoes which sell for $1.05 a bag and some

small ones which sell for $.75 a bag. If he mixes them together he will have 100 bags which he will sell for $85.50. How many bags of each size did he have?

7. A chemist has 18 qt of 90% solution of sulfuric acid. How much additional water must be used to have a 22% acid solution?

8. An alloy of lead and tin weighs 78 lb and contains 67% tin. How much lead must be melted in to make an alloy containing 60% lead?

9. A nurse is told to dilute with water 3 oz of a certain mixture which contains 80% medicine. She is to make a mixture which contains 20% medicine. How much water must the nurse add?

10. How much water must a chemist add to a pint of 90% solution of hydrochloric acid to dilute it to an 80% solution?

11. How many quarts of permanent antifreeze must be added to 3 gal of a 10% solution to make it a 25% solution?

12. Milk testing 2% butterfat is mixed with milk testing 4% butterfat to make 10 gal testing 3.25% butterfat. How much of each is used?

13. Suppose a science instructor has 5 qt of 25% sulfuric acid solution. He would like to obtain a sulfuric acid solution of 35% acid by adding a solution of 75% acid to his original solution. How much of the more concentrated acid must be added to achieve the desired concentration?

GLOSSARY

Addition method: If $a = b$ and $c = d$, then $a + c = b + d$.

Comparison method: If $x = m$ and $x = n$, then $m = n$.

Substitution method: If $x = m$, then m can be substituted for x in the second equation in the system.

CHAPTER REVIEW PROBLEMS

1. Check each of the following ordered number pairs (x, y) to see whether it satisfies the equation $2x - 3y = -7$.

    a. $(0, \frac{3}{7})$ **b.** $(-\frac{7}{2}, 0)$ **c.** $(-3, \frac{1}{3})$ **d.** $(-\frac{1}{2}, 2)$

2. **a.** On a coordinate system, graph the following system of equations:
$$5x - y = 4$$
$$2 + y = -3x.$$

    b. Read the solution of the system you graphed in 2a.

Chapter Review Problems

c. Solve the system given in 2a using one of the methods which does not involve a graph. Check your solution. Why is it best to use an algebraic method to solve this system?

3. a. Using the comparison method, solve the following system of equations:

$$x = \frac{y}{2} + 1$$
$$x = 2y - 3.$$

b. Explain why the comparison method is convenient to use for this system.

4. a. Using the substitution method, solve the following system of equations:

$$x = y - 2$$
$$5x - y = 1.$$

b. Explain why the substitution method is convenient to use for this system.

5. a. Using the addition method, solve the following system of equations:

$$5x - 3y = 7$$
$$x + 3y = 5.$$

b. Explain why the addition method is convenient to use for this system.

6. Solve each system of equations using any method you wish.

a. $3x + y = 5$
   $x - y = 1$

b. $x = -y$
   $10x - y = 33$

c. $\frac{x+y}{2} = \frac{1}{4}$
   $x - y = 0$

d. $3x = -2y$
   $\frac{6x + 10y}{3} = 2y$

e. $15x + 4y = 11$
   $x + y = 0$

f. $x + y = 2x - y$
   $2x + 3y = 7$

g. $x + y = 3$
   $x - y = -4$

h. $5x = y - 5$
   $x + y = -1$

i. $3x + y = 25$
   $\frac{x + 5y}{5} = -1$

j. $3x + 5y = -15$
   $5y - x = 31$

k. $x + y = 100$
   $50x - y = 100x - 149$

l. $2(x + 2y) = 2$
   $3(y - 2x) = -21$

m. $\dfrac{2y - x}{2} = -2$

$\dfrac{3(x + y)}{-5} = 3$

n. $3(x - y) = y - 16$
$2(2y - 5x) = y + 12$

7. Solve each of the following problems.
   a. The sum of two numbers is 16. Their difference is 18. What are the two numbers?
   b. If one of two numbers is multiplied by 12, a number which is 16 less than the second number is obtained. If the second number is 20 times the first number, what are the two numbers?
   c. The sum of two numbers is 21. Three times the smaller number subtracted from the larger number gives 3. What are the two numbers?
   d. A number is one-third of the sum of itself and another number. What are the two numbers if the first number is 1 less than twice the second number?
   e. A line segment 25 in. long is divided into three parts. Part A is three times as long as Part B, and the length of Part B is three-fifths of the length of Part C. How long is each part?
   f. The perimeter of an isosceles triangle is 29 in. If one leg is 4 in. longer than the base, determine the length of each side of the triangle.
   g. The perimeter of a rectangle is 21.4 in. Twice the width is .2 in. less than the length. What are the dimensions of the rectangle?
   h. A rectangle and a square have equal perimeters. The length of the rectangle is 2 in. more than a side of the square. The length of the rectangle is 4 in. more than its width, and the area of the square is equal to 9 sq in. What are the dimensions of each figure?
   i. A square and a rectangle have the same perimeter. The rectangle's dimensions are: length, 27 ft; width, 12 ft. How long is each side of the square?
   j. The width of a rectangle is 20% less than the length. Compute the width and the length if the perimeter is 342 ft.
   k. The difference of the measures of two complementary angles is 1°. What is the measure of each of the two angles?
   l. In a triangle, the sum of the measures of two angles is 51°. If the measure of one of these angles is decreased by 11°, its measure will equal the measure of the other angle. What is the measure of each of the three angles of the triangle?

Chapter Review Problems

m. Two bicyclists start out from the same point on a circular track with a radius of 60 ft. Bicyclist A rides at a speed of 9 mph and bicyclist B at a speed of 12 mph. If they started out riding in opposite directions, how long would it take them to meet?

n. A merchant made a mixture of 150 lb of tea worth $109.50 by mixing tea worth $1.25 per lb, with tea worth $.65 per lb. How many pounds of each kind did he use?

o. How much skimmed milk (milk without butter fat) must a farmer add to 100 qt of milk testing 4.6% butterfat so as to obtain milk testing 3.2% butterfat?

p. A grocer mixes a certain number of pounds of $.15 candy with some $.12 candy. He obtains 32 lb of mixed candy which sells for $4.44. How many pounds of each kind of candy did he use?

q. A first aid treatment for extensive burns calls for a solution which is 5% baking soda. How much water must be added to 12 oz of a 15% solution of baking soda and water so that the new solution will be only 5% baking soda?

r. A grocer bought some pecans at $.60 per lb and some walnuts for $.45 per lb. He mixed them and obtained 90 lb which he sold at $.55 per lb. How many pounds of pecans were in the mixture?

s. Brine is a solution of salt and water. If a tub contains 50 lb of a 5% solution of brine, how much water must evaporate to change it to an 8% solution?

12

Roots

and Radical Equations

12.1 Square Root

Let us consider the following equation:

$$x^2 = 36.$$

The solution set of this equation is $\{6, -6\}$, since "$6^2 = 36$" and "$(-6)^2 = 36$" are both true statements. Furthermore, there are only two numbers each of which squared gives 36.

Thus the equation $x^2 = 36$, or its equivalent equation $x^2 - 36 = 0$, has two solutions: 6 and -6, which are additive inverses. Since $6 \times 6 = 36$, we call 6 a *square root* of 36. Also, since $(-6)(-6) = 36$, we call -6 a *square root* of 36. Thus, 36 has two square roots: one a *positive* number, the other a *negative* number. To distinguish between the two square roots, we call the *positive* square root of 36

the *principal* square root of 36.

A short name for the principal square root of 36 is $\sqrt{36}$. Since the other square root of 36 is the additive inverse of the principle square root, it is $-\sqrt{36}$, or simply -6. Thus,

$$(\sqrt{36})^2 = 36 \qquad (-\sqrt{36})^2 = 36.$$

12.1 Square Root

Agreement: Whenever we use $\sqrt{x}$ $[x \geq 0]$, we mean the *principal* square root of x.

To make the meaning of $\sqrt{x}$ precise, we state the following definition.

DEFINITION 12.1 Given a real number $x \geq 0$, $\sqrt{x}$ is a nonnegative number such that $\sqrt{x} \times \sqrt{x} = x$.

We made sure that in $\sqrt{x}$, x be nonnegative. The following argument will show why this is necessary.

$\sqrt{x}$ is a real number which multiplied by itself results in x.

Suppose x is negative. Then some number multiplied by itself gives a negative number for the product.

We know that, given a real number, only one of the following is true.

(1) It is positive.
(2) It is negative.
(3) It is 0.

We also know the following.

(4) The product of a pair of positive numbers is a positive number.
(5) The product of a pair of negative numbers is a positive number.
(6) $0 \times 0 = 0$.

Therefore, there is no real number which multiplied by itself results in a product which is a negative number. Thus, in $\sqrt{x}$, x must be nonnegative.

Now consider the following equation:

$$x^3 = 8.$$

In solving this equation, we are seeking a real number x, such that

$$x \times x \times x = 8.$$

As we examine the set of real numbers, we can see that the solution set is $\{2\}$, since

$$2 \times 2 \times 2 = 8$$

or

$$2^3 = 8.$$

We know that -2 is not a solution of $x^3 = 8$ because $(-2)^3 = -8$.

The number 2 is called the *third root*, or the *cube root* of 8. We shall write $\sqrt[3]{8}$ as an abbreviation for the cube root of 8. Thus,

$$\sqrt[3]{8} = 2$$

and

$$\sqrt[3]{8} \times \sqrt[3]{8} \times \sqrt[3]{8} = 8.$$

Similarly, $\sqrt[3]{27} = 3$, $\sqrt[3]{125} = 5$, and $\sqrt[3]{1} = 1$.

Now consider the equation

$$x^4 = 81.$$

The real number solution set is $\{3, -3\}$ since

$$3^4 = 81$$

and

$$(-3)^4 = 81.$$

We use the symbol $\sqrt[4]{81}$ to mean the positive, or *principal fourth root* of 81. Thus,

$$\sqrt[4]{81} = 3$$

and

$$-\sqrt[4]{81} = -3.$$

In general, $\sqrt[n]{x}$, called the *n*th *root of x*, is the real number such that

$$\underbrace{\sqrt[n]{x} \times \sqrt[n]{x} \times \ldots \times \sqrt[n]{x}}_{\sqrt[n]{x} \text{ used } n \text{ times as a factor,}} = x$$

where *n* is a natural number greater than 2. (In $\sqrt{x}$, we assume that $n = 2$.) If there are two real *n*th roots of *x*, we shall agree that $\sqrt[n]{x}$ means the *positive* *n*th root.

The following is the standard vocabulary used in connection with expressions of the form $\sqrt[n]{x}$. We call

$\sqrt[n]{x}$ a *radical*

$\sqrt[n]{}$ a *radical sign*

x in $\sqrt[n]{x}$ a *radicand*

n in $\sqrt[n]{x}$ an *index*.

We have shown that in $\sqrt{x}$, *x* must be a *nonnegative* real number. But must *x* be nonnegative in the case of $\sqrt[3]{x}$? Let us examine $\sqrt[3]{-8}$, for example. It should be a number such that

$$\sqrt[3]{-8} \times \sqrt[3]{-8} \times \sqrt[3]{-8} = -8.$$

We know such a real number; it is -2, since

$$(-2)(-2)(-2) = -8.$$

12.1 Square Root

Thus,
$$\sqrt[3]{-8} = -2.$$

Each of the following statements is true.

(1) $\sqrt{-9}$ is not a real number, because no real number squared is equal to -9.
(2) $\sqrt[3]{-27}$ is a real number; it is -3, because $(-3)^3 = -27$.
(3) $\sqrt[4]{-16}$ is not a real number, because there is no real number resulting in -16 when raised to the fourth power.
(4) $\sqrt[5]{-32}$ is a real number; it is -2, because $(-2)^5 = -32$.

We can generalize the situation by stating the following property.

PROPERTY OF RADICALS $\sqrt[n]{x}$ is a real number for each *even* natural number n and each real number $x \geq 0$.

$\sqrt[n]{x}$ is a real number for each *odd* natural number n and each real number x.

EXERCISE 12.1

1. Tell the solution set of each of the following equations.
   a. $x^2 = 9$
   b. $x^2 = 16$
   c. $x^2 = 121$
   d. $x^2 = 225$
   e. $x^2 = 0$
   f. $x^2 - 64 = 0$
   g. $x^2 - 100 = 0$
   h. $x^2 - 3 = 0$
   i. $x^2 - 12 = 13$
   j. $x^2 - 64 = 36$

2. For each of the following, tell what is the radical, the radical sign, the radicand, and the index.
   a. $\sqrt[3]{-27}$
   b. $\sqrt[5]{4}$
   c. $\sqrt[4]{0}$
   d. $\sqrt{10}$

3. Tell which of the following are true and which are false.
   a. $(-\sqrt{2})^2 = -2$
   b. $(\sqrt{2})^2 = 2$
   c. $(\sqrt{3})^2 = 9$
   d. $(-\sqrt{5})^2 = 5$
   e. $(-\sqrt{6})^2 = (\sqrt{6})^2$
   f. $(\sqrt{5} \times 0)^2 = 0^2$
   g. $(\sqrt{2} \times \sqrt{3})^2 = 6$
   h. $(-\sqrt{2} \times \sqrt{3})^2 = -6$
   i. $[(-\sqrt{2})(-\sqrt{3})]^2 = 6$
   j. $-(\sqrt{2})^2 = -2$

4. For each of the following, give an integer name.
   a. $\sqrt{100}$
   b. $\sqrt[3]{-27}$
   c. $\sqrt[6]{64}$
   d. $\sqrt[7]{-128}$
   e. $\sqrt[5]{-243}$
   f. $\sqrt[9]{-1}$

5. Tell which of the following are real numbers and which are not.
   a. $\sqrt{-1}$ b. $\sqrt[3]{-1}$ c. $\sqrt[6]{-1}$ d. $\sqrt[9]{-1}$

12.2 Products of Roots

We shall now use the various properties of real numbers to work with expressions involving square roots. Let us first observe a pattern in the following examples.

$$\sqrt{4} \times \sqrt{9} = 2 \times 3 = 6 \qquad \sqrt{4 \times 9} = \sqrt{36} = 6$$
$$\text{Therefore, } \sqrt{4} \times \sqrt{9} = \sqrt{4 \times 9}.$$

$$\sqrt{4} \times \sqrt{16} = 2 \times 4 = 8 \qquad \sqrt{4 \times 16} = \sqrt{64} = 8$$
$$\text{Therefore, } \sqrt{4} \times \sqrt{16} = \sqrt{4 \times 16}.$$

$$\sqrt{4} \times \sqrt{25} = 2 \times 5 = 10 \qquad \sqrt{4 \times 25} = \sqrt{100} = 10$$
$$\text{Therefore, } \sqrt{4} \times \sqrt{25} = \sqrt{4 \times 25}.$$

$$\sqrt{4} \times \sqrt{36} = 2 \times 6 = 12 \qquad \sqrt{4 \times 36} = \sqrt{144} = 12$$
$$\text{Therefore, } \sqrt{4} \times \sqrt{36} = \sqrt{4 \times 36}.$$

Since $\sqrt{0} = 0$, we also have $\sqrt{0} \times \sqrt{0} = \sqrt{0 \times 0}$.

We shall now state the theorem suggested by this pattern and prove it.

THEOREM 12.1 Product of Square Roots For all real numbers x and y where $x \geq 0$ and $y \geq 0$,

$$\sqrt{xy} = \sqrt{x} \times \sqrt{y}.$$

Proof. Since $\sqrt{x} \geq 0$ and $\sqrt{y} \geq 0$, we know that $\sqrt{x} \times \sqrt{y} \geq 0$; that is, the product of two nonnegative numbers is nonnegative.

$$(\sqrt{x} \times \sqrt{y})^2 = (\sqrt{x} \times \sqrt{y})(\sqrt{x} \times \sqrt{y})$$
$$= (\sqrt{x} \times \sqrt{x})(\sqrt{y} \times \sqrt{y})$$
$$= xy.$$

We have shown that xy is the square of the nonnegative number $\sqrt{x} \times \sqrt{y}$. By the definition of a radical, we know that xy is also the square of the nonnegative number $\sqrt{xy}$.

$$(\sqrt{xy})^2 = \sqrt{xy} \times \sqrt{xy} = xy.$$

Thus, since xy is the square of the nonnegative number $\sqrt{x} \times \sqrt{y}$, and also of the nonnegative number $\sqrt{xy}$, it follows that

$$\sqrt{xy} = \sqrt{x} \times \sqrt{y}.$$

Theorem 12.1 can be extended to the nth root.

12.2 Products of Roots

THEOREM 12.2 $\sqrt[n]{xy} = \sqrt[n]{x} \times \sqrt[n]{y}$.

Note that in Theorem 12.2, n is a natural number greater than 2. If n is even, then both x and y must be nonnegative. If n is odd, then x and y can be any real numbers.

Study the following examples which illustrate the use of Theorem 12.1 in *simplifying radicals*.

Example 1 $\sqrt{48} = \sqrt{16 \times 3} = \sqrt{16} \times \sqrt{3} = 4\sqrt{3}$.

Example 2 $\sqrt{t^9} = \sqrt{t^8 \times t} = \sqrt{t^8} \times \sqrt{t} = t^4\sqrt{t}$ $[t \geq 0]$.

Note that simplifying a radical means expressing it in the form $a\sqrt{b}$, where a is a rational number and b is not divisible by a perfect square such as 4, 9, 16, ...

We can also use Theorem 12.1 in simplifying products of radicals.

Example 3 $\sqrt{55} \times \sqrt{5} = \sqrt{11 \times 5} \times \sqrt{5}$
$= \sqrt{11} \times \sqrt{5} \times \sqrt{5}$
$= \sqrt{11} \times 5$
$= 5\sqrt{11}$.

Example 4 $\sqrt{7} \times \sqrt{6} \times \sqrt{77} \times \sqrt{22}$
$= \sqrt{7} \times (\sqrt{2} \times \sqrt{3}) \times (\sqrt{7} \times \sqrt{11}) \times (\sqrt{2} \times \sqrt{11})$
$= (\sqrt{7} \times \sqrt{7}) \times (\sqrt{2} \times \sqrt{2}) \times (\sqrt{11} \times \sqrt{11}) \times \sqrt{3}$
$= 7 \times 2 \times 11 \times \sqrt{3}$
$= 154\sqrt{3}$.

Now we illustrate the use of Theorem 12.2 in simplifying radicals.

Example 5 $\sqrt[3]{16} = \sqrt[3]{8 \times 2} = \sqrt[3]{8} \times \sqrt[3]{2} = 2\sqrt[3]{2}$.

Example 6 $\sqrt[4]{30,000} = \sqrt[4]{10,000} \times \sqrt[4]{3} = 10\sqrt[4]{3}$.

Example 7 $\sqrt[3]{-54} \times \sqrt[3]{4} = \sqrt[3]{-27 \times 2} \times \sqrt[3]{4}$
$= \sqrt[3]{-27} \times \sqrt[3]{2} \times \sqrt[3]{4}$
$= -3\sqrt[3]{2} \times 4$
$= -3\sqrt[3]{8}$
$= -3 \times 2$
$= -6$.

EXERCISE 12.2

1. Simplify.
   - a. $\sqrt{98}$
   - b. $\sqrt{44}$
   - c. $\sqrt{99}$
   - d. $\sqrt{242}$
   - e. $\sqrt{450}$
   - f. $\sqrt{243}$
   - g. $\sqrt{80}$
   - h. $\sqrt{63}$
   - i. $\sqrt{4} \times \sqrt{3}$
   - j. $\sqrt{12} \times \sqrt{9}$
   - k. $\sqrt{7} \times \sqrt{11}$
   - l. $\sqrt{50}$
   - m. $\sqrt{18}$
   - n. $3\sqrt{2} \times 3\sqrt{32}$
   - o. $3\sqrt{5} \times 3\sqrt{32}$
   - p. $\sqrt{18} \times \sqrt{2}$
   - q. $\sqrt{26} \times \sqrt{52}$
   - r. $\sqrt{256} \times \sqrt{28}$
   - s. $\sqrt{22} \times \sqrt{11}$
   - t. $\sqrt{36} \times \sqrt{7}$
   - u. $\sqrt{8} \times \sqrt{3}$
   - v. $\sqrt{27} \times \sqrt{3}$
   - w. $\sqrt{75} \times \sqrt{3}$
   - x. $\sqrt{10} \times \sqrt{30}$
   - y. $5\sqrt{7} \times 3\sqrt{2}$
   - z. $7\sqrt{5} \times \sqrt{6}$
   - a'. $\sqrt{5} \times \sqrt{15}$
   - b'. $\sqrt{125}$
   - c'. $\sqrt{3} \times \sqrt{6}$
   - d'. $\sqrt{30} \times \sqrt{10} \times \sqrt{3}$
   - e'. $\sqrt{27} \times \sqrt{12} \times \sqrt{18} \times \sqrt{2}$
   - f'. $4\sqrt{2} \times 5\sqrt{3}$
   - g'. $\sqrt{72} \times \sqrt{2}$
   - h'. $\sqrt{12} \times \sqrt{6} \times \sqrt{11}$
   - i'. $\sqrt{22} \times \sqrt{33} \times \sqrt{6}$
   - j'. $\sqrt{42} \times \sqrt{7} \times \sqrt{6}$
   - k'. $\sqrt{17} \times \sqrt{34} \times \sqrt{5}$
   - l'. $8 \times \sqrt{7} \times 3 \times \sqrt{2}$
   - m'. $5\sqrt{6} \times 5\sqrt{6}$
   - n'. $(-7\sqrt{2})(10\sqrt{3})$
   - o'. $(-5\sqrt{3})(-4\sqrt{18})$
   - p'. $(-7\sqrt{5})(-2\sqrt{45})$

2. Simplify. If the replacement set of the variables is not the set of all real numbers, state the restriction.

Example 1 $\sqrt{t^6} = t^3$ $[t \geq 0]$.
Note that $\sqrt{t^6}$ is the principal square root of t^6. Thus, t^3 must be nonnegative, and so t must also be nonnegative.

Example 2 $\sqrt{m^{12}} = m^6$.
Note that for all m, m^6 and m^{12} are nonnegative. Therefore m may be any real number.

Example 3 $\sqrt{18c^8} = \sqrt{18} \times \sqrt{c^8}$
$\phantom{Example 3 \ \ \sqrt{18c^8}} = 3\sqrt{2} \times c^4$
$\phantom{Example 3 \ \ \sqrt{18c^8}} = 3c^4\sqrt{2}.$

12.2 Products of Roots

Since c^8 and c^4 are nonnegative numbers for each value of c, c can be any real number.

Example 4
$$\sqrt{a^7} = \sqrt{a^6 \times a}$$
$$= \sqrt{a^6} \times \sqrt{a}$$
$$= a^3\sqrt{a} \quad [a \geq 0].$$

We know that a must be nonnegative, because for $a < 0$, $a^7 < 0$ and the square root of a negative number does not exist in the set of real numbers.

a. $\sqrt{m^4}$
b. $\sqrt{c^8}$
c. $\sqrt{t^{16}}$
d. $\sqrt{n^{28}}$
e. $\sqrt{4a^4}$
f. $\sqrt{16b^{16}}$
g. $\sqrt{81m^{20}}$
h. $\sqrt{200x^8}$
i. $\sqrt{x^{11}}$
j. $\sqrt{m^9}$
k. $\sqrt{c^3}$
l. $\sqrt{y^{31}}$
m. $\sqrt{t^{17}}$
n. $\sqrt{81a^5}$
o. $\sqrt{18a^9}$
p. $\sqrt{250m^{17}}$
q. $\sqrt{a^4b^8c^{12}}$
r. $\sqrt{(x-y)^4}$
s. $\sqrt{a^{12}m^8p^{20}}$
t. $\sqrt{.25t^5}$
u. $\sqrt{5x} \times \sqrt{5x^2}$
v. $\sqrt{2} \times \sqrt{18b^2}$
w. $\sqrt{3m^3} \times \sqrt{6m^4n}$
x. $\sqrt{a} \times \sqrt{2a} \times \sqrt{8a^2}$

3. Simplify.

a. $\sqrt[3]{7} \times \sqrt[3]{7} \times \sqrt[3]{7}$
b. $\sqrt[3]{24}$
c. $\sqrt[3]{-24}$
d. $\sqrt[5]{32}$
e. $\sqrt[5]{64}$
f. $\sqrt[4]{32}$
g. $\sqrt[4]{40,000}$
h. $\sqrt[4]{80,000}$
i. $\sqrt[3]{-108}$
j. $\sqrt[3]{-16}$

4. Simplify. If the replacement set of the variables is not the set of all real numbers, state the restriction.

Example 1 $\sqrt[3]{8x^4} = \sqrt[3]{8x^3 \times x} = \sqrt[3]{8x^3} \times \sqrt[3]{x} = 2x\sqrt[3]{x}$.
x can be any real number.

Example 2 $\sqrt[4]{16x^3y^4z^4} = \sqrt[4]{16y^4z^4} \times \sqrt[4]{x^3} = 2yz\sqrt[4]{x^3}$
$$[x \geq 0, y \geq 0, z \geq 0].$$

a. $\sqrt[3]{8x^3}$
b. $\sqrt[3]{8x^5}$

c. $\sqrt[3]{48r^4}$
d. $\sqrt[4]{a^9}$
e. $\sqrt[3]{-8x^2y^5}$
f. $\sqrt[3]{-48a^2b^4}$
g. $\sqrt[4]{a^8b^{12}c^7}$
h. $\sqrt[3]{2x} \times \sqrt[3]{4x^2}$
i. $\sqrt[3]{9x} \times \sqrt[3]{9x^3}$
j. $\sqrt[3]{-54m^7n^4}$

12.3 Simplifying Expressions; Rationalizing Denominators

As we learn how to solve certain types of equations, we will encounter expressions such as the following:

$$\frac{8 + \sqrt{12}}{2}.$$

Using the distributive property and Theorem 12.1, we can simplify such expressions.

$$\frac{8 + \sqrt{12}}{2} = \frac{2 \times 4 + 2\sqrt{3}}{2}$$
$$= \frac{2(4 + \sqrt{3})}{2}$$
$$= 4 + \sqrt{3}.$$

Thus, a simpler name for $\frac{8 + \sqrt{12}}{2}$ is $4 + \sqrt{3}$.

Study the following examples.

Example 1 $\quad \frac{12 + 6\sqrt{7}}{3} = \frac{3(4 + 2\sqrt{7})}{3}$
$$= 4 + 2\sqrt{7}.$$

Example 2 $\quad \frac{6 - 3\sqrt{20}}{12} = \frac{6 - 3(2\sqrt{5})}{12}$
$$= \frac{6 - 6\sqrt{5}}{12}.$$
$$= \frac{6(1 - \sqrt{5})}{6 \times 2}$$
$$= \frac{1 - \sqrt{5}}{2}.$$

When dealing with fractional expressions, we frequently encounter irrational numbers in the denominator. For example, $\frac{1}{\sqrt{2}}$ is such an expression. If we wish to compute a rational number approximation to $\frac{1}{\sqrt{2}}$,

12.3 Simplifying Expressions; Rationalizing Denominators

we discover that it would be less cumbersome if we did not have the irrational number in the denominator. Recall that decimal numerals for irrational numbers are nonterminating and nonrepeating.

Let us see what the computations look like using 1.414 as a three-decimal place approximation to $\sqrt{2}$.

$$\frac{1}{\sqrt{2}} \doteq \frac{1}{1.414} \text{ or } \frac{1000}{1414}$$

$$
\begin{array}{r}
.707 \\
1414 \overline{\smash{\big)}\ 1000.000} \\
\underline{989\ 8} \\
10\ 200 \\
\underline{9\ 898} \\
302
\end{array}
$$

Thus, $\frac{1}{\sqrt{2}} \doteq .71$, correct to two decimal places.

We can avoid these computations by obtaining an expression equivalent to $\frac{1}{\sqrt{2}}$, which has a rational number in the denominator.

$$\frac{1}{\sqrt{2}} = \frac{1 \times \sqrt{2}}{\sqrt{2} \times \sqrt{2}} = \frac{\sqrt{2}}{2}.$$

Using $\frac{\sqrt{2}}{2}$ in place of $\frac{1}{\sqrt{2}}$ provides us with an equivalent expression which is much more convenient for computations. The following proves the point:

$$\frac{\sqrt{2}}{2} \doteq \frac{1.414}{2}$$
$$\doteq .707$$
$$\doteq .71.$$

The process of writing $\frac{1}{\sqrt{2}}$ as $\frac{\sqrt{2}}{2}$ is called *rationalizing the denominator*. Study the following examples which illustrate this process.

Example 1 $\quad \frac{5}{\sqrt{5}} = \frac{5\sqrt{5}}{\sqrt{5} \times \sqrt{5}} = \frac{5\sqrt{5}}{5} = \sqrt{5}.$

Example 2 $\dfrac{7}{\sqrt{2}} = \dfrac{7\sqrt{2}}{\sqrt{2} \times \sqrt{2}} = \dfrac{7\sqrt{2}}{2}.$

Example 3 $\dfrac{15}{\sqrt{12}} = \dfrac{15}{2\sqrt{3}} = \dfrac{15\sqrt{3}}{2\sqrt{3} \times \sqrt{3}} = \dfrac{5 \times 3\sqrt{3}}{2 \times 3} = \dfrac{5\sqrt{3}}{2}.$

Now observe the pattern displayed in the following examples involving quotients of square roots.

$$\sqrt{\tfrac{9}{4}} = \tfrac{3}{2}; \quad \text{also} \quad \dfrac{\sqrt{9}}{\sqrt{4}} = \tfrac{3}{2}; \quad \text{therefore,} \quad \sqrt{\tfrac{9}{4}} = \dfrac{\sqrt{9}}{\sqrt{4}}.$$

$$\sqrt{\tfrac{25}{16}} = \tfrac{5}{4}; \quad \text{also} \quad \dfrac{\sqrt{25}}{\sqrt{16}} = \tfrac{5}{4}; \quad \text{therefore,} \quad \sqrt{\tfrac{25}{16}} = \dfrac{\sqrt{25}}{\sqrt{16}}.$$

$$\sqrt{\tfrac{25}{4}} = \tfrac{5}{2}; \quad \text{also} \quad \dfrac{\sqrt{25}}{\sqrt{4}} = \tfrac{5}{2}; \quad \text{therefore,} \quad \sqrt{\tfrac{25}{4}} = \dfrac{\sqrt{25}}{\sqrt{4}}.$$

This pattern suggests the following theorem.

THEOREM 12.3 $\sqrt{\dfrac{x}{y}} = \dfrac{\sqrt{x}}{\sqrt{y}} \quad [x \geq 0,\ y > 0].$

Proof. We will start by showing that $\sqrt{y} \times \sqrt{\dfrac{x}{y}} = \sqrt{x}.$

$$\sqrt{y} \times \sqrt{\dfrac{x}{y}} = \sqrt{y \times \dfrac{x}{y}}$$
$$= \sqrt{x}.$$

Thus, $\sqrt{y} \times \sqrt{\dfrac{x}{y}} = \sqrt{x}.$ Now we multiply both sides of the equation by $\dfrac{1}{\sqrt{y}}.$

$$\dfrac{1}{\sqrt{y}}\left(\sqrt{y} \times \sqrt{\dfrac{x}{y}}\right) = \dfrac{1}{\sqrt{y}} \times \sqrt{x}$$

$$\left(\dfrac{1}{\sqrt{y}} \times \sqrt{y}\right) \times \sqrt{\dfrac{x}{y}} = \dfrac{\sqrt{x}}{\sqrt{y}}.$$

Thus, $\sqrt{\dfrac{x}{y}} = \dfrac{\sqrt{x}}{\sqrt{y}},$

which was to be proved.

12.3 Simplifying Expressions; Rationalizing Denominators

The following three examples illustrate the use of Theorem 12.3.

Example 1 $\quad \dfrac{\sqrt{55}}{\sqrt{11}} = \sqrt{\dfrac{55}{11}} = \sqrt{5}.$

Example 2 $\quad \sqrt{\dfrac{5}{6}} = \dfrac{\sqrt{5}}{\sqrt{6}} = \dfrac{\sqrt{5} \times \sqrt{6}}{\sqrt{6} \times \sqrt{6}} = \dfrac{\sqrt{30}}{6}.$

Example 3 $\quad \sqrt{\dfrac{98}{3}} = \dfrac{\sqrt{98}}{\sqrt{3}} = \dfrac{\sqrt{49}\sqrt{2}}{\sqrt{3}}$

$\qquad\qquad = \dfrac{7\sqrt{2}}{\sqrt{3}} = \dfrac{7\sqrt{2} \times \sqrt{3}}{\sqrt{3} \times \sqrt{3}} = \dfrac{7\sqrt{6}}{3}.$

Notice that in Examples 2 and 3 we rationalized the denominators. We do not consider an expression to be in simplest form unless we have rationalized its denominator.

Theorem 12.4 is an extension of Theorem 12.3 to the *n*th root.

THEOREM 12.4 $\quad \sqrt[n]{\dfrac{x}{y}} = \dfrac{\sqrt[n]{x}}{\sqrt[n]{y}}.$

Using the Property of Radicals, specify restrictions on *n*, *x*, and *y* in Theorem 12.4.

Study the examples below in which Theorem 12.4 is applied.

Example 1 $\quad \dfrac{\sqrt[3]{24}}{\sqrt[3]{3}} = \sqrt[3]{\dfrac{24}{3}} = \sqrt[3]{8} = 2.$

Example 2 $\quad \sqrt[3]{\dfrac{-27}{x^6}} = \dfrac{\sqrt[3]{-27}}{\sqrt[3]{x^6}} = \dfrac{-3}{x^2} \quad [x \neq 0].$

EXERCISE 12.3

1. Simplify.

   a. $\dfrac{9 + 12\sqrt{5}}{6}$ b. $\dfrac{8 - 6\sqrt{7}}{4}$

   c. $\dfrac{5 + \sqrt{27}}{-8}$ d. $\dfrac{10 - \sqrt{200}}{7}$

   e. $\dfrac{3 + 3\sqrt{7}}{3}$ f. $\dfrac{10 - 25\sqrt{2}}{5}$

   g. $\dfrac{2 + \sqrt{8}}{2}$ h. $\dfrac{12 + \sqrt{12}}{2}$

i. $\dfrac{-8 + 14\sqrt{2}}{4}$ j. $\dfrac{-9 + 15\sqrt{7}}{-6}$

k. $\dfrac{-8 - 4\sqrt{3}}{-4}$ l. $\dfrac{-21 - \sqrt{18}}{3}$

m. $\dfrac{6 - \sqrt{32}}{2}$ n. $\dfrac{12 + \sqrt{72}}{-6}$

2. Write equivalent expressions which do not contain parentheses and simplify.
   a. $\sqrt{3}(\sqrt{3} + \sqrt{5})$ b. $2(\sqrt{6} + 7)$
   c. $\sqrt{3}(\sqrt{2} - \sqrt{5})$ d. $-\sqrt{7}(\sqrt{5} - \sqrt{6})$
   e. $\sqrt{5}(3\sqrt{5} + 2\sqrt{5})$ f. $\sqrt{3}(\sqrt{12} + \sqrt{27})$

3. Simplify.
   a. $\dfrac{2\sqrt{18} + 3\sqrt{36}}{-3}$ b. $\dfrac{3\sqrt{75} - 4\sqrt{50}}{5}$
   c. $\dfrac{-5\sqrt{150} + 2\sqrt{50}}{5}$ d. $\dfrac{9\sqrt{3} - 2\sqrt{9}}{-3}$
   e. $\dfrac{10\sqrt{3} + \sqrt{10{,}000}}{-10}$ f. $\dfrac{2\sqrt{14} - \sqrt{36}}{-2}$

4. Write equivalent expressions which do not contain parentheses and simplify.
   a. $-\sqrt{2}(\sqrt{10} - \sqrt{20})$ b. $\sqrt{2}(\sqrt{15} + \sqrt{21})$
   c. $\sqrt{11}(\sqrt{22} - \sqrt{33})$ d. $-\sqrt{15}(\sqrt{3} + \sqrt{27})$
   e. $-\sqrt{12}(\sqrt{2} + \sqrt{3})$ f. $2\sqrt{5}(\sqrt{10} - 3\sqrt{5})$

5. Using the following approximations, compute to two decimal places.

   $\sqrt{2} \doteq 1.414 \qquad \sqrt{3} \doteq 1.732 \qquad \sqrt{5} \doteq 2.236$

   Example 1 $\sqrt{12} = 2\sqrt{3} \doteq 2 \times 1.732 = 3.464 \doteq 3.46.$

   Example 2 $\dfrac{1}{\sqrt{3}} = \dfrac{1 \times \sqrt{3}}{\sqrt{3} \times \sqrt{3}} = \dfrac{\sqrt{3}}{3} \doteq \dfrac{1.732}{3} \doteq .577 \doteq .58.$

   a. $\sqrt{50}$ b. $\dfrac{2}{\sqrt{3}}$ c. $\sqrt{18} + \sqrt{8}$

6. Tell which of the following are true and which are false.
   a. $\dfrac{1}{\sqrt{5}} = \dfrac{\sqrt{5}}{5}$ b. $\dfrac{3}{\sqrt{3}} = 3$
   c. $\dfrac{2}{\sqrt{2}} = \sqrt{2}$ d. $\sqrt{5 \times 6} = \sqrt{5} \times \sqrt{6}$

12.3 Simplifying Expressions; Rationalizing Denominators

e. $\sqrt{(-5)(-6)} = \sqrt{-5} \times \sqrt{-6}$ f. $\sqrt[3]{(-8)(-27)} = \sqrt[3]{-8} \times \sqrt[3]{-27}$

g. $\sqrt{\dfrac{12}{2}} = \dfrac{\sqrt{12}}{\sqrt{2}}$ h. $\sqrt[3]{\dfrac{-5}{-3}} = \dfrac{\sqrt[3]{-5}}{\sqrt[3]{-3}}$

i. $\sqrt{\dfrac{-5}{-3}} = \dfrac{\sqrt{-5}}{\sqrt{-3}}$ j. $\sqrt{\dfrac{0}{5}} = \dfrac{\sqrt{0}}{\sqrt{5}}$

k. $\dfrac{5}{\sqrt{5}} = \dfrac{\sqrt{5}}{1}$ l. $\sqrt{0 \times 0} = \sqrt{0} \times \sqrt{0}$

7. Simplify.

a. $\dfrac{\sqrt{27}}{\sqrt{3}}$ b. $\dfrac{\sqrt{45}}{\sqrt{3}}$ c. $\dfrac{\sqrt{28}}{\sqrt{2}}$

d. $\dfrac{\sqrt{98}}{\sqrt{7}}$ e. $\dfrac{\sqrt{7}}{\sqrt{3}}$ f. $\dfrac{\sqrt{16}}{\sqrt{25}}$

g. $\dfrac{\sqrt{121}}{\sqrt{49}}$ h. $\dfrac{1}{\sqrt{6}}$ i. $\dfrac{1}{\sqrt{7}}$

j. $\dfrac{1}{\sqrt{9}}$ k. $\dfrac{1}{\sqrt{3}}$ l. $\sqrt{\dfrac{1}{3}}$

m. $\dfrac{1}{\sqrt{17}}$ n. $\sqrt{\dfrac{1}{17}}$ o. $\dfrac{2}{\sqrt{2}}$

p. $\dfrac{11}{\sqrt{11}}$ q. $\dfrac{5\sqrt{7}}{\sqrt{5}}$ r. $\dfrac{9}{\sqrt{3}}$

s. $\dfrac{10}{\sqrt{5}}$ t. $\dfrac{10\sqrt{3}}{\sqrt{5}}$ u. $\dfrac{\sqrt{48}}{\sqrt{27}}$

v. $\dfrac{5}{\sqrt{5}}$ w. $\dfrac{12}{\sqrt{12}}$ x. $\dfrac{27}{\sqrt{27}}$

y. $\sqrt{\dfrac{49}{8}}$ z. $\sqrt{\dfrac{125}{5}}$ a'. $\dfrac{\sqrt{2} \times \sqrt{3}}{\sqrt{6}}$

b'. $\dfrac{\sqrt{5} \times \sqrt{14}}{\sqrt{7} \times \sqrt{2}}$ c'. $\dfrac{\sqrt{5} \times \sqrt{14}}{\sqrt{7} \times \sqrt{5}}$ d'. $\dfrac{2}{\sqrt{2}} \times \dfrac{7}{\sqrt{7}}$

e'. $\dfrac{5\sqrt{17}}{\sqrt{5}} \times \dfrac{10}{\sqrt{10}}$ f'. $\dfrac{443}{\sqrt{443}}$ g'. $\dfrac{\sqrt{6} \times \sqrt{8}}{\sqrt{28}}$

h'. $\dfrac{\sqrt{12} \times \sqrt{27}}{\sqrt{18}}$ i'. $\dfrac{\sqrt{14} \times \sqrt{28}}{\sqrt{7}}$ j'. $\dfrac{\sqrt{21} \times \sqrt{35}}{\sqrt{7} \times \sqrt{5}}$

k'. $\dfrac{\sqrt{51} \times \sqrt{17}}{\sqrt{3} \times \sqrt{2}}$ l'. $\dfrac{\sqrt{35} \times \sqrt{15}}{\sqrt{3} \times \sqrt{5}}$ m'. $\dfrac{\sqrt{80} \times \sqrt{36}}{\sqrt{3} \times \sqrt{8}}$

n'. $\dfrac{\sqrt[3]{21}}{\sqrt[3]{3}}$ o'. $\dfrac{\sqrt[3]{48}}{\sqrt[3]{12}}$ p'. $\sqrt[3]{-\dfrac{8}{27}}$

q'. $\sqrt[3]{\dfrac{125}{8}}$ r'. $\dfrac{\sqrt[3]{64}}{\sqrt[3]{8}}$ s'. $\dfrac{\sqrt[3]{64}}{\sqrt[3]{16}}$

t'. $\dfrac{\sqrt[5]{64}}{\sqrt[5]{2}}$ u'. $\dfrac{\sqrt[3]{-16}}{\sqrt[3]{-54}}$ v'. $\dfrac{\sqrt[3]{2}\sqrt[3]{4}}{\sqrt[3]{-8}}$

12.4 Exponents of the Form $\dfrac{1}{n}$

In the properties involving exponents, so far we have limited ourselves to exponents which are integers. Consider the Product of Powers Property.

$$x^m \times x^n = x^{m+n}.$$

An example of this property is

$$3^2 \times 3^3 = 3^{2+3} = 3^5.$$

Now suppose we are seeking an exponent n such that

$$3^n \times 3^n = 3.$$

If we assume that the Product of Powers Property holds, we have

$$3^n \times 3^n = 3^{n+n} = 3^{2n}.$$

If $3^n \times 3^n = 3$, then $3^{2n} = 3^1$. We can conclude that the exponents, $2n$ and 1, are equal. Thus,

$$2n = 1$$
$$n = \tfrac{1}{2}.$$

Thus, the exponent we were seeking is $\tfrac{1}{2}$, and

$$3^{\tfrac{1}{2}} \times 3^{\tfrac{1}{2}} = 3.$$

How shall we define the exponent $\tfrac{1}{2}$? Since we also know that

$$\sqrt{3} \times \sqrt{3} = 3 \quad \text{and} \quad (-\sqrt{3})(-\sqrt{3}) = 3,$$

it seems reasonable to define $3^{\tfrac{1}{2}}$ as either $\sqrt{3}$ or $-\sqrt{3}$. We shall choose the principal square root, and agree that

$$3^{\tfrac{1}{2}} = \sqrt{3}.$$

Using a similar argument, we make the following definition.

DEFINITION 12.2 For all nonnegative real numbers x, $x^{\tfrac{1}{2}} = \sqrt{x}$.

Generalizing further, we define $x^{\tfrac{1}{n}}$ as follows:

DEFINITION 12.3 $x^{\frac{1}{n}} = \sqrt[n]{x}$.

In the last definition, n is a natural number greater than 1. If n is odd, then x may be any real number. If n is even, then x must be nonnegative.

According to the definition of $x^{\frac{1}{n}}$, we can conclude the following:

$$x^{\frac{1}{3}} = \sqrt[3]{x} \qquad\qquad x^{\frac{1}{4}} = \sqrt[4]{x} \quad [x \geq 0]$$
$$(-8)^{\frac{1}{3}} = \sqrt[3]{-8} = -2 \qquad 81^{\frac{1}{4}} = \sqrt[4]{81} = 3$$

12.5 Rational-number Exponents

In the preceding section we considered exponents of the form $\frac{1}{n}$, where n is a natural number greater than 1. We now extend our study of exponents to include all rational numbers. Study the examples below and observe the pattern emerging from them. Again we will assume that the Product of Powers Property holds for exponents of the form $\frac{1}{n}$.

$$4^{\frac{1}{3}} \times 4^{\frac{1}{3}} = 4^{\frac{1}{3}+\frac{1}{3}} = 4^{\frac{2}{3}}.$$

Also,

$$\sqrt[3]{4} \times \sqrt[3]{4} = \sqrt[3]{4 \times 4} = \sqrt[3]{4^2}.$$

Since $4^{\frac{1}{3}} = \sqrt[3]{4}$, it follows that

$$4^{\frac{2}{3}} = \sqrt[3]{4^2}.$$

Consider a second example:

$$5^{\frac{1}{4}} \times 5^{\frac{1}{4}} \times 5^{\frac{1}{4}} = 5^{\frac{1}{4}+\frac{1}{4}+\frac{1}{4}} = 5^{\frac{3}{4}}.$$

Also,

$$\sqrt[4]{5} \times \sqrt[4]{5} \times \sqrt[4]{5} = \sqrt[4]{5 \times 5 \times 5} = \sqrt[4]{5^3}.$$

Since $5^{\frac{1}{4}} = \sqrt[4]{5}$, it follows that

$$5^{\frac{3}{4}} = \sqrt[4]{5^3}.$$

These two examples suggest the following definition of a rational-number exponent.

DEFINITION 12.4 $x^{\frac{m}{n}} = \sqrt[n]{x^m}$.

For this definition, m and n are natural numbers greater than 1. If n is odd, then x may be any real number. If n is even, then x^m must be nonnegative.

It can be shown that all the properties which hold for integer exponents also hold for rational-number exponents. Assuming these properties, study the following examples.

Example 1 $(125y^9)^{\frac{1}{3}} = (125)^{\frac{1}{3}} \times (y^9)^{\frac{1}{3}}$
$= \sqrt[3]{125} \times y^{\frac{9}{3}}$
$= 5y^3.$

Example 2 $(36x^{-2})^{-\frac{1}{2}} = (36)^{-\frac{1}{2}} \times (x^{-2})^{-\frac{1}{2}}$ $[x > 0]$
$= \dfrac{1}{36^{\frac{1}{2}}} \times x$
$= \dfrac{x}{\sqrt{36}}$
$= \dfrac{x}{6}.$

Now observe the following examples which suggest another pattern.
$$\sqrt[3]{8^2} = \sqrt[3]{64} = 4 \quad \text{and} \quad (\sqrt[3]{8})^2 = (2)^2 = 4;$$
hence,
$$\sqrt[3]{8^2} = (\sqrt[3]{8})^2.$$
$$\sqrt[4]{16^2} = \sqrt[4]{256} = 4 \quad \text{and} \quad (\sqrt[4]{16})^2 = (2)^2 = 4;$$
hence,
$$\sqrt[4]{16^2} = (\sqrt[4]{16})^2.$$

The theorem suggested by these examples is the following:

THEOREM 12.5 $\sqrt[n]{x^m} = (\sqrt[n]{x})^m.$

(m and n are natural numbers, $n \geq 2$, x is a nonnegative real number.)

Proof. $\sqrt[n]{x^m} = x^{\frac{m}{n}}$
$= x^{\frac{1}{n} \times m}$
$= (x^{\frac{1}{n}})^m$
$= (\sqrt[n]{x})^m.$

Thus, $\sqrt[n]{x^m} = (\sqrt[n]{x})^m.$

The following two examples illustrate the use of Theorem 12.5.

Example 1 $81^{\frac{3}{4}} = \sqrt[4]{81^3} = (\sqrt[4]{81})^3 = 3^3 = 27.$

Example 2 $(-8)^{\frac{5}{3}} = \sqrt[3]{(-8)^5} = (\sqrt[3]{-8})^5 = (-2)^5 = -32.$

12.5 Rational-number Exponents

EXERCISE 12.4

1. Label the following with T for true and F for false.
   a. $5^{\frac{1}{2}} = \sqrt{5}$
   b. $(-3)^{\frac{1}{3}} = \sqrt[3]{-3}$
   c. $4^{\frac{1}{6}} - \sqrt[4]{6}$
   d. In $\sqrt[3]{-4}$, 3 is the radicand.
   e. In $\sqrt[3]{-4}$, -4 is the radicand.
   f. In $\sqrt[5]{2}$, 5 is the index.
   g. In $\sqrt{3}$, 2 is the index.
   h. $5^{\frac{1}{2}}$ is a radical.
   i. $\sqrt[12]{-10}$ is not a real number.
   j. $\sqrt[13]{-10}$ is not a real number.
   k. $\sqrt[20]{5}$ is not a real number.
   l. $\sqrt[19]{5}$ is not a real number.
   m. $2^{\frac{1}{2}} \times 2^{\frac{1}{3}} = 2^{\frac{2}{3}}$
   n. $(-6)^{\frac{1}{5}}(-6)^{\frac{1}{5}}(-6)^{\frac{1}{5}} = (-6)^{\frac{3}{5}}$

2. For each of the following, write an equivalent symbol which contains no fractional exponents. Express your answers in simplest form. If the replacement set for a variable is not the set of real numbers, state the necessary restriction.
   a. $x^{\frac{1}{3}}$
   b. $32^{\frac{1}{5}}$
   c. $\dfrac{6x}{6^{\frac{1}{2}}}$
   d. $(5y)^{\frac{1}{2}}$
   e. $5y^{\frac{1}{2}}$
   f. $\dfrac{x^2}{x^{\frac{1}{2}}}$
   g. $(81x^3)^{\frac{1}{4}}$
   h. $24^{\frac{1}{2}} + 54^{\frac{1}{2}}$
   i. $(-\frac{1}{8})^{\frac{1}{3}}$
   j. $(-8)^{\frac{1}{3}}$

3. Tell which of the following are true for all real-number replacements of the variables.
   a. $x^{\frac{1}{3}} = \sqrt{x^3}$
   b. $\dfrac{x}{y} = \dfrac{\sqrt{x}}{\sqrt{y}}$ $[y \neq 0]$
   c. $x^{\frac{4}{3}} = \sqrt[4]{x^3}$ $[x \geq 0]$
   d. $x^{\frac{2}{3}} = \sqrt[3]{x^2}$
   e. $x^{\frac{2}{5}} \times x^{\frac{2}{5}} = x^{\frac{4}{25}}$
   f. $x^{\frac{1}{3}} \times x^{\frac{1}{3}} = x^{\frac{2}{3}}$

4. For each of the following, write an equivalent symbol using a radical sign.
   a. $z^{\frac{1}{3}}$
   b. $r^{\frac{2}{3}}$
   c. $d^{\frac{3}{4}}$
   d. $2x^{\frac{1}{2}}$
   e. $3y^{\frac{4}{3}}$
   f. $r^{\frac{2}{5}}$

5. For each of the following, write an equivalent symbol using fractional exponents.
   a. $\sqrt{3}$
   b. $\sqrt[3]{11}$
   c. $\sqrt[5]{x^2}$
   d. $\sqrt[4]{r^3}$
   e. $\sqrt[7]{t^3}$
   f. $\sqrt[5]{x^8}$
   g. $\sqrt[3]{(x+y)^2}$
   h. $\sqrt{(r-t)^3}$
   i. $\sqrt[3]{(a+b+c)^4}$

6. Give the simplest name involving no exponents for each of the following:

a. 3^{-3} b. 53^0 c. $4^{-\frac{1}{2}}$

d. 3^{-1} e. $4^{-\frac{3}{2}}$ f. $(-3)^{-2}$

g. $\dfrac{4^{-2}}{6^0}$ h. 3×3^{-3} i. $16 \times 8^{-\frac{1}{3}}$

j. $\dfrac{6}{4^{-2}}$ k. $16^{\frac{1}{2}} \times 4^{-\frac{1}{2}}$ l. $\dfrac{3}{25^{-\frac{1}{2}}}$

m. $25^{-\frac{5}{2}}$ n. $\left(\frac{3}{4}\right)^{-2}$ o. $\left(\frac{25}{16}\right)^{-\frac{1}{2}}$

p. 32×2^{-5} q. $5^{-1} \times 5^{-3} \times 5^4$ r. $\dfrac{7^{-2}}{7^{-3}}$

s. $125^{\frac{2}{3}}$ t. $27^{\frac{4}{3}}$ u. $8^{-\frac{2}{3}}$

7. Simplify. Do not use a negative exponent or a radical sign.

a. $x^{\frac{3}{2}} \times x^{\frac{1}{2}}$ b. $y^{\frac{1}{3}} \times y^{\frac{1}{4}}$ c. $d^{\frac{4}{3}} \times d^{\frac{1}{2}}$

d. $r^{\frac{7}{3}} \times r^{-\frac{1}{3}}$ e. $\dfrac{3y}{y^{\frac{2}{3}}}$ f. $\dfrac{s^3}{s^{\frac{1}{3}}}$

g. $\dfrac{r^{\frac{5}{6}}}{r^{\frac{1}{3}}}$ h. $\dfrac{p^{\frac{1}{3}}}{p^{\frac{1}{2}}}$ i. $\dfrac{x^{\frac{1}{4}}}{x^{\frac{1}{8}}}$

8. Simplify.

a. $(x^4 y^2)^{\frac{3}{2}}$ b. $\left(\dfrac{125 y^4}{216 y^{-1}}\right)^{\frac{1}{3}}$

c. $\left(\dfrac{9x^{-3}}{4z^2}\right)^{\frac{1}{2}}$ d. $\dfrac{x^{-2} y^5}{x^5 y^{-2}}$

e. $(x^{\frac{3}{4}} y^{\frac{4}{3}})^3$ f. $(r^{\frac{2}{3}} s^{\frac{1}{3}})^3$

g. $\dfrac{(z^4 y^{\frac{3}{4}})^4}{r^{\frac{5}{4}}}$ h. $(x^{-3})^{-\frac{1}{3}}$

i. $(16 r^2 t^3)^{\frac{3}{4}}$ j. $(a^{-2} \times b^{-3})^{-\frac{1}{6}}$

9. Express each of the following in simplest form using only positive exponents and no radical signs.

a. $\dfrac{-x^{-2}}{2}$ b. $\dfrac{r^{-2} s^{-3}}{t^2}$

c. $\dfrac{x^{-\frac{4}{3}}}{x}$ d. $\dfrac{r^{-\frac{3}{4}}}{3}$

e. $\dfrac{7}{-6s^{-2}}$ f. $\dfrac{(x+y)^{-2}}{3}$

g. $\dfrac{5}{(r-t)^{-3}}$ h. $rs^{-3} + r^{-2}s$

i. $3^{-1} c y^{-2}$ j. $2^{-4} d x^{-3} y^{-4}$

12.6 Square Root and Absolute Value

Let us examine the symbol
$$\sqrt{x^2}.$$
On first thought we might think that $\sqrt{x^2}$ is equal to x. This is certainly true if x is nonnegative. But what if x is negative? For example, what is $\sqrt{(-3)^2}$ equal to? Agreeing that raising to a power comes first, we have
$$\sqrt{(-3)^2} = \sqrt{9} = 3 \quad \text{and} \quad 3 = |-3|.$$
Also, $\quad \sqrt{(-7)^2} = \sqrt{49} = 7 \quad \text{and} \quad 7 = |-7|.$

The patterns above suggest that if $x < 0$, $\sqrt{x^2} = |x|$.
Is this also true if $x \geq 0$? Examine the following:
$$\sqrt{5^2} = \sqrt{25} = 5 \quad \text{and} \quad 5 = |5|$$
$$\sqrt{0^2} = \sqrt{0} = 0 \quad \text{and} \quad 0 = |0|.$$
These patterns suggest the following definition.

DEFINITION 12.5 $\forall_x \ \sqrt{x^2} = |x|.$

Study the following examples.

Example 1 $\sqrt{(-2)^2} = |-2| = 2.$

Example 2 $\sqrt{x^2 y^2} = |xy|.$

Note that we no longer need to restrict the replacement set for x and y.

Example 3 $\sqrt{(x-1)^2} = |x-1|.$

Example 4 $\sqrt{a^4} = \sqrt{(a^2)^2} = |a^2| = a^2.$

Note that $a^2 \geq 0$ whether a is negative or nonnegative.

12.7 Radical Equations

If an equation contains a variable under a radical sign, we call it a *radical equation*. For example,
$$\sqrt{x} - 1 = 2$$
is a radical equation. To determine solutions of radical equations, we will make use of the following:

PROPERTY OF nth POWER For all real numbers x and y and for each natural number n, if $x = y$, then $x^n = y^n$.

Here are two examples in which this property is illustrated.

$$\text{If } 4 = 3 + 1, \text{ then } 4^3 = (3 + 1)^3.$$
$$\text{If } -2 = 3 - 5, \text{ then } (-2)^5 = (3 - 5)^5.$$

Some radical equations are solved below. Study these solutions and observe what caution you must exercise when solving radical equations.

Example 1 $\sqrt{x} - 1 = 2$
$\sqrt{x} = 3$
$(\sqrt{x})^2 = 3^2$
$x = 9.$

The solution set *seems* to be $\{9\}$.

Check.

$\sqrt{x} - 1$	2
$\sqrt{9} - 1$	2
$3 - 1$	
2	

Thus, the solution set is $\{9\}$.

Example 2 $\sqrt{y + 3} = \sqrt{3y - 5}$
$(\sqrt{y + 3})^2 = (\sqrt{3y - 5})^2$
$y + 3 = 3y - 5$
$8 = 2y$
$y = 4.$

The solution set *seems* to be $\{4\}$.

Check.

$\sqrt{y + 3}$	$\sqrt{3y - 5}$
$\sqrt{4 + 3}$	$\sqrt{3 \times 4 - 5}$
$\sqrt{7}$	$\sqrt{12 - 5}$
$\sqrt{7}$	$\sqrt{7}$

Thus, the solution set is $\{4\}$.

Example 3 $2\sqrt{z} + 5 = 1$
$2\sqrt{z} = -4$
$(2\sqrt{z})^2 = (-4)^2$
$4z = 16$
$z = 4.$

12.7 Radical Equations

The solution set *seems* to be {4}.

Check.

$$\begin{array}{c|c} 2\sqrt{z}+5 & 1 \\ \hline 2\sqrt{4}+5 & 1 \\ 2\times 2+5 & \\ 9 & \end{array}$$

Since $9=1$ is false, {4} is *not* the solution set.
Thus, the solution set of $2\sqrt{z}+5=1$ in the set of real numbers is ϕ.

Note that the equation $2\sqrt{z}=-4$ and its equivalent equation $\sqrt{z}=-2$ require that the principal square root of a number be -2. Since the principal square root of any nonnegative real number is a nonnegative number, this equation has no real-number solution.

Also note that the technique used in solving radical equations leads to new equations which may or may not be equivalent to the equations with which we started. In the first two examples, we obtained equivalent equations, but in the third example, the final equation ($z=4$) was not equivalent to the original equation ($2\sqrt{z}+5=1$).

Sometimes this technique leads to a so-called *quadratic equation*. For example,

$$\sqrt{x}+x=1$$
$$\sqrt{x}=1-x$$
$$(\sqrt{x})^2=(1-x)^2$$
$$x=1-2x+x^2$$
$$x^2-3x+1=0.$$

The last equation is a *quadratic* equation because it is of the form $ax^2+bx+c=0$, where a, b, and c are real numbers. We will learn to solve such equations in Chapter 14.

EXERCISE 12.5

1. Give a number name which does not contain a radical sign ($\sqrt{}$). Use absolute value notation where necessary.
   - a. $\sqrt{(-5)^2}$
   - b. $\sqrt{(-\frac{1}{2})^2}$
   - c. $\sqrt{3^2}$
   - d. $-\sqrt{36}$
   - e. $(-\sqrt{25})^2$
   - f. $\sqrt{2\frac{1}{4}}$
   - g. $\sqrt{(-.7)^2}$
   - h. $-\sqrt{(-\frac{1}{3})^2}$
   - i. $-\sqrt{0.09}$
   - j. $\sqrt{25x^2}$
   - k. $\sqrt{9a^2b^2}$
   - l. $\sqrt{16y^4}$

m. $\sqrt{81a^4b^2}$

n. $\sqrt{\dfrac{4x^2}{y^2}}$

o. $\sqrt{(-3cd)^2}$

p. $\sqrt{(-2a)^4}$

q. $\sqrt{(x-y)^2}$

r. $\sqrt{(x-y)^4}$

2. Determine the solution set of each equation.

a. $\sqrt{n} = 2$

b. $\sqrt{x} - 1 = 1$

c. $2\sqrt{y} + 4 = 20$

d. $\sqrt{2p} = 10$

e. $\sqrt{2r-4} = 2$

f. $\sqrt{\dfrac{x-1}{3}} = 2$

g. $\dfrac{\sqrt{x+1}}{3} = 4$

h. $\sqrt{3s-2} = 25$

i. $\sqrt{2y+3} = \sqrt{y+5}$

j. $\sqrt{x} + 1 = 0$

k. $4\sqrt{z} - 1 = 24$

l. $9\sqrt{s} - 40 = 18$

m. $5 + 4\sqrt{t} = 30$

n. $\sqrt{1-2z} = \sqrt{5z+15}$

o. $\sqrt{\dfrac{2n+1}{3}} = \sqrt{\dfrac{n+3}{2}}$

p. $\sqrt[3]{2x} = 2$

q. $\sqrt{\dfrac{2}{s+3}} = \sqrt{\dfrac{1}{2s+1}}$

r. $\sqrt[3]{s} + 1 = 4$

s. $\sqrt[3]{2y} - 1 = 3$

t. $\sqrt[3]{3t} + 2 = -1$

u. $\sqrt[3]{m} + 1 = -2$

v. $\sqrt[3]{x} = \sqrt[3]{x}$

3. Determine a negative number whose square decreased by 2 is equal to 1.

4. A number is multiplied by 2 and 5 is added to the product; the square root of the result is equal to 3. What was the original number?

5. A number is increased by 1; the square root of the sum is divided by 2, resulting in 2. What was the original number?

GLOSSARY

Cube root: Example: $\sqrt[3]{-8} = -2$, because $(-2)(-2)(-2) = -8$.

Index: In $\sqrt[n]{x}$, n is the index.

Principal square root: The positive number whose square is equal to a given number. Example: 3 is the principal square root of 9, since $3 > 0$ and $3^2 = 9$; we write $\sqrt{9} = 3$.

Radical: $\sqrt[n]{x}$ is a radical.

Radical equation: An equation which contains a variable under a radical sign.

Radical sign: $\sqrt[n]{\ }$ is a radical sign (n is a natural number, $n \geq 2$).
Radicand: In $\sqrt[n]{x}$, x is the radicand.
Square root: A square root of a given number is a number which when multiplied by itself results in the given number. Example: $\sqrt{64} = 8$, because $8 \times 8 = 64$.

CHAPTER REVIEW PROBLEMS

1. Simplify.
   a. $2s^{\frac{1}{2}} \times s^{\frac{3}{2}}$
   b. $x^{\frac{1}{3}} \div x^{\frac{1}{3}}$
   c. $r^{\frac{1}{3}} \times r^{\frac{1}{6}}$
   d. $x^{-3} \times x^7$
   e. $12y^{-4} \div (3y^{-2})$
   f. $\dfrac{r^{-\frac{2}{3}}}{r^{\frac{1}{3}}}$
   g. $\dfrac{7x^3}{\dfrac{1}{x^2}}$
   h. $(36y^6x^2)^{\frac{1}{2}}$

2. Give equivalent expressions containing only positive exponents.
   a. $\dfrac{4x^2}{x^{-5}}$
   b. $\dfrac{5}{x^{-2}}$
   c. $\dfrac{x^{-2}}{x^{-4}}$
   d. $\dfrac{ab^{-2}}{m^2n^{-4}}$
   e. $\dfrac{a^{-3}b^{-4}}{x^{-2}y^{-3}}$
   f. $\dfrac{(x-y)^{-2}(x+y)^{-3}}{(x-y)^4(x+y)^{-7}}$

3. Give equivalent expressions containing fractional exponents.
   a. $\sqrt{r}$
   b. $\sqrt[5]{s}$
   c. $\sqrt[3]{x^2}$
   d. $\sqrt[5]{(x-3)^2}$
   e. $\sqrt[3]{y^2}$
   f. $\sqrt[4]{r^3}$
   g. $\sqrt{x^5}$
   h. $\sqrt[3]{(y-x)^7}$
   i. $\sqrt[7]{(y-x)^3}$
   j. $\sqrt{x-y}$

4. Tell which of the following are true and which are false.
   a. $(1.2^3)^2 = (1.2)^5$
   b. $\sqrt[3]{2} = 2^{-3}$
   c. $\left(\dfrac{3}{5}\right)^3 = \dfrac{9}{15}$
   d. $\left(\dfrac{1}{2}\right)^{-1} = 2$
   e. $\sqrt{\dfrac{3}{5}} = \dfrac{\sqrt{15}}{5}$
   f. $\sqrt[5]{-32} = -2$
   g. $\sqrt[3]{-8} = 2$
   h. $(16)^{\frac{1}{4}} = 2$
   i. $(2^2)^2 = 8$
   j. $\left(\dfrac{2}{3}\right)^3 = \dfrac{8}{27}$

k. $\sqrt{45} = 3\sqrt{15}$
l. $\sqrt[3]{27} = 9$
m. $(16)^{\frac{5}{4}} = 64$
n. $\sqrt{6} \times \sqrt{3} = 3\sqrt{2}$
o. $\dfrac{\sqrt{56}}{\sqrt{14}} = 2$
p. $\sqrt{27} = 3\sqrt{3}$
q. $5^{\frac{2}{3}} = \sqrt{5^3}$
r. $(-2)^{\frac{1}{2}} = \sqrt{2}$
s. $\left(\dfrac{1}{2}\right)^{-2} = 4$
t. $\dfrac{\sqrt{39}}{\sqrt{3}} = \sqrt{13}$
u. $\sqrt{72} = 6\sqrt{2}$
v. $\dfrac{2}{\sqrt{2}} = \dfrac{2\sqrt{2}}{2} = \sqrt{2}$
w. $\dfrac{4}{\sqrt{7}} = \dfrac{4\sqrt{7}}{7}$
x. $\sqrt{\dfrac{7}{9}} = \dfrac{3}{\sqrt{7}}$

5. Give equivalent expressions containing radical signs.
   a. $r^{\frac{2}{3}}$
   b. $[(2x)^2]^{\frac{1}{3}}$
   c. $\left(\dfrac{3t}{2}\right)^{\frac{4}{3}}$
   d. $(-5x)^{-\frac{1}{3}}$

6. Rationalize each denominator and simplify wherever possible.
   a. $\dfrac{3}{\sqrt{2}}$
   b. $\dfrac{5}{\sqrt{7}}$
   c. $\dfrac{3}{\sqrt{2}} \times \dfrac{5}{\sqrt{3}}$
   d. $\dfrac{\sqrt{2}}{\sqrt{5} \times \sqrt{8}}$
   e. $\dfrac{\sqrt{5} \times \sqrt{15}}{\sqrt{30} \times \sqrt{6}}$
   f. $\dfrac{\sqrt{75}}{\sqrt{125}}$

7. Simplify.
   a. $\dfrac{36xy^3z^2}{\frac{1}{3}y^{-2}z^2}$
   b. $\dfrac{7x^5}{x^{-3}}$
   c. $(81x^3y^4)^{\frac{1}{2}}$
   d. $r^{\frac{3}{2}} \times r^{\frac{2}{3}}$
   e. $\dfrac{x^{-6}}{\frac{1}{x^2}}$
   f. $\dfrac{r}{r^{-2}}$

8. Determine the real-number solution set.
   a. $\sqrt{x} = 1$
   b. $\sqrt{x} = 3$
   c. $3\sqrt{x} = 12$
   d. $\sqrt{x} = -5$
   e. $\sqrt{2x} = 1$
   f. $\sqrt{3y - 2} = -1$
   g. $\sqrt{z - 10} = 6$
   h. $\sqrt{n + 1} = \sqrt{3n - 7}$
   i. $\sqrt{\dfrac{2 - 3m}{3}} = \sqrt{\dfrac{m + 1}{4}}$
   j. $\sqrt{\dfrac{2}{3t - 1}} = \sqrt{\dfrac{3}{t + 4}}$

13

Polynomials; Factoring

13.1 Multiplication of Two Binomials

The distributive and commutative properties are two key properties on which much of the work with polynomials is based. To illustrate how these properties are used in multiplying the binomials $x + 2$ and $x + 5$, study the following development.

$$\boxed{(x + 2)}\ (x + 5) = \boxed{(x + 2)} \times x + \boxed{(x + 2)} \times 5$$
$$= x^2 + 2x + 5x + 10$$
$$= x^2 + 7x + 10.$$

Here are some more examples for your study.

Example 1 $\boxed{(x - 6)}\ (x + 4) = \boxed{(x - 6)} \times x + \boxed{(x - 6)} \times 4$
$$= x^2 - 6x + 4x - 24$$
$$= x^2 - 2x - 24.$$

Example 2 $\boxed{(3x - 7)}\ (2x - 5) = \boxed{(3x - 7)} \times 2x + \boxed{(3x - 7)} \times (-5)$
$$= 6x^2 - 14x - 15x + 35$$
$$= 6x^2 - 29x + 35.$$

Example 3 $\boxed{(2x+3y)}\ (3x-4y)$

$= \boxed{(2x+3y)} \times 3x + \boxed{(2x+3y)} \times (-4y)$
$= (2x) \times (3x) + (3y) \times (3x) + (2x) \times (-4y) + (3y) \times (-4y)$
$= 6x^2 + 9xy - 8xy - 12y^2$
$= 6x^2 + xy - 12y^2.$

13.2 Multiplication of Binomials of the Form x + a

If you examine thoughtfully the examples in the preceding section, you should discover a pattern. For example, in the first example we used, we found that $(x+2)(x+5)$ and $x^2 + 7x + 10$ are equivalent expressions. We refer to $x+2$ and $x+5$ as *factors* of $x^2 + 7x + 10$, because

$$(x+2)(x+5) = x^2 + 7x + 10.$$

This is similar to the situation in arithmetic where we call 5 and 3 factors of 15, because $5 \times 3 = 15$.

The product of $x+2$ and $x+5$ is shown below with arrows drawn to indicate how the coefficients in the product, $x^2 + 7x + 10$, are related to the coefficients of the factors, $x+2$ and $x+5$.

$$2 \times 5 = 10$$
$$(x+2)(x+5) = x^2 + 7x + 10$$
$$2 + 5 = 7$$

The same is done for the product of $y-3$ and $y+6$. For convenience we use $y + (-3)$ as an equivalent expression to $y-3$.

$$-3 \times 6 = -18$$
$$[y+(-3)](y+6) = y^2 + 3y - 18$$
$$-3 + 6 = 3$$

Here is one more example showing that the same pattern holds.

$$(-2) \times (-5) = 10$$
$$[z+(-2)][z+(-5)] = z^2 - 7z + 10$$
$$(-2) + (-5) = -7$$

It appears that the pattern

$$(x+a)(x+b) = x^2 + (a+b)x + a \times b$$

holds for all replacements of a and b by real-number names. We now state the pattern as a theorem.

13.2 Multiplication of Binomials of the Form $x + a$

THEOREM 13.1 $(x + a)(x + b) = x^2 + (a + b)x + a \times b$.

Proof. $\boxed{(x + a)} \ (x + b) = \boxed{(x + a)} \times x + \boxed{(x + a)} \times b$
$= x^2 + a \times x + x \times b + a \times b$
$= x^2 + a \times x + b \times x + a \times b$
$= x^2 + (a + b)x + a \times b$.

The following examples show a direct application of Theorem 13.1 when multiplying two binomials of the form $x + a$.

Example 1 $(x + 3)(x + 39) = x^2 + (3 + 39) \times x + 3 \times 39$
$= x^2 + 42x + 117$.

Example 2 $(x - 15)(x + 20) = [x + (-15)](x + 20)$
$= x^2 + (-15 + 20)x + (-15) \times 20$
$= x^2 + 5x - 300$.

Example 3 $(x - 7)(x - 9) = [x + (-7)][x + (-9)]$
$= x^2 + [-7 + (-9)]x + (-7) \times (-9)$
$= x^2 + (-16)x + 63$
$= x^2 - 16x + 63$.

Theorem 13.1 enables us to write down quickly the product of two factors of the form $x + a$. Streamlining this multiplication still further, we can write Examples 2 and 3 as follows:

$(x - 15)(x + 20) = x^2 + (-15 + 20)x - 15 \times 20$
$= x^2 + 5x - 300;$

$(x - 7)(x - 9) = x^2 - (7 + 9)x + 7 \times 9$
$= x^2 - 16x + 63.$

EXERCISE 13.1

1. Determine the products, using the distributive properties.
   - a. $(x + 4)(x + 2)$
   - b. $(y + 7)(y + 9)$
   - c. $(x - 2)(x + 4)$
   - d. $(v - 3)(v - 6)$
   - e. $(y + 7)(y - 3)$
   - f. $(z - 8)(z + 5)$
   - g. $(x - 3)(x + 4)$
   - h. $(2x + 3)(x + 4)$
   - i. $(3x + 5)(x + 2)$
   - j. $(x + 7)(4x + 5)$
   - k. $(6y + 4)(3y - 1)$
   - l. $(5t - 2)(3t + 1)$
   - m. $(3r - 4)(2r - 3)$
   - n. $(7t - 3)(6t + 4)$

o. $(5x - 2)(x + 1)$
p. $(2x + 4)(x - 3)$
q. $(3x - 1)(x + 2)$
r. $(x - 5)(2x + 3)$
s. $(2x - 1)(3x + 4)$
t. $(6x - 7)(x + 3)$
u. $(2x + 2)(2x + 2)$
v. $(3x - 2)(3x + 2)$
w. $(5x + 2)(5x + 2)$
x. $(6x + 3)(6x - 3)$
y. $(-5x - 1)(x - 3)$
z. $(3x + 3)(3x - 3)$
a'. $(-4x + 1)(2x - 1)$
b'. $(3x + 7)(x - 1)$
c'. $(9x + 7)(9x - 7)$
d'. $(-2x + 3)(-2x - 3)$
e'. $(2x^2 + 1)(3x^2 + 4)$
f'. $(4x^2 - 3)(5x^2 + 5)$
g'. $(6x^2 - 1)(7x^2 - 1)$
h'. $(3 - 2a^2)(3a^2 - 1)$
i'. $(4 - 6x^2)(1 - 5x^2)$
j'. $(3 - 5c^2)(-2 - 3c^2)$
k'. $(2x + y)(3x + 2y)$
l'. $(3x - y)(4x + 3y)$
m'. $(5x - 2y)(5x + 2y)$
n'. $(3x - 7y)(3x - 7y)$
o'. $(5x + 4a)(4x - a)$
p'. $(-2n + 3m)(-5n - 6m)$

2. Determine the products by the use of Theorem 13.1.
   a. $(x + 1)(x + 2)$
   b. $(u + 3)(u + 6)$
   c. $(n + 7)(n + 11)$
   d. $(y + 2)(y + 25)$
   e. $(c + 12)(c + 8)$
   f. $(a + 9)(a + 9)$
   g. $(c - 1)(c + 3)$
   h. $(m - 6)(m + 15)$
   i. $(p + 5)(p - 4)$
   j. $(r + 2)(r - 50)$
   k. $(m - 2)(m - 3)$
   l. $(d - 3)(d - 20)$
   m. $(n - 11)(n - 10)$
   n. $(r - 2)(r - 105)$
   o. $(y - 5)(y - 5)$
   p. $(t - 2)(t - 2)$
   q. $(w - 5)(w + 5)$
   r. $(a + 10)(a - 10)$
   s. $(s - 12)(s + 12)$
   t. $(h - 20)(h + 20)$
   u. $(k - 20)(k - 20)$
   v. $(w + 20)(w + 20)$
   w. $(x - 1.5)(x + 3)$
   x. $(t + 2.5)(t + 4)$
   y. $(u - \frac{1}{4})(u - \frac{1}{2})$
   z. $(r - 2.6)(r - 2.5)$

3. Prove: $(x - a)(x + b) = x^2 + (-a + b)x - ab$.
4. Prove: $(x - a)(x - b) = x^2 - (a + b)x + ab$.
5. Determine the products, using the distributive properties.
   a. $(x^2 + 2)(4x^2 + 1)$
   b. $(2x^2 + 1)(4x^2 + 3)$
   c. $(3x^2 - 2)(5x + 2)$
   d. $(4x^2 - 1)(4x^2 + 1)$
   e. $(9x^2 + 4)(9x^2 - 4)$
   f. $(2x^2 - 5)(2x^2 - 5)$

13.3 Special Products

g. $(3x + y)(4x + y)$
h. $(4a - b)(3a + b)$
i. $(6a^2 - c)(4a + c)$
j. $(3x^2 - n)(4x - n)$
k. $(9x + a^2)(3x - a)$
l. $(-m^2 + 3x)(m + 5)$
m. $(x + a + b)(x - 3)$
n. $(2x - c - m)(x^2 + 5)$
o. $(x^2 + 2x + 3)(x - 4)$
p. $(3n^2 + x + a)(x + 2)$
q. $(y^2 + y + 3)(y - 1)$
r. $(4m^2 + n + c)(c + 3)$

13.3 Special Products

We have learned to multiply two binomials by using the distributive property. There are some binomials which, when multiplied, reveal a special pattern. The following examples illustrate this.

Example 1 $(x + 3)^2 = (x + 3)(x + 3)$
$= x^2 + (3 + 3)x + 3 \times 3$
$= x^2 + 6x + 9.$

Example 2 $(5n + 6)^2 = (5n + 6) \times (5n + 6)$
$= (5n + 6) \times (5n) + (5n + 6) \times 6$
$= 25n^2 + 30n + 30n + 36$
$= 25n^2 + 60n + 36.$

The pattern displayed in the examples above is the following. For all polynomials P and Q,

$$(P + Q)^2 = P^2 + 2PQ + Q.$$

Notice that we use P and Q as variables which can be replaced by polynomials. In the preceding examples we used the following replacements.

In Example 1, P was replaced by x and Q was replaced by 3.

In Example 2, P was replaced by $5n$ and Q was replaced by 6.

Replacing P by $2y^2$ and Q by 3, we would obtain

$(2y^2 + 3)^2 = (2y^2)^2 + 2 \times (2y^2) \times 3 + 3^2$
$= 4y^4 + 12y^2 + 9.$

Now consider the following examples of squaring binomials where subtraction is involved. We use the definition of subtraction to change to addition. Then we apply the pattern for the square of a sum.

Example 1 $(x-3)^2 = [x+(-3)]^2$
$= x^2 + 2x(-3) + (-3)^2$
$= x^2 - 6x + 9.$

Example 2 $(4x-5y)^2 = [4x+(-5y)]^2$
$= (4x)^2 + 2(4x)(-5y) + (-5y)^2$
$= 16x^2 - 40xy + 25y^2.$

The following is the pattern displayed by the examples above. For all polynomials P and Q,

$$(P-Q)^2 = P^2 - 2PQ + Q^2.$$

This pattern can actually be derived from the pattern for the sum as follows:

$$(P-Q)^2 = [P+(-Q)]^2$$
$$= P^2 + 2P(-Q) + (-Q)^2$$
$$= P^2 - 2PQ + Q^2.$$

We have stated that P and Q can be replaced by polynomials. The following example shows the application of both patterns, where P is replaced by $x - y$ and Q by 4.

$$[(x-y)+4]^2 = (x-y)^2 + 2 \times (x-y) \times 4 + 4^2$$
$$= x^2 - 2xy + y^2 + 8x - 8y + 16.$$

Now we give two examples which reveal another pattern.

Example 1 $(x+2)(x-2) = (x+2) \times x + (x+2) \times (-2)$
$= x^2 + 2x - 2x - 4$
$= x^2 - 4.$

Example 2 $(3x-2)(3x+2) = (3x-2) \times 3x + (3x-2) \times 2$
$= 9x^2 - 6x + 6x - 4$
$= 9x^2 - 4.$

The two examples show the following pattern. For all polynomials P and Q,

$$(P+Q)(P-Q) = P^2 - Q^2.$$

We can apply this pattern directly and obtain products quickly, as illustrated below.

13.3 Special Products

Example $[3x + (2a + 1)][3x - (2a + 1)] = (3x)^2 - (2a + 1)^2$
$= 9x^2 - (4a^2 + 4a + 1)$
$= 9x^2 - 4a^2 - 4a - 1.$

EXERCISE 13.2

1. Tell which of the following are true for all real-number replacements of the variables.
   a. $(x + 1)^2 = x^2 + x + 1$
   b. $(2x + 1)^2 = x^2 + 4x + 1$
   c. $(3x - 2)^2 = 9x^2 - 6x + 4$
   d. $(2x - 3)^2 = 4x^2 - 12x - 9$
   e. $(4x - 1)^2 = 16x^2 - 8x + 1$
   f. $(x + 2)(x - 2) = x^2 - 4$
   g. $(2x + 1)(2x - 1) = 2x^2 - 1$
   h. $(3x + 2)(3x - 2) = 9x^2 + 4$
   i. $[(x + 1) + a][(x + 1) - a] = (x + 1)^2 - a^2$
   j. $[(x + y) + (a + b)][(x + y) - (a + b)] = (x + y)^2 - (a + b)^2$

2. Determine the products.
   a. $(2x + 5)(2x - 5)$
   b. $(3c + 2)^2$
   c. $(5k - 1)^2$
   d. $(c - n)(c + n)$
   e. $(n^2 + 1)(n^2 - 1)$
   f. $(-4 + x)(4 + x)$
   g. $(-3 + 4x)^2$
   h. $(m^2 + c)(m^2 - c)$
   i. $(5x + 3y)^2$
   j. $(2x - 5y)^2$
   k. $(2c - k)(2c + k)$
   l. $(x + \sqrt{2})(x - \sqrt{2})$
   m. $(4n - \sqrt{5})(4n + \sqrt{5})$
   n. $(x + \sqrt{3})^2$
   o. $(2y - \sqrt{5})^2$
   p. $(\sqrt{5} - \sqrt{3})(\sqrt{5} + \sqrt{3})$
   q. $[(x+y)+z] \times [(x+y)+z]$
   r. $[(x+y)+z] \times [(x+y)-z]$
   s. $[a+(b+c)] \times [a-(b+c)]$
   t. $[(3x + 1) - a]^2$
   u. $[x + (a + b)]^2$
   v. $[x - (a + b)]^2$
   w. $[(a+b)+c] \times [(a+b)-c]$
   x. $[(a + b) + (c + d)]^2$
   y. $[(a + b) - (c + d)]^2$
   z. $[(a + b) + (c + d)] \times [(a + b) - (c + d)]$

3. Multiply and simplify.
   a. $(x + 1)^2 + (x + 1)(x - 2)$

b. $(3a - 2)^2 + (2a + 1)^2$
c. $(3x + 1)(3x - 1) - (2x + 1)^2$
d. $(5c + 3)^2 + (5c - 3)^2$

13.4 Factoring Trinomials of the Form $x^2 + px + q$

Theorem 13.1, which states that

$$(x + a)(x + b) = x^2 + (a + b)x + a \times b,$$

is quite helpful in factoring trinomials. Our problem now will be to start with a trinomial and find two factors whose product is that trinomial. That is, we start with

$$x^2 + (a + b)x + a \times b$$

and will have to find

$$(x + a) \quad \text{and} \quad (x + b)$$

such that

$$x^2 + (a + b)x + a \times b = (x + a)(x + b).$$

Suppose we have the trinomial

$$x^2 + 13x + 22.$$

To find its factors, the problem reduces to finding replacements for a and b for which

$$a + b = 13 \quad \text{and} \quad a \times b = 22.$$

To do this, we shall resort mostly to trial-and-error and common sense. Some intelligent guessing will help. What two integers have the sum 13 and product 22? Answer: 11 and 2. Thus,

$$x^2 + 13x + 22 = (x + 11)(x + 2).$$

Example 1 What are the factors of $m^2 - 5m + 4$?
   To answer this, we must answer the question, "What two numbers have the sum -5 and the product 4?"
   Answer: -1 and -4.
   Therefore, $m^2 - 5m + 4 = (m - 1)(m - 4)$.

Example 2 Factor $y^2 + 10y - 24$.
   We need to answer the question, "What two numbers have the sum 10 and the product -24?
   Answer: -2 and 12.
   Therefore, $y^2 + 10y - 24 = (y - 2)(y + 12)$.

13.4 Factoring Trinomials of the Form $x^2 + px + q$

Example 3 Factor $m^2 - 6m - 7$.
What two numbers have the sum -6 and the product -7?
Answer: 1 and -7.
Therefore, $m^2 - 6m - 7 = (m + 1)(m - 7)$.

Example 4 Factor $s^2 - \frac{1}{6}s - \frac{1}{6}$.
What two numbers have the sum $-\frac{1}{6}$ and the product $-\frac{1}{6}$? (It takes a little more searching here!)
Answer: $-\frac{1}{2}$ and $\frac{1}{3}$.
Therefore, $s^2 - \frac{1}{6}s - \frac{1}{6} = (s - \frac{1}{2})(s + \frac{1}{3})$.

Example 5 Factor $n^4 + 7n^2 + 12$.
At first glance this does not appear to be of the form $x^2 + px + q$. But writing the trinomial as $(n^2)^2 + 7(n^2) + 12$ reveals that it is of this form. Now we seek two numbers whose sum is 7 and whose product is 12. They are 3 and 4.
Therefore, $n^4 + 7n^2 + 12 = (n^2 + 3)(n^2 + 4)$.

Example 6 Factor $m^2n^2 + 6mn + 8$.
Observe that $(mn)^2 + 6(mn) + 8$ is equivalent to the given trinomial.
Verify that $m^2n^2 + 6mn + 8 = (mn + 2)(mn + 4)$.

The special products which we considered in the previous section are helpful in factoring some polynomials. Recall that

$$(P + Q)^2 = P^2 + 2PQ + Q^2$$
$$(P - Q)^2 = P^2 - 2PQ + Q^2$$
$$(P + Q)(P - Q) = P^2 - Q^2.$$

The following examples illustrate the use of these patterns in factoring.

Example 1 Factor $x^2 + 8x + 16$.
An equivalent expression is

$$x^2 + 2 \times 4x + 4^2.$$

Thus, $x^2 + 8x + 16 = (x + 4)^2$.

Example 2 Factor $m^2 - 6m + 9$.
An equivalent expression is

$$m^2 - 2 \times 3m + 3^2.$$

Thus, $m^2 - 6m + 9 = (m - 3)^2$.

Example 3 Factor $9x^2 - 16y^2$.

An equivalent expression is

$$(3x)^2 - (4y)^2.$$

Thus, $9x^2 - 16y^2 = (3x + 4y)(3x - 4y)$.

EXERCISE 13.3

1. Tell which of the following are correct factorizations and which are not.

   a. $x^2 - 1 = (x + 1)(x - 1)$
   b. $x^2 - 2 = (x + 2)(x - 2)$
   c. $x^2 + 3x + 1 = (x + 1)(x + 2)$
   d. $x^4 - 1 = (x^2 - 1)(x^2 + 1)$
   e. $x^2 - 4x - 3 = (x + 1)(x - 3)$
   f. $x^2 - 8x + 7 = (x - 1)(x - 7)$
   g. $x^4 + 7x^2 + 12x = (x^2 + 3)(x^2 + 4)$
   h. $x^9 - 1 = (x^3 - 1)(x^3 + 1)$
   i. $x^2 + 6x + 9 = (x + 3)(x + 3)$
   j. $x^2 - 2x + 1 = (x - 1)(x + 1)$

2. Factor each of the following.

   a. $x^2 + 5x + 6$ **b.** $p^2 + 10p + 16$
   c. $z^2 + 15z + 36$ **d.** $a^2 + 10a + 9$
   e. $n^2 + 9n + 14$ **f.** $s^2 + 16s + 15$
   g. $t^2 + 2t - 3$ **h.** $x^2 + x - 30$
   i. $r^2 + r - 12$ **j.** $k^2 - 15k + 56$
   k. $x^2 - 5x - 14$ **l.** $y^2 + 14y + 48$
   m. $z^2 - 7z - 44$ **n.** $t^2 + 8t - 9$
   o. $r^2 - 14r + 48$ **p.** $x^2 + 14x + 49$
   q. $y^2 + 6y - 16$ **r.** $t^2 - 6t - 27$
   s. $b^2 + b - 72$ **t.** $c^2 + 3c - 10$
   u. $u^2 + 4u - 45$ **v.** $m^2 - m - 6$
   w. $n^2 - 3n - 4$ **x.** $p^2 - 5p - 14$
   y. $v^2 - 5v - 50$ **z.** $d^2 - 5d - 24$
   a'. $c^2 - 9c - 10$ **b'.** $e^2 - 3e + 2$
   c'. $g^2 - 15g + 14$ **d'.** $h^2 - 15h + 56$
   e'. $w^2 - 13w + 22$ **f'.** $k^2 - 17k + 30$

g'. $m^2 - 6m + 9$
h'. $n^2 - 10n + 25$
i'. $y^2 - 8y + 16$
j'. $r^2 + 6r + 9$
k'. $t^2 + 10t + 21$
l'. $n^2 + 8n - 9$
m'. $z^2 - 5z - 24$
n'. $t^2 - 4t + 4$
o'. $x^2 + 10x + 25$
p'. $x^2 + 8x - 33$
q'. $r^2 - 7r + 6$
r'. $r^2 - 8r + 16$
s'. $x^2 + x - 12$
t'. $z^2 - z - 30$
u'. $k^2 + k - 56$
v'. $r^2 - 49$
w'. $s^2 - 4s - 21$
x'. $t^2 - 36$
y'. $x^2 - 7x - 18$
z'. $y^2 - 25$
a''. $r^2 - 100$
b''. $a^2 - 625$
c''. $t^2 - 196$
d''. $h^2 - 289$
e''. $n^2 - 144$
f''. $n^2 - 1225$
g''. $n^2 - 2025$
h''. $s^2 - 2500$
i''. $x^2 + 10x + 25$
j''. $a^2 - 4a + 4$
k''. $m^2 + 22m + 121$
l''. $k^2 + 30k + 225$
m''. $s^2 + 24s + 144$
n''. $h^2 - 16h + 64$
o''. $r^2 - 28r + 196$
p''. $u^2 + 40u + 400$
q''. $b^2 - 50b + 625$
r''. $c^2 + 60c + 900$
s''. $4x^2 - 25y^2$
t''. $9a^2 - 100b^2$
u''. $81m^2 - 121n^2$
v''. $.01p^2 - .04r^2$
w''. $.09u^2 - .16v^2$
x''. $\frac{1}{4}t^2 - \frac{1}{9}s^2$
y''. $\frac{4}{9}c^2 - \frac{9}{16}d^2$
z''. $\frac{16}{9}r^2 - \frac{1}{25}d^2$

13.5 Factoring Trinomials of the Form $ax^2 + bx + c$

In order to learn to factor polynomials like $3x^2 + 13x + 4$, let us further examine multiplication of binomials. Observe the following scheme, where the arrows draw our attention to the relationship between coefficients.

$$3x \times x = 3x^2$$
$$(3x + 1)(x + 4) = 3x^2 + 13x + 4$$
$$1 \times 4 = 4$$

The diagram shows how we obtain $3x^2$ and 4 in the product. Now examine the following diagram.

$$(3x + 1) \quad (x + 4) = 3x^2 + 13x + 4$$
$$1 \times x = x \quad x + 12x = 13x$$
$$3x \times 4 = 12x$$

Now, if we are faced with the problem of factoring $3x^2 + 13x + 4$, we shall look for

(1) two monomials whose product is $3x^2$ [$3x$ and x],
(2) two numbers whose product is 4,
(3) but the two numbers must be such that one of them multiplied by x plus the other multiplied by $3x$ results in $13x$ [1 and 4, since $1 \times x + 4 \times 3x = 13x$].

Before doing some examples, let us agree that the coefficients a, b, and c in $ax^2 + bx + c$ are integers and that resulting binomial factors have integers for their coefficients. Thus, we will consider *factoring over the set of integers*. This is the case with the example $3x^2 + 13x + 4 = (3x + 1)(x + 4)$. The coefficients in the trinomial are 3, 13, and 4, which are integers. The coefficients in the first binomial factor $(3x + 1)$ are 3 and 1; and the coefficients in the second binomial factor $(x + 4)$ are 1 and 4. These are also integers.

Example 1 Factor $3y^2 + 10y + 8$.

Two monomials with integral coefficients whose product is $3y^2$ are $3y$ and y. Let us write down parts of the binomial factors:

$$(3y + ?)(y + ?).$$

Now what two integers will give 8 for their product? Let us try 1 and 8.

$$(3y + 1)(y + 8) = 3y^2 + 25y + 8$$
We need 10 here.

Let us try the factors $(3y + 8)$ and $(y + 1)$.

$$(3y + 8)(y + 1) = 3y^2 + 11y + 8$$
We need 10 here.

So, 1 and 8 are not the numbers we are seeking.

Let us try 2 and 4.

$$(3y + 2)(y + 4) = 3y^2 + 14y + 8$$
We need 10 here.

13.5 Factoring Trinomials of the Form $ax^2 + bx + c$

What about the factors $(3y + 4)$ and $(y + 2)$?

$$(3y + 4)(y + 2) = 3y^2 + 10y + 8.$$

We have finally succeeded in obtaining the factors we need.

Example 2 Factor $6x^2 + 7x + 2$.

After trying several alternatives on the side, we find that

$$6x^2 + 7x + 2 = (2x + 1)(3x + 2).$$

We verify the factorization by performing multiplication.

$$\begin{aligned}(2x + 1)(3x + 2) &= 6x^2 + 4x + 3x + 2 \\ &= 6x^2 + 7x + 2.\end{aligned}$$

Example 3 Factor $15a^2 + 7a - 2$.

Several tries lead us to see that

$$15a^2 + 7a - 2 = (5a - 1)(3a + 2).$$

To check, we multiply.

$$\begin{aligned}(5a - 1)(3a + 2) &= 15a^2 + 10a - 3a - 2 \\ &= 15a^2 + 7a - 2.\end{aligned}$$

Example 4 Factor $9m^2 - 30m + 24$.

We find that

$$9m^2 - 30m + 24 = (3m - 4)(3m - 6).$$

To check:

$$\begin{aligned}(3m - 4)(3m - 6) &= 9m^2 - 18m - 12m + 24 \\ &= 9m^2 - 30m + 24.\end{aligned}$$

Example 5 Factor $-6x^2 + 14x - 4$.

A few tries lead to the conclusion that $-6x^2 + 14x - 4$ can be factored as

$$(-3x + 1)(2x - 4)$$

or

$$(3x - 1)(-2x + 4).$$

To check:

$$\begin{aligned}(3x - 1)(-2x + 4) &= -6x^2 + 12x + 2x - 4 \\ &= -6x^2 + 14x - 4.\end{aligned}$$

Some trinomials of degrees higher than 2 can be factored by the same methods because they pattern like quadratic binomials. Consider the following two examples.

Example 1   Factor $12x^4 + 13x^2 + 3$.
Replacing x^2 by y, we obtain
$$12y^2 + 13y + 3.$$
Its factorization is
$$(4y + 3)(3y + 1).$$
Now replacing y by x^2, we have
$$12x^4 + 13x^2 + 3 = (4x^2 + 3)(3x^2 + 1).$$

Example 2   Factor $6x^6 - x^3 - 2$.
Since $x^6 = (x^3)^2$, we may replace x^3 by y and obtain
$$6y^2 - y - 2.$$
Its factorization is
$$(2y + 1)(3y - 2).$$
Now replacing y by x^3, we have
$$6x^6 - x^3 - 2 = (2x^3 + 1)(3x^3 - 2).$$

In all of the examples given so far, we have considered only trinomials with one variable. The following example illustrates a method of factoring trinomials with more than one variable.

Example 3   Factor $6x^2 + xy - 2y^2$.
We can obtain $6x^2$ by multiplying, for example, x by $6x$. And we can obtain $-2y^2$ by multiplying $-y$ by $2y$. Let us try the two factors, $x - y$ and $6x + 2y$:
$$(x - y)(6x + 2y) = 6x^2 - 6xy + 2xy - 2y^2$$
$$= 6x^2 - 4xy - 2y^2.$$
We did not obtain the desired middle term. Another way to factor $6x^2$ over the set of integers is by using the factors $2x$ and $3x$. Let us try $2x - y$ and $3x + 2y$:
$$(2x - y)(3x + 2y) = 6x^2 - 3xy + 4xy - 2y^2$$
$$= 6x^2 + xy - 2y^2.$$
Thus, we have the desired factorization.

13.5 Factoring Trinomials of the Form $ax^2 + bx + c$

EXERCISE 13.4

1. Factor each trinomial over the set of integers.
   - a. $2n^2 + 5n + 2$
   - b. $3a^2 + 5a + 2$
   - c. $5x^2 + 11x + 2$
   - d. $3s^2 - 8s + 4$
   - e. $2t^2 + 11t + 12$
   - f. $6m^2 + 11m + 5$
   - g. $5u^2 + 6u + 1$
   - h. $6x^2 + 5x + 1$
   - i. $6c^2 + 11c + 3$
   - j. $6r^2 + 13r + 6$
   - k. $10v^2 + 17v + 3$
   - l. $9d^2 + 46d + 5$
   - m. $12h^2 + 13h + 3$
   - n. $21k^2 + 10k + 1$
   - o. $15p^2 + 11p + 2$
   - p. $2x^2 + 3x - 2$
   - q. $3a^2 + 20a - 7$
   - r. $4m^2 + 13m - 12$
   - s. $5r^2 + 11r - 12$
   - t. $6s^2 + 29s - 5$
   - u. $6t^2 + t - 2$
   - v. $3x^2 - 14x - 5$
   - w. $5n^2 - 19n - 4$
   - x. $6u^2 - 11u - 7$
   - y. $6y^2 - 13y - 5$
   - z. $2a^2 + 5a - 3$
   - a'. $13m^2 + 11m - 2$
   - b'. $6d^2 - 5d - 21$
   - c'. $10x^2 + 19x - 15$
   - d'. $20n^2 + n - 12$
   - e'. $2k^2 - 5k + 2$
   - f'. $5h^2 - 17h + 6$
   - g'. $7s^2 - 31s + 12$
   - h'. $6v^2 - 7v + 2$
   - i'. $21y^2 - 29y + 10$
   - j'. $10p^2 - 27p + 18$
   - k'. $12r^2 - 13r + 3$
   - l'. $20a^2 - 56a + 15$
   - m'. $6x^2 - 31x + 40$
   - n'. $6u^2 - 23u + 7$
   - o'. $15v^2 - 51v + 18$
   - p'. $6n^2 - 35n + 50$
   - q'. $-x^2 - 5x + 14$
   - r'. $-3m^2 + 7m - 2$
   - s'. $-2a^2 - 5a - 2$
   - t'. $-15t^2 + 11t + 12$
   - u'. $-12s^2 - 25s + 7$
   - v'. $x^2 + 11x + 24$
   - w'. $2y^2 + 15y + 7$
   - x'. $12a^2 + 25a + 12$
   - y'. $3m^2 + 20m - 7$
   - z'. $15t^2 - 26t + 8$
   - a''. $18m^2 + 61m - 7$
   - b''. $4k^2 - 32k + 55$
   - c''. $12s^2 - 37s + 28$
   - d''. $10 + 11y + y^2$
   - e''. $14c + 8c^2 - 9$
   - f''. $-2x^2 + 12 - 5x$
   - g''. $21n^2 + 21 - 58n$
   - h''. $-12y^2 + 14 - 13y$
   - i''. $3x^4 + 5x^2 + 2$
   - j''. $4x^4 + 9x^2 + 2$

k''. $6x^4 + 13x^2 - 5$ l''. $8x^4 - 14x^2 + 3$
m''. $2x^6 + 5x^3 - 3$ n''. $2x^2 + 7xy + 3y^2$
o''. $3a^2 + 5ab - 2b^2$ p''. $12m^2 + 5mn - 3n^2$
q''. $12a^2 - 23ax + 5x^2$ r''. $4u^2 - 15tu - 4t^2$

2. Factor each trinomial over the set of integers. In each case find three binomial factors.

    a. $2a^3 + 3a^2b - 2ab^2 - 3b^3$ b. $x^4 - y^4$

13.6 Complete Factorization

In factoring integers, we say that an integer is factored *completely* if it is shown as a product of prime factors only. For example,

$$2 \times 5 \times 7$$

is a *complete factorization* of 70. Note that 2, 5, and 7 are prime numbers. This factorization is also called *prime factorization*. When factoring a positive integer, we need not bother with negative factors. However, when factoring a negative integer one of the factors may have to be negative. For example, a complete factorization of -70 is $-2 \times 5 \times 7$.

We have seen that to factor a polynomial over the integers means to show it as a product of polynomials over the integers. A polynomial, of course, may have more than one such factorization. For example, five such factorizations of $4x^2 - 16y^2$ are:

(1) $(2x - 4y)(2x + 4y)$
(2) $4(x - 2y)(x + 2y)$
(3) $4(x^2 - 4y^2)$
(4) $2(2x - 4y)(x + 2y)$
(5) $2(x - 2y)(2x + 4y)$.

Notice that we do not consider such trivial factorizations as $1(4x^2 - 16y^2)$ or $-1(-4x^2 + 16y^2)$. It is easy to see that any polynomial P has the four trivial factors, $1, -1, P$ and $-P$. To describe other factors of P, we make the following definition.

DEFINITION 13.1 If P, Q, and R are nonzero polynomials over the integers such that $P \times Q = R$, and neither P nor Q is 1 or -1, then P and Q are *proper factors* of R.

For example, 3, $x - y$, and $x + y$ are proper factors of $3(x^2 - y^2)$. $3(x - y)(x + y)$ is the correct factorization of $3(x^2 - y^2)$, since

$$3(x - y)(x + y) = 3(x^2 - y^2).$$

13.6 Complete Factorization

Thus, 3, $x - y$, and $x + y$ are the proper factors of $3(x^2 - y^2)$.

If a polynomial has proper factors, then we say that it is *factorable*. For example, $5x + 10y$ is factorable over the integers. Its proper factors are 5 and $x + 2y$, since

$$5(x + 2y) = 5x + 10y.$$

Observe that $x + 2y$ is not factorable over the integers.

DEFINITION 13.2 A *prime polynomial* is a polynomial over the integers, other than 1, 0, or -1, which is nonfactorable over the integers.

DEFINITION 13.3 If P, a factorable polynomial over the integers, is written as

$$T \times U \times V \times \ldots$$

where $T, U, V, \ldots$ are prime polynomials, then P is *completely factored*. $T \times U \times V \ldots$ is a *complete*, or *prime factorization* of P.

The following examples illustrate prime factorization of polynomials.

Example 1   Factor $16x + 8y$ completely.

$$16x + 8y = 8(2x + y).$$

Therefore, a complete factorization of $16x + 8y$ is $8(2x + y)$. Notice that we do not bother to factor 8, since factorization of integers is trivial.

Note that $2(8x + 4y)$ is not a prime factorization of $16x + 8y$, because $8x + 4y$ is factorable over the integers.

$$8x + 4y = 4(2x + y).$$

Example 2   Factor $-3m + 12n$ completely.

$$-3m + 12n = -3(m - 4n)$$
$$= 3(-m + 4n).$$

Both $-3(m - 4n)$ and $3(-m + 4n)$ are complete factorizations of $-3m + 12n$. Either is acceptable.

Example 3   Factor $6x^2 + 16x - 6$ completely.

$$6x^2 + 16x - 6 = 2(3x^2 + 8x - 3)$$
$$= 2(x + 3)(3x - 1).$$

Observe that none of the following three factorizations of $6x^2 + 16x - 6$ is a complete factorization.

(1) $2(3x^2 + 8x - 3)$
(2) $(2x + 6)(3x - 1)$
(3) $(x + 3)(6x - 2)$.

Whenever you are asked to factor a polynomial over the integers, you are to factor it completely. This means that you are to give its complete, or prime factorization.

13.7 Factoring Techniques

We have had quite a bit of experience by now in factoring a variety of polynomials. Experience is the only sure way to become proficient in this skill. There are some techniques which we must be alert to. For example, in the polynomial

$$3xy + 4xz,$$

we notice that $3xy$ and $4xz$ have a *common factor*.

We can use a distributive property and obtain the following:

$$3xy + 4xz = x(3y + 4z).$$

Thus we have a binomial $3xy + 4xz$ and an expression which is equivalent to it and which is the product of x and $3y + 4z$. Therefore, x and $3y + 4z$ are factors of $3xy + 4xz$.

Here are a few more examples of the same kind. Study them thoughtfully.

Example 1 $3rt^2 + 7ts = t(3rt) + t \times 7s$
$\phantom{\text{Example 1 } 3rt^2 + 7ts } = t(3rt + 7s)$.

Neither of the factors t and $3rt + 7s$ can be factored any further and therefore $t(3rt + 7s)$ is a complete factorization of $3rt^2 + 7ts$.

Example 2 $2xy^2 + 5x^2y + 3xt = x(2y^2) + x(5xy) + x(3t)$
$\phantom{\text{Example 2 } 2xy^2 + 5x^2y + 3xt} = x(2y^2 + 5xy + 3t)$.

Example 3 $6rt + 3ts = 3t(2r) + 3t \times s$
$\phantom{\text{Example 3 } 6rt + 3ts} = 3t(2r + s)$.

Example 4 $5xy - 10x^2z = 5x \times y - 5x \times 2xz$
$\phantom{\text{Example 4 } 5xy - 10x^2z} = 5x(y - 2xz)$.

13.7 Factoring Techniques

The four examples above suggest one question you should ask yourself when factoring a polynomial: does it have a common factor?

Some polynomials may be factored by noting terms which have a common *binomial* factor.

For example, to factor the expression
$$a(x - y) + b(x - y),$$
we apply a distributive property to yield
$$a(x - y) + b(x - y) = (a + b)(x - y).$$
Thus,
$$(a + b) \text{ and } (x - y)$$
are factors of $a(x - y) + b(x - y)$.

The examples below illustrate the use of this technique in a variety of polynomials.

Example 1 Factor $mx + my + px + py$.
We observe that this expression is equivalent to
$$m(x + y) + p(x + y),$$
which in turn is equivalent to
$$(m + p)(x + y).$$
Thus, $mx + my + px + py = (m + p)(x + y)$.

Example 2 Factor $bx + by - cx - cy$.
We observe that this expression is equivalent to
$$b(x + y) - c(x + y),$$
which in turn is equivalent to
$$(b - c)(x + y).$$
Thus, $bx + by - cx - cy = (b - c)(x + y)$.

Example 3 Factor $4x^2 - 81y^2$.
Recall that the difference of squares $P^2 - Q^2$ factors as $(P - Q)(P + Q)$. In the expression above P^2 is replaced by $4x^2$ and Q^2 by $81y^2$. We have then
$$4x^2 - 81y^2 = (2x)^2 - (9y)^2$$
$$= (2x - 9y)(2x + 9y).$$
Thus, $4x^2 - 81y^2 = (2x - 9y)(2x + 9y)$.

Example 4 Factor $x^2 - 4x + 4 - 25y^2$.

We observe that the first three terms may be grouped together and viewed as a perfect square, and that the last term is also a perfect square.

$(x^2 - 4x + 4) - 25y^2$ is equivalent to $(x - 2)^2 - (5y)^2$. Thus, we have the expression in the form of the difference of two squares. Hence,

$$(x - 2)^2 - (5y)^2 = [(x - 2) - (5y)][(x - 2) + (5y)]$$
$$= (x - 5y - 2)(x + 5y - 2).$$

Sometimes quite a bit of regrouping and simplifying is needed to get an expression into the form of the difference of two squares. Study each of the following examples rather diligently, since they are somewhat more difficult.

Example 1 Factor $r^2 - x^2 + t^2 - 2rt + 2xy - y^2$.

We regroup to produce an equivalent expression:

$$(r^2 - 2rt + t^2) - x^2 + 2xy - y^2.$$

Next we obtain

$$(r^2 - 2rt + t^2) - (x^2 - 2xy + y^2).$$

Since

$$r^2 - 2rt + t^2 = (r - t)^2$$

and

$$x^2 - 2xy + y^2 = (x - y)^2,$$

we have

$$(r^2 - 2rt + t^2) - (x^2 - 2xy + y^2) = (r - t)^2 - (x - y)^2.$$

The latter is in the form of the difference of two squares $P^2 - Q^2$, where $P = r - t$ and $Q = x - y$. Since $P^2 - Q^2 = (P - Q)(P + Q)$, we have

$$(r - t)^2 - (x - y)^2 = [(r - t) - (x - y)][(r - t) + (x - y)]$$
$$= (r - t - x + y)(r - t + x - y).$$

Thus, $r^2 - x^2 + t^2 - 2rt + 2xy - y^2 = (r - t - x + y)(r - t + x - y)$.

Binomials and trinomials can frequently be factored by first finding equivalent expressions in the form of the difference of two squares.

Example 2 Factor $x^4 + x^2 + 1$.

13.7 Factoring Techniques

Note that the expression is nearly a perfect square. We know that $x^4 + 2x^2 + 1$ is a perfect square, and we are aware that zero is the only number which may be added without changing the value of the expression. Since $x^2 + (-x^2)$ is equivalent to zero, it may be added to get the equivalent expression

$$x^4 + x^2 + 1 = x^4 + x^2 + 1 + [(x^2) + (-x^2)].$$

The last expression is equivalent to

$$\begin{aligned}(x^4 + 2x^2 + 1) - x^2 &= (x^2 + 1)^2 - x^2 \\ &= [(x^2 + 1) + x][(x^2 + 1) - x] \\ &= (x^2 + x + 1)(x^2 - x + 1).\end{aligned}$$

Thus, $x^2 + x + 1$ and $x^2 - x + 1$ are factors of $x^4 + x^2 + 1$ and we have

$$x^4 + x^2 + 1 = (x^2 + x + 1)(x^2 - x + 1).$$

Example 3 Factor $x^4 - 12x^2y^2 + 16y^4$.

Observe first that

$$x^4 - 8x^2y^2 + 16y^4$$

is a perfect square, since it is equal to $(x^2 - 4y^2)^2$.

Of course, this last expression is not equivalent to the given expression. To obtain an equivalent expression we must add $-4x^2y^2$.

$$(x^4 - 8x^2y^2 + 16y^4) - (4x^2y^2) = x^4 - 12x^2y^2 + 16y^4.$$

We now have the difference of two squares,

$$(x^2 - 4y^2)^2 - (2xy)^2,$$

which is equivalent to

$$[(x^2 - 4y^2) + 2xy][(x^2 - 4y^2) - 2xy]$$

or

$$(x^2 + 2xy - 4y^2)(x^2 - 2xy - 4y^2).$$

Thus

$$x^4 - 12x^2y^2 + 16y^4 = (x^2 + 2xy - 4y^2)(x^2 - 2xy - 4y^2).$$

Sometimes we encounter expressions involving cubes which can be factored. To see this, consider the following multiplication.

$$\begin{aligned}(x + y)(x^2 - xy + y^2) &= (x + y) \times x^2 - (x + y) \times xy + (x + y) \times y^2 \\ &= x^3 + x^2y - x^2y - xy^2 + xy^2 + y^3 \\ &= x^3 + y^3.\end{aligned}$$

Thus, $x^3 + y^3$ has as its factors

$$x + y \text{ and } x^2 - xy + y^2$$

or

$$x^3 + y^3 = (x + y)(x^2 - xy + y^2).$$

Now consider another multiplication which is similar to the one above.

$$\begin{aligned}(x - y)(x^2 + xy + y^2) &= (x - y) \times x^2 + (x - y) \times xy + (x - y) \times y^2 \\ &= x^3 - x^2y + x^2y - xy^2 + xy^2 - y^3 \\ &= x^3 - y^3.\end{aligned}$$

Thus, $x^3 - y^3$ has as its factors

$$x - y \text{ and } x^2 + xy + y^2$$

or

$$x^3 - y^3 = (x - y)(x^2 + xy + y^2).$$

The last two examples are helpful in understanding how to factor expressions which are the sum or difference of two cubes. The next two examples illustrate how to factor such expressions.

Example 1 Factor $8x^3 + 27$.

We observe that $8x^3 + 27$ is of the form $P^3 + Q^3$.

$$\begin{aligned}8x^3 + 27 &= (2x)^3 + (3)^3 \\ &= (2x + 3)[(2x)^2 - (2x)(3) + (3)^2] \\ &= (2x + 3)(4x^2 - 6x + 9).\end{aligned}$$

Example 2 Factor $x^6 - y^6$.

We note that $x^6 - y^6$ is of the form $P^3 - Q^3$.

$$\begin{aligned}x^6 - y^6 &= (x^2)^3 - (y^2)^3 \\ &= (x^2 - y^2)[(x^2)^2 + (x^2)(y^2) + (y^2)^2] \\ &= (x^2 - y^2)(x^4 + x^2y^2 + y^4) \\ &= (x - y)(x + y)(x^4 + x^2y^2 + y^4).\end{aligned}$$

Any expression which is the sum or the difference of two cubes may be factored by this method.

EXERCISE 13.5

1. Tell which of the following are true and which are false.

   a. 2×11 is the prime factorization of 22.

13.7 Factoring Techniques

b. 4×8 is a factorization of 32.
c. 4×8 is the prime factorization of 32.
d. $x + y$ is a proper factor of $x^2 - y^2$.
e. $(x + y)^2$ is a proper factor of $(x + y)^3$.
f. $(x + y)^2(x + y)$ is a prime factorization of $(x + y)^3$.
g. $x^2 - y^2$ is a prime polynomial.
h. $x + y$ is a prime polynomial.
i. 1 is a factor of every polynomial.
j. $7x + 9y$ is factorable.

2. Why is $\frac{1}{2}x + y$ not a polynomial over the integers?

3. Give the prime factorization of each of the following integers.
   a. 62
   b. 42
   c. 165
   d. 51

4. Is $3\left(\frac{x}{3} + y\right)$ a factorization of $x + 3y$ over the integers? Why or why not?

5. Give the complete factorization of each of the following polynomials.
   a. $3x^2 + 3x - 6$
   b. $-8a^2 - 28a - 12$
   c. $4y^2 - 14y + 6$
   d. $x^2 - y^2 + x + y$
   e. $x^2 + xy - 2y^2 + 2x + 4y$
   f. $9m^2 - 3m - 6$
   g. $9s^2 - 9t^2$
   h. $6a^2 + 10ab - 4b^2$
   i. $a^2 - b^2 - ac - bc$
   j. $2m^2 + 5mn + 3n^2 + m + n$

6. Factor each of the following.
   a. $3rt + 6s$
   b. $4cr + 5cz$
   c. $6xt + 9ts$
   d. $3xy^2 + 5xy$
   e. $7rt + 9t^2y$
   f. $4xy + 8x^2r$
   g. $5xy + 10xz + 15x^2y$
   h. $13 + 39x$
   i. $5x^2 + 7x$
   j. $y^3 + 3y^2 + 2y$
   k. $25y^2z + 10y^3x$
   l. $2xy + 4x^2y + 6x^2y^2$
   m. $2xy + 5xt + 6t$
   n. $7x - 3xy$
   o. $14xy + 15xy^2 + 13x^2z$
   p. $5 - 15t$
   q. $6 - 9xy + 3z$
   r. $3x^3 - 5x^2 - 2xy^2$
   s. $5z^2y + 10yx^2 - 15zx$
   t. $16rt - 4tx$
   u. $8x^2y^3 - 2xy$
   v. $9rt^3 - 3rt^2$
   w. $xyz^2 - x^2yz + x^2yz^2$
   x. $3xy - 4yz^2 + 4yx^2$
   y. $9rs^2 - 6st$
   z. $14xy^2 + 7x^2y - 21x^2y$

a'. $2x^2 + 4x - 6$
c'. $4r^2 + 24r + 36$
e'. $cr + dr + ct + dt$
g'. $x^4 + 3x^2 - 4$
i'. $y^2 - 6y$
k'. $(r + s)^2 + 2(r + s) + 1$
m'. $16x^2 - 81y^2$
o'. $y^5 - 4y$
q'. $7x^2 + 42xy + 63y^2$
s'. $y(x + 2) + r(x + 2)$
u'. $9x^4 + 7x^2 + 2$
w'. $9 - 81z^4$
y'. $3x^3 + 24$
a''. $4z^2 - (x + 3)^2$
c''. $z^4 - 4z^2 - 5$
e''. $4 - x^2 - 2xy - y^2$
g''. $9x^4 + 11x^2 + 2$
i''. $x^2 + y - x - y^2$
k''. $(x^2 - 6x + 9) - z^2$
m''. $15 + (x + y) - 6(x + y)^2$
o''. $x^4 - y^4$
q''. $s^2 - (t^2 + 4t + 4)$
s''. $cz^2 - c - z^2 + 1$
u''. $9x^3 - 18x^2r + 9rx^2$

b'. $3x^2 + 3x - 90$
d'. $5m^2 + 40m - 45$
f'. $16r^2 - 9t^2 + 4r - 3t$
h'. $8y^3 - 125x^3$
j'. $x^2 - 4x + 4 - 25y^2$
l'. $x^6 + 8$
n'. $9cy^2 - d^2c$
p'. $\frac{1}{8}y^3 - \frac{1}{27}x^3$
r'. $2x^2 - 9y - 18 + yx^2$
t'. $r^2 - 2rt + t^2 - x^2 + 2xy - y^2$
v'. $r^2t - rt^2$
x'. $8y^3 - 1$
z'. $r^2st - r^3 - s^2t + rs$
b''. $x^2 + 2ry - y^2 - r^2$
d''. $c^2r - r + c^2t - t$
f''. $cr + cs - dr - ds$
h''. $r^3 + 3r - (r + 3)$
j''. $tw + vx - vw - tx$
l''. $x^2 - y^2 + x + y$
n''. $81x^4 + 64y^4$
p''. $r(y - x) + t(x - y)$
r''. $x^4 - 7x^2 - 8$
t''. $w^2 - r^2 - z^2 + 2rz$
v''. $y^8 - z^8$

13.8 Use of Factoring to Simplify Rational Expressions

The techniques of factoring which we studied in the previous section are helpful in simplifying some rational expressions. Study each of the examples which follow to see how these techniques are put to use.

Example 1 $\dfrac{6x^2y^6}{12x^3y^4} = \dfrac{(6x^2y^4)y^2}{(6x^2y^4)(2x)} = \dfrac{y^2}{2x}$ $[x \neq 0, y \neq 0]$.

The expression $\dfrac{y^2}{2x}$ is in *simplest form*. In general, we say that a ra-

13.8 Use of Factoring to Simplify Rational Expressions

tional expression is in simplest form if the numerator and denominator do not have a common factor other than 1 or -1.

Notice that we have specified that 0 for x and 0 for y are the nonpermissible values, since division by 0 is undefined.

Example 2 $\quad \dfrac{8x+12}{4x-20} = \dfrac{4(2x+3)}{4(x-5)} = \dfrac{2x+3}{x-5}.$

$\dfrac{2x+3}{x-5}$ is in simplest form. 5 for x is a nonpermissible value.

Example 3 $\quad \dfrac{6xy-21x^2}{3x^2y+9x} = \dfrac{3x(2y-7x)}{3x(xy+3)} = \dfrac{2y-7x}{xy+3}.$

0 for x is a nonpermissible value. Also all values of x and y for which $xy = -3$ are nonpermissible, since they result in 0 in the denominator.

Example 4 $\quad \dfrac{x^2-1}{2x^2+x-3} = \dfrac{(x-1)(x+1)}{(x-1)(2x+3)} = \dfrac{x+1}{2x+3}.$

The nonpermissible values of x are 1 and $-\frac{3}{2}$.

Example 5 $\quad \dfrac{2x^2-3x-2}{3x^2-5x-2} = \dfrac{(2x+1)(x-2)}{(3x+1)(x-2)} = \dfrac{2x+1}{3x+1}.$

The nonpermissible values of x are $-\frac{1}{3}$ and 2.

Example 6 $\quad \dfrac{3(2x-y)+5(2x-y)}{6x-3y} = \dfrac{(2x-y)(3+5)}{3(2x-y)} = \dfrac{8}{3}.$

The nonpermissible values of x and y are all values for which $2x-y=0$, that is, $2x=y$.

Example 7 $\quad \dfrac{x^3+y^3}{5(x+y)} = \dfrac{(x+y)(x^2-xy+y^2)}{5(x+y)} = \dfrac{x^2-xy+y^2}{5}.$

All values such that $x=-y$ are nonpermissible.

Example 8 $\quad \dfrac{x^3-y^3}{x^2-y^2} = \dfrac{(x-y)(x^2+xy+y^2)}{(x-y)(x+y)} = \dfrac{x^2+xy+y^2}{x+y}.$

All values such that $x=y$ or $x=-y$ are nonpermissible.

13.9 Rationalizing Binomial Denominators

In this section we will be concerned with the use of the distributive property in working with binomials which contain radicals. The examples below show how we may simplify certain binomials.

Example 1 $6\sqrt{3} + 8\sqrt{3} = (6 + 8)\sqrt{3}$
$= 14\sqrt{3}.$

Example 2 $3\sqrt{48} + 4\sqrt{12} = 3(4\sqrt{3}) + 4(2\sqrt{3})$
$= 12\sqrt{3} + 8\sqrt{3}$
$= (12 + 8)\sqrt{3}$
$= 20\sqrt{3}.$

In Examples 1 and 2, we were able to simplify a binomial so that it could be expressed as a monomial.

We can also use the distributive property to find products involving radicals.

Example 3 $\sqrt{3}(2 + \sqrt{3}) = \sqrt{3} \times 2 + \sqrt{3} \times \sqrt{3}$
$= 2\sqrt{3} + 3.$

Example 4 $(2 + \sqrt{3})(\sqrt{5} + 4)$
$= (2 + \sqrt{3})\sqrt{5} + (2 + \sqrt{3})4$
$= 2\sqrt{5} + (\sqrt{3} \times \sqrt{5}) + (2 \times 4) + (\sqrt{3} \times 4)$
$= 2\sqrt{5} + \sqrt{15} + 8 + 4\sqrt{3}.$

Note that binomials which contain radicals are treated like any other binomial.

Examples 5 and 6 illustrate how to square expressions involving radicals. We will use the following property:

$$(ab)^2 = a^2 \times b^2.$$

Example 5 $(2\sqrt{3})^2 = 2^2 \times (\sqrt{3})^2 = 4 \times 3 = 12.$

Example 6 $(-3\sqrt{5})^2 = (-3)^2 \times (\sqrt{5})^2 = 9 \times 5 = 45.$

We may apply the special patterns in multiplying binomials to binomials containing radicals. We will be especially concerned with the pattern

$$(P + Q)(P - Q) = P^2 - Q^2.$$

13.9 Rationalizing Binomial Denominators

The following examples show how we use this pattern.

Example 7 $(1 + \sqrt{2})(1 - \sqrt{2}) = 1^2 - (\sqrt{2})^2$
$= 1 - 2$
$= -1.$

Example 8 $(2\sqrt{3} - 3\sqrt{5})(2\sqrt{3} + 3\sqrt{5}) = (2\sqrt{3})^2 - (3\sqrt{5})^2$
$= 4 \times 3 - 9 \times 5$
$= 12 - 45$
$= -33.$

Note that in Examples 7 and 8, each product is a rational number. Once we know how to determine such products, we are able to *rationalize binomial denominators*. Examples 9 and 10 illustrate this.

Example 9 $\dfrac{5}{4 + \sqrt{7}} = \dfrac{5(4 - \sqrt{7})}{(4 + \sqrt{7})(4 - \sqrt{7})}$
$= \dfrac{5(4 - \sqrt{7})}{16 - 7}$
$= \dfrac{5(4 - \sqrt{7})}{9}.$

It is important to observe two things about this example.

(1) Multiplying both the numerator and the denominator by the same number results in another name for the same number. Therefore,
$$\dfrac{5}{4 + \sqrt{7}} = \dfrac{5(4 - \sqrt{7})}{9}.$$

(2) Multiplying $4 + \sqrt{7}$ by $4 - \sqrt{7}$ results in a rational number, namely 9. Note that each of $4 + \sqrt{7}$ and $4 - \sqrt{7}$ is an irrational number.

Example 10 $\dfrac{-4}{1 - \sqrt{5}} = \dfrac{-4(1 + \sqrt{5})}{(1 - \sqrt{5})(1 + \sqrt{5})}$
$= \dfrac{-4(1 + \sqrt{5})}{1 - 5}$
$= \dfrac{-4(1 + \sqrt{5})}{-4}$
$= 1 + \sqrt{5}.$

EXERCISE 13.6

1. Find equivalent expressions by factoring and simplifying. In each case, replacements of the variables which lead to undefined symbols are excluded.

   a. $\dfrac{3a + 6}{9a + 3}$
   b. $\dfrac{7m + 14}{14m - 7}$
   c. $\dfrac{4 - 6t}{2t - 8}$
   d. $\dfrac{2xy + 4x}{6x^2 + 8x^2y}$
   e. $\dfrac{16a^2b^2 - 4ab}{a^2b^2 + ab}$
   f. $\dfrac{27mn + 9m^2}{18m^2 + 9m}$
   g. $\dfrac{25a^2 + 5a}{5a - 10ab}$
   h. $\dfrac{7m^2n^2 - 28mn}{7mn}$
   i. $\dfrac{k^2 + 2k - 3}{2k^2 + 11k + 15}$
   j. $\dfrac{2s^2 + 11s - 6}{-2s^2 + 7s - 3}$
   k. $\dfrac{3u^2 - 13u - 10}{6u^2 + u - 2}$
   l. $\dfrac{-a^2 + 4a + 21}{-2a^2 + 15a - 7}$

2. Factor and simplify.

   a. $\dfrac{7(x - 5y) + 2(x - 5y)}{18(x - 5y)}$
   b. $\dfrac{3(5a - b) - (5a - b)}{8(5a - b)}$
   c. $\dfrac{4(x^2 - 1) - 2(1 - x^2)}{6(x^2 - 1)}$
   d. $\dfrac{x^4 + x^2 + 1}{3(x^2 - x + 1)}$
   e. $\dfrac{x^4 - 12x^2y^2 + 16y^4}{3(x^2 - 2xy - 4y^2)}$
   f. $\dfrac{3(x^2 - y^2)}{12(x + y)}$
   g. $\dfrac{4(x^3 + y^3)}{20(x - y)}$
   h. $\dfrac{-5(x^3 - y^3)}{20(x - y)}$
   i. $\dfrac{4(x^3 - y^3)}{-12(x^2 + xy + y^2)}$
   j. $\dfrac{-(x^3 + y^3)}{-(x^2 - xy + y^2)}$

3. Determine equivalent expressions as illustrated in the preceding examples.

   a. $5\sqrt{2} + 7\sqrt{2}$
   b. $16\sqrt{7} - 5\sqrt{7}$
   c. $3\sqrt{11} - 12\sqrt{11}$
   d. $3\sqrt{6} + \sqrt{6} + 7\sqrt{6}$
   e. $2\sqrt{3} + 5\sqrt{3} + 6\sqrt{7} - 2\sqrt{7}$
   f. $4\sqrt{7} + 2\sqrt{5} + 3\sqrt{7} + 4\sqrt{5}$
   g. $3\sqrt{6} - \sqrt{6}$
   h. $2\sqrt{2} + \sqrt{2}$
   i. $\sqrt{8} + \sqrt{2}$
   j. $3\sqrt{8} + 7\sqrt{2}$
   k. $\sqrt{48} + \sqrt{96}$
   l. $\sqrt{3}(\sqrt{6} - \sqrt{12})$
   m. $-\sqrt{5}(\sqrt{3} - \sqrt{5})$
   n. $\sqrt{7}(2\sqrt{7} - 3\sqrt{2})$
   o. $-2\sqrt{11}(4 - \sqrt{11})$
   p. $\sqrt{5}(\sqrt{5} + \sqrt{7})$

13.9 Rationalizing Binomial Denominators

q. $\sqrt{5}(\sqrt{5} - \sqrt{10})$
r. $\sqrt{2}\left(\dfrac{2\sqrt{5}}{\sqrt{2}} + \dfrac{2\sqrt{7}}{\sqrt{8}}\right)$
s. $\dfrac{\sqrt{20}}{\sqrt{2}}(\sqrt{10} + \sqrt{5})$
t. $\tfrac{1}{2}(4\sqrt{5} - 10\sqrt{6})$
u. $\dfrac{\sqrt{90}}{\sqrt{3}}(\sqrt{30} - \sqrt{2})$
v. $\dfrac{\sqrt{21}}{\sqrt{3}}(\sqrt{5} + \sqrt{7})$
w. $3(\sqrt{2} + 4)$
x. $-5(7 + \sqrt{2})$
y. $(3\sqrt{2})^2$
z. $(2\sqrt{3})^2$
a'. $(6\sqrt{11})^2$
b'. $(-5\sqrt{10})^2$
c'. $(-2\sqrt{14})^2$
d'. $(\tfrac{1}{4}\sqrt{8})^2$
e'. $(\tfrac{2}{5}\sqrt{2\tfrac{1}{2}})^2$
f'. $(1.2\sqrt{10})^2$
g'. $(1.1\sqrt{11})^2$
h'. $(\tfrac{2}{3}\sqrt{\tfrac{1}{3}})^2$
i'. $(2\sqrt{2})^2(3\sqrt{3})^2$
j'. $(-\sqrt{5})^2(-\sqrt{6})^2$
k'. $(-2\sqrt{3})^2(4\sqrt{7})^2$
l'. $(-\sqrt{3})^2(-5\sqrt{2})^2$
m'. $(-\sqrt{2})^2(-\tfrac{1}{2}\sqrt{5})^2$
n'. $(-\tfrac{1}{3}\sqrt{3})^2(\tfrac{1}{2}\sqrt{2})^2$

4. Compute the products.
   a. $(\sqrt{2} + \sqrt{3})(\sqrt{2} - \sqrt{3})$
   b. $(\sqrt{7} + \sqrt{3})(\sqrt{7} - \sqrt{3})$
   c. $(\sqrt{6} + \sqrt{8})(\sqrt{6} - \sqrt{8})$
   d. $(3\sqrt{5} + \sqrt{2})(3\sqrt{5} - \sqrt{2})$
   e. $(4\sqrt{7} - \sqrt{5})(4\sqrt{7} + \sqrt{5})$

5. Rationalize each denominator.
   a. $\dfrac{1}{2 + \sqrt{3}}$
   b. $\dfrac{1}{3 - \sqrt{2}}$
   c. $\dfrac{-5}{\sqrt{5} - 1}$
   d. $\dfrac{-3}{\sqrt{7} - 2}$
   e. $\dfrac{-4}{-\sqrt{5} - 8}$
   f. $\dfrac{2}{\sqrt{3} + \sqrt{2}}$
   g. $\dfrac{-3}{\sqrt{2} - \sqrt{5}}$
   h. $\dfrac{\sqrt{2}}{\sqrt{7} - \sqrt{5}}$
   i. $\dfrac{-1}{2\sqrt{2} + 2\sqrt{5}}$
   j. $\dfrac{-7}{-3\sqrt{5} + 2\sqrt{7}}$

6. Compute the products.
   a. $(5\sqrt{3} + \sqrt{10})(\sqrt{3} - 4)$
   b. $(\tfrac{1}{2}\sqrt{2} + \tfrac{1}{3}\sqrt{3})(\tfrac{1}{4}\sqrt{2} + \tfrac{1}{5}\sqrt{3})$

c. $(2\sqrt{5} + \sqrt{3})(3\sqrt{5} + 2\sqrt{2})$
d. $(\sqrt{7} - \sqrt{2})(\sqrt{3} + \sqrt{11})$
e. $(2 + \sqrt{3})^2$
f. $(3 + \sqrt{2})^2$
g. $(4 - \sqrt{3})^2$
h. $(\sqrt{7} - 2)^2$
i. $(-2 - \sqrt{3})^2$

13.10 Polynomials

We have already worked with polynomials. We know, for example, that each of the following is a polynomial.

$$2x^3 + 3x^2 - x + 3 \qquad 4x^5 - 3x^2 + x + 2$$
$$4x^4 - x^2 + 3x - 4 \qquad -x + 3$$
$$3x^2 - 5x + 1 \qquad -2$$

To make this concept precise, we state the following definitions.

DEFINITION 13.4 Any expression of the general form

$$a_r x^r + a_{r-1} x^{r-1} + \cdots + a_1 x + a_0$$

with $a_r \neq 0$ is called a *polynomial* of degree r in x.

[*Note:* $a_0, a_1, \ldots, a_r$ are read "a sub-zero, a sub-one, $\ldots, a$ sub-r."]

The letter variables $a_0, a_1, \ldots, a_r$ are replaceable by real-number names, and r is a positive integer. Of course, for the polynomial to be of degree r, a_r cannot be zero.

DEFINITION 13.5 $a_r, a_{r-1}, \ldots, a_1, a_0$ in the polynomial

$$a_r x^r + a_{r-1} x^{r-1} + \cdots + a_1 x + a_0$$

are called *coefficients*.

The degree of the polynomial

$$2x^3 + 3x^2 - x + 3$$

is three. Here a_3 is replaced by 2, a_2 by 3, a_1 by -1, and a_0 by 3.

The degree of the polynomial $4x^4 - x^2 + 3x - 4$ is four, and the replacements for the coefficients are 4 for a_4, 0 for a_3, -1 for a_2, 3 for a_1, and -4 for a_0.

The degree of the polynomial -2 is zero, since $-2 = -2x^0$.

We shall use the symbols $P(x)$, $Q(x)$, and so on as symbols for polynomials in x. If $P(x)$ stands for a polynomial in x, $P(3)$ is the number which results when we replace all occurrences of x in the polynomial with 3. For example, if

$$P(x) = 2x^3 + 3x^2 - x + 3,$$

then

$$P(3) = 2(3)^3 + 3(3)^2 - 3 + 3 = 81$$
$$P(2) = 2(2)^3 + 3(2)^2 - 2 + 3 = 29$$
$$P(-1) = 2(-1)^3 + 3(-1)^2 - (-1) + 3 = 5.$$

In adding polynomials, we make use of the properties of operations. Study the examples below to see how it is done.

Example 1 $(4x^5 - 3x^2 + x + 2) + (-x + 3)$
$$= 4x^5 - 3x^2 + x + 2 - x + 3$$
$$= 4x^5 - 3x^2 + (x - x) + (2 + 3)$$
$$= 4x^5 - 3x^2 + 5.$$

Example 2 $(3x^2 - 5x + 1) - (-x + 3)$
$$= 3x^2 - 5x + 1 + x - 3$$
$$= 3x^2 + (-5x + x) + (1 - 3)$$
$$= 3x^2 - 4x - 2.$$

In multiplying polynomials we also make use of our properties.

Example 3 $(-x + 3)(3x^2 - 5x + 1)$
$$= (-x + 3) \times 3x^2 - (-x + 3) \times 5x + (-x + 3) \times 1$$
$$= -3x^3 + 9x^2 + 5x^2 - 15x - x + 3$$
$$= -3x^3 + 14x^2 - 16x + 3.$$

13.11 Dividing Polynomials

When dividing one number by another, we can show the result in different ways. For example, in dividing 7 by 2, we can show the result as follows:

$$\tfrac{7}{2} = 3, \text{ remainder } 1 \quad \text{or} \quad 7 = 3 \times 2 + 1.$$

Here are a few more examples shown in the same way.

$$\tfrac{9}{4} = 2, \text{ remainder } 1 \quad \text{or} \quad 9 = 4 \times 2 + 1$$
$$\tfrac{11}{3} = 3, \text{ remainder } 2 \quad \text{or} \quad 11 = 3 \times 3 + 2$$

$\frac{14}{5} = 2$, remainder 4   or $14 = 5 \times 2 + 4$

$\frac{16}{3} = 5$, remainder 1   or $16 = 3 \times 5 + 1$

$\frac{17}{6} = 2$, remainder 5   or $17 = 6 \times 2 + 5$.

To show the pattern in the previous examples, we may write

$$\frac{c}{d} = q, \text{ remainder } r \quad \text{or} \quad c = dq + r.$$

Observe that it is true for each example that

$$0 \le r < d.$$

When dividing polynomials, the same pattern appears.

$$\frac{c(x)}{d(x)} = q(x), \text{ remainder } r(x)$$

or

$$c(x) = d(x) \times q(x) + r(x).$$

The degree of $r(x)$ is less than the degree of $d(x)$. Of course, $r(x)$ may be equal to 0 or be of degree 0; that is, $r(x)$ may be equal to some nonzero real number.

Before considering a method for dividing a polynomial by a polynomial, let us examine the case of division of a polynomial by a monomial. Read each step in the following example of division of $3x^3 + 2x^2 - 5x + 1$ by x.

```
     3x³ divided by x
     │       2x² divided by x
     │       │      -5x divided by x
     ↓       ↓       ↓
     3x² + 2x  -  5
 x)3x³ + 2x² - 5x + 1
   3x³ ←─────────────── 3x² multiplied by x
       2x² - 5x + 1 ←── 3x³ subtracted from 3x³ + 2x² - 5x + 1
       2x² ←─────────── 2x multiplied by x
           -5x + 1 ←─── 2x² subtracted from 2x² - 5x + 1
           -5x ←─────── -5 multiplied by x
                 1 ←─── -5x subtracted from -5x + 1.
```

Notice that 1 is the remainder. Its degree in x is 0, since $1 \times x^0 = 1 \times 1 = 1$. This is less than the degree of x, which is 1, since $x = x^1$.

We check our division by multiplying the quotient $3x^2 + 2x - 5$ by the divisor x and adding the remainder 1 to the product.

$$x(3x^2 + 2x - 5) = 3x^3 + 2x^2 - 5x.$$

13.11 Dividing Polynomials

Adding 1, we obtain $3x^3 + 2x^2 - 5x + 1$, which is the original dividend.

To divide one polynomial by another, it is important to arrange both the dividend and divisor in order of descending powers of the variable. For example, to divide $x - 12 + x^2$ by $-3 + x$, we write the dividend as $x^2 + x - 12$ and the divisor as $x - 3$.

$$
\begin{array}{r}
\overset{\overset{\displaystyle x^2 \text{ divided by } x}{\downarrow}\;\;\overset{\displaystyle 4x \text{ divided by } x}{\downarrow}}{x + 4} \\
x - 3 \overline{\smash{)}\, x^2 + x - 12} \\
\underline{x^2 - 3x}\longleftarrow x \text{ multiplied by } x - 3 \\
4x - 12 \longleftarrow x^2 - 3x \text{ subtracted from } x^2 + x - 12 \\
\underline{4x - 12}\longleftarrow 4 \text{ multiplied by } x - 3 \\
0 \longleftarrow 4x - 12 \text{ subtracted from } 4x - 12.
\end{array}
$$

The quotient is $x + 4$ and the remainder is 0. Thus, we have

$$x^2 + x - 12 = (x - 3)(x + 4).$$

To check division, we multiply.

$$(x - 3)(x + 4) = x^2 - 3x + 4x - 12$$
$$= x^2 + x - 12.$$

Study the following two examples.

Example 1

$$
\begin{array}{r}
x^2 + x - 3 \\
x + 5 \overline{\smash{)}\, x^3 + 6x^2 + 2x - 4} \\
\underline{x^3 + 5x^2} \\
x^2 + 2x - 4 \\
\underline{x^2 + 5x} \\
-3x - 4 \\
\underline{-3x - 15} \\
11
\end{array}
$$

The quotient is $x^2 + x - 3$ and the remainder is 11. Thus, $x^3 + 6x^2 + 2x - 4 = (x^2 + x - 3)(x + 5) + 11$.

Check. $(x^2 + x - 3)(x + 5) = x^3 + 5x^2 + x^2 + 5x - 3x - 15$
$$= x^3 + 6x^2 + 2x - 15$$
and $x^3 + 6x^2 + 2x - 15 + 11 = x^3 + 6x^2 + 2x - 4$.

Example 2 In dividing $(3x^3 + 2x - 5)$ by $(x + 2)$, we need to be aware that the coefficient of x^2 in $3x^3 + 2x - 5$ is 0.

$$\begin{array}{r}3x^2 - 6x + 14\\ x+2\overline{\smash{)}3x^3 + 0x^2 + 2x - 5}\\ \underline{3x^3 + 6x^2}\\ -6x^2 + 2x\\ \underline{-6x^2 - 12x}\\ 14x - 5\\ \underline{14x + 28}\\ -33\end{array}$$

Thus,

$$3x^3 + 2x - 5 = (3x^2 - 6x + 14)(x + 2) + (-33).$$

Check. $(3x^2 - 6x + 14)(x + 2) = 3x^3 + 6x^2 - 6x^2 - 12x + 14x + 28$
$= 3x^3 + 2x + 28$
and $\quad 3x^3 + 2x + 28 + (-33) = 3x^3 + 2x - 5.$

EXERCISE 13.7

1. Tell the degree of each of the following polynomials.
   a. $5x^2 + x + 1$ b. $-3x^4 + x^3 + 2$
   c. $2 + x^2 - x$ d. $-2x + 5x^6 + 1$
   e. 5 f. $2 - x$
   g. $16 - x^3 + x - 5x^7$ h. $5x - 2 + 17x^2$

2. What is the coefficient of x^3 in each of the following polynomials?
   a. $-4x^3 + x - 1$ b. $x - x^2 - x^3$ c. $x^2 + x + 1$

3. What will be the coefficient of x^4 after the multiplication is performed in $(2x^2 + 1)(-7x^2 - 1)$?

4. If $P(x) = 5x^3 - 4x^2 + 2x - 1$, $Q(x) = 2x^2 - 3x + 3$, and $F(x) = 5x - 3$, determine the following.
   a. $P(-1); \quad P(0); \quad P(3)$ b. $Q(0); \quad Q(1); \quad Q(-1)$
   c. $F(2); \quad F(-3); \quad F(-2)$ d. $P(x) + Q(x)$
   e. $F(x) + Q(x)$ f. $F(x) + P(x)$
   g. $P(x) - F(x)$ h. $P(x) - Q(x)$
   i. $F(x) \times Q(x)$ j. $Q(x) \times P(x)$

5. Divide.
   a. $(x^2 + 7x + 18) \div (x + 9)$
   b. $(2r^3 - 5r^2 - 3r + 3) \div (r + 2)$
   c. $(x^2 - 8x + 9) \div (x - 4)$

d. $(x^3 - y^3) \div (x - y)$
   e. $(x^3 - 64) \div (x + 4)$
   f. $(9r^3 + 11r + 8) \div (3r - 2)$
   g. $(4y^3 - 8y^2 - 9y + 7) \div (2y - 3)$
   h. $(r^6 - 13r^3 + 42) \div (r^3 - 7)$
   i. $(t^2 - 17t + 60) \div (t - 12)$
   j. $(-15x^3 - 19x^2 - 30x - 8) \div (3x^2 + 5x - 4)$

GLOSSARY

Coefficient: In the polynomial $a_r x^r + a_{r-1} x^{r-1} + \ldots + a_1 x + a_0$, a_r, a_{r-1}, $\ldots$, a_1, a_0 are coefficients.

Degree of a polynomial: In a polynomial with one variable, the greatest exponent used with the variable whose coefficient is not zero.

Factor: One of the numbers or expressions used in a product. Examples: 2 and 3 are factors of 6, because $2 \times 3 = 6$. The factors of $x(x + 2)$ are x and $x + 2$.

CHAPTER REVIEW PROBLEMS

1. Determine the products.
   a. $3(x + 3)$
   b. $5(m - 7)$
   c. $\sqrt{3}(\sqrt{6} + 3)$
   d. $(n + 3)(n + 4)$
   e. $(s - 4)(s + 6)$
   f. $(t - 7)(t - 2)$
   g. $(2u + 1)(3u + 4)$
   h. $(3u - 1)(4u + 2)$
   i. $(4k - 3)(7k - 4)$
   j. $(-2y + 1)(y - 6)$
   k. $(\sqrt{5} + 6)(\sqrt{5} - 6)$
   l. $(2\sqrt{3} + \sqrt{2})(\sqrt{3} - \sqrt{2})$
   m. $(5\sqrt{6})^2$
   n. $-3(\sqrt{3})^2$
   o. $(-3\sqrt{3})^2$
   p. $(2a - m)^2$
   q. $(3t + s)^2$
   r. $(4n - 3x)^2$
   s. $(2m + 7)(2m - 7)$
   t. $(3x^2 + 4y^2)(3x^2 - 4y^2)$
   u. $(-7c + 3a)^2$
   v. $[(x - 2) + a]^2$
   w. $[(2y + m) - n]^2$
   x. $[(x - y) + (a - b)]^2$

2. Factor.
   a. $3x^2 + 6x$
   b. $3\sqrt{11} - 6\sqrt{11}$
   c. $t^2 + 2t + 1$
   d. $6m^2 - m - 1$

e. $a(x - y) + b(x - y)$
f. $7(2x + 5) + 3(2x + 5)$
g. $9y^2 - 16z^2$
h. $\sqrt[3]{24} - \sqrt[3]{16}$
i. $9m^2 - 30mn + 25n^2$
j. $(3t + s)(m + n) + (3t + s)(m - n)$
k. $6x^2 + 7x + 2$
l. $4x^2 + x - 3$
m. $3k^2 - 17k + 20$
n. $-2t^2 - 5t - 3$

3. Factor and simplify. In each case, replacements of the variables which lead to undefined symbols are excluded.

a. $\dfrac{4a + 6}{12a + 14}$
b. $\dfrac{5x - 10}{10x - 5}$

c. $\dfrac{3xy + 9x}{6xy - 3x}$
d. $\dfrac{x^2 - 2x - 3}{2x^2 + 3x + 1}$

e. $\dfrac{6k^2 + 7k - 3}{3k^2 + 2k - 1}$
f. $\dfrac{s^2 + 4s + 3}{s^2 + 3s + 2}$

g. $\dfrac{nk + 3n + k + 3}{nm + 4n + m + 4}$
h. $\dfrac{ac + 4a - 2c - 8}{ak - a - 2k + 2}$

4. Rationalize each denominator and simplify.

a. $\dfrac{1}{2 + \sqrt{5}}$
b. $\dfrac{2}{3 - \sqrt{7}}$

c. $\dfrac{-3}{2\sqrt{2} + \sqrt{3}}$
d. $\dfrac{-2}{\sqrt{7} - \sqrt{5}}$

5. Tell which of the following are true for all permissible replacements of the variables.

a. $\dfrac{2x + 12}{4x - 2} = \dfrac{x + 6}{x - 1}$
b. $\dfrac{x + 7}{x - 1} = -7$

c. $\dfrac{ax + ay}{bx + by} = \dfrac{a}{b}$
d. $\dfrac{x^2 - 1}{x - 1} = x - 1$

e. $\dfrac{n^2 - 1}{n^2 + 2n + 1} = \dfrac{n - 1}{n + 1}$
f. $\dfrac{k^2 + k + 1}{k^2 + 2k - 3} = \dfrac{2}{3}$

6. Given the polynomials

$$P(x) = 2x^3 - x^2 + x - 2$$
$$Q(x) = x - 1,$$

determine each of the following.

a. $P(1)$
b. $Q(\sqrt{2})$
c. $P(x) + Q(x)$
d. $P(x) - Q(x)$
e. $P(x) \times Q(x)$
f. $P(x) \div Q(x)$

Chapter Review Problems

7. Given the polynomials

$$P(x) = 4x^5 - x^4 + x^2 - 2 \qquad R(x) = x + 1$$
$$Q(x) = x^2 - 1 \qquad S(x) = x - 1,$$

determine each of the following.

- **a.** $P(1)$
- **b.** $Q(-3)$
- **c.** $R(-\frac{7}{9})$
- **d.** $S(\sqrt{2})$
- **e.** $P(x) \times Q(x)$
- **f.** $P(x) \div Q(x)$
- **g.** $Q(x) \times R(x)$
- **h.** $Q(2) \times S(-3)$
- **i.** $\dfrac{Q(-9)}{S(-2)}$
- **j.** $P(1) \times Q(-1) \times R(0) \times S(2)$
- **k.** $P(x) \div R(x)$
- **l.** $P(x) \div S(x)$

14

Quadratic Equations

14.1 Solving Quadratic Equations

In this chapter we shall study equations which are *quadratic*. Every equation which can be written in the form

$$ax^2 + bx + c = 0,$$

where a, b, and c are real numbers and $a \neq 0$, is called a *quadratic equation*. The following are examples of quadratic equations.

Example 1 $\quad 3x^2 - 5x + 2 = 0$.
   Here $a = 3$, $b = -5$, and $c = 2$.

Example 2 $\quad 3 - x^2 = 2x$.
   An equivalent equation of the form $ax^2 + bx + c = 0$ is

$$-x^2 - 2x + 3 = 0,$$

where $a = -1$, $b = -2$, and $c = 3$.

Example 3 $\quad x^2 = 7x$.
   An equivalent equation is $x^2 - 7x = 0$. Here $a = 1$, $b = -7$, and $c = 0$.

14.1 Solving Quadratic Equations

To learn how to solve quadratic equations, we first consider the case where $c = 0$. For example, $x^2 + 7x = 0$ in such an equation. We use our knowledge of factoring in solving this equation.

Example 1 Solve: $x^2 + 7x = 0$
$x(x + 7) = 0.$

It is stated that the product of x and $x + 7$ is equal to 0. Recall that this is true only if x is 0 or $x + 7$ is 0. Thus, the last equation is equivalent to

$$x = 0 \quad \text{or} \quad x + 7 = 0.$$

The solution set of the last sentence is $\{0, -7\}$. Thus, the solution set of $x^2 + 7x = 0$ is $\{0, -7\}$.

Check.

$x^2 + 7x$	0	$x^2 + 7x$	0
$0^2 + 7 \times 0$	0	$(-7)^2 + 7(-7)$	0
$0 + 0$		$49 + (-49)$	
0		0	

Now let us consider a quadratic equation in which b is 0.

Example 2 Solve: $x^2 - 16 = 0.$

An equivalent equation is

$$x^2 = 16.$$

It is easy to see that we are seeking a number whose square is 16. There are two such numbers, 4 and -4, since

$$4^2 = 16 \quad \text{and} \quad (-4)^2 = 16.$$

Thus, the solution set of $x^2 - 16 = 0$ is $\{4, -4\}$.

The equation can also be solved by factoring.

$$x^2 - 16 = 0$$
$$(x - 4)(x + 4) = 0$$
$$x - 4 = 0 \quad \text{or} \quad x + 4 = 0$$
$$x = 4 \quad \text{or} \quad x = -4.$$

Again we see that the solution set is $\{4, -4\}$.

Now consider an equation of the form $ax^2 + bx + c = 0$, where no coefficient is 0. We first examine an equation which can be easily factored.

Example 3 Solve: $x^2 + 8x + 12 = 0$
$(x + 6)(x + 2) = 0.$

Thus, the given quadratic equation is equivalent to

$$x + 6 = 0 \quad \text{or} \quad x + 2 = 0.$$

The solution set of the last sentence is $\{-6, -2\}$. Thus, the solution set of $x^2 + 8x + 12 = 0$ is $\{-6, -2\}$. We check to be sure that this is so.

Check. -6 for x: -2 for x:

$x^2 + 8x + 12$	0	$x^2 + 8x + 12$	0
$(-6)^2 + 8(-6) + 12$	0	$(-2)^2 + 8(-2) + 12$	0
$36 + (-48) + 12$		$4 + (-16) + 12$	
$-12 + 12$		$-12 + 12$	
0		0	

Here is one more example is which we use the factoring method.

Example 4 Solve: $3x^2 - 12 = 0$
$3(x^2 - 4) = 0$
$3(x - 2)(x + 2) = 0.$

Now we apply the Multiplication Property for Equations and multiply both sides of the equation by $\frac{1}{3}$.

$$\tfrac{1}{3} \times 3(x - 2)(x + 2) = \tfrac{1}{3} \times 0$$
$$(x - 2)(x + 2) = 0.$$

The last equation is equivalent to

$$x - 2 = 0 \quad \text{or} \quad x + 2 = 0.$$

The solution set of this sentence is $\{2, -2\}$.

Check these two solutions in $3x^2 - 12 = 0$.

EXERCISE 14.1

1. Tell what replacements are made for a, b, and c in $ax^2 + bx + c = 0$ to obtain each equation.

　　a. $x^2 + 4x + 1 = 0$　　　　　　b. $2x^2 - 10x + 5 = 0$

14.2 Patterns in Squaring Binomials

c. $-3x^2 + 6 = 0$
d. $-x^2 - 5x = 0$
e. $3x - x^2 - 4 = 0$
f. $-5 - 7x + x^2 = 0$
g. $x - 16x^2 = 0$
h. $2 - 3x^2 = 0$

2. For each equation write an equivalent equation of the form $ax^2 + bx + c = 0$.

a. $3x^2 = x + 2$
b. $2x = 5x^2 - 1$
c. $-9 = -3x^2 + 2x$
d. $2(x^2 + 1) = 10x$
e. $x^2 = 8x$
f. $3(1 - 4x) = x(5 + x)$

3. Solve each of the following equations.

a. $x^2 - 4 = 0$
b. $x^2 - 25 = 0$
c. $x^2 = 121$
d. $2x^2 = 72$
e. $3x^2 = 48$
f. $4x^2 = 400$
g. $5x^2 - 500 = 0$
h. $7x^2 - 700 = 0$
i. $x^2 + 10x = 0$
j. $x^2 - 17x = 0$
k. $x^2 + 3x = 0$
l. $2x^2 + 6x = 0$
m. $-3x^2 - 21x = 0$
n. $6x^2 = 9x$
o. $5x^2 = -7x$
p. $4x = 15x^2$

4. Solve each equation by factoring.

a. $x^2 + 5x + 6 = 0$
b. $x^2 + 10x + 16 = 0$
c. $x^2 + 15x + 36 = 0$
d. $x^2 + 2x - 3 = 0$
e. $x^2 + x - 30 = 0$
f. $x^2 + 12x - 45 = 0$
g. $x^2 - x - 6 = 0$
h. $x^2 - 5x - 50 = 0$
i. $x^2 - 9x - 10 = 0$
j. $x^2 - 5x - 24 = 0$
k. $x^2 - 17x = -30$
l. $x^2 = 15x - 56$
m. $14 - 15x = -x^2$
n. $3x - 2 = x^2$
o. $x^2 = 5x + 14$
p. $33 = x^2 + 8x$
q. $12 = x^2 + x$
r. $-x^2 - x = -56$
s. $-48 = x^2 + 14x$
t. $9 - 8x = x^2$

14.2 Patterns in Squaring Binomials

Let us square a few binomials and look for a pattern.

$$(x + 3)^2 = x^2 + 6x + 9$$
$$(x + 5)^2 = x^2 + 10x + 25$$
$$(x - 4)^2 = x^2 - 8x + 16$$

$$(x - 7)^2 = x^2 - 14x + 49$$
$$(x - \tfrac{1}{2})^2 = x^2 - x + \tfrac{1}{4}.$$

Now take a good look at the examples above. The following pattern correctly describes all of the following examples.

$$(x + a)^2 = x^2 + 2 \times a \times x + a^2.$$

Trinomials such as

$$x^2 + 6x + 9 \quad \text{and} \quad x^2 + 10x + 25$$

are called *perfect squares* because they result from squaring binomials.

Now suppose we are given a binomial such as

$$x^2 + 12x$$

and we are seeking a third term which we will add to $x^2 + 12x$ so that it becomes a perfect square trinomial. By examining the pattern above, we can see that if we take $\tfrac{1}{2} \times 12$, or 6, and square it, we obtain the desired third term, 36. Now adding 36 to our binomial, we have

$$x^2 + 12x + 36$$

which is a perfect square, since

$$(x + 6)^2 = x^2 + 12x + 36.$$

In general, if we are given the binomial

$$x^2 + bx,$$

then $\left(\dfrac{b}{2}\right)^2$ is the third term which will form a perfect square. The following is true for all replacements of the variables:

$$\left(x + \frac{b}{2}\right)^2 = x^2 + bx + \left(\frac{b}{2}\right)^2.$$

14.3 Perfect Squares and Quadratic Equations

Suppose we are asked to find the solution set of the equation

$$(y + 5)^2 = 36.$$

You may reason as follows: Some number, $y + 5$, squared gives 36. There are two such numbers: 6 and -6. Therefore,

$$y + 5 = 6 \quad \text{or} \quad y + 5 = -6$$
$$y = 1 \quad \text{or} \quad y = -11.$$

So, the solution set of $(y + 5)^2 = 36$ should be $\{1, -11\}$.

14.3 Perfect Squares and Quadratic Equations

Check.   1 for y: -11 for y:

$(y + 5)^2$	36
$(1 + 5)^2$	36
6^2	
36	

$(y + 5)^2$	36
$(-11 + 5)^2$	36
$(-6)^2$	
36	

Note that

$$(y + 5)^2 = 36$$
$$y^2 + 10y + 25 = 36$$
$$y^2 + 10y - 11 = 0$$

are all equivalent equations.

Suppose that, instead of being given the equation $(y + 5)^2 = 36$, we were given its equivalent equation, $y^2 + 10y - 11 = 0$. We probably would factor the latter as follows:

$$(y - 1)(y + 11) = 0$$

and immediately find the solution set to be $\{1, -11\}$.

There is another way to attack this problem. Suppose we start with $y^2 + 10y - 11 = 0$ and arrive at $(y + 5)^2 = 36$. One way to do this would be to reason along these lines:

$$y^2 + 10y - 11 = 0.$$

Now let us use the Addition Property for Equations and add 11 to each side of the equation:

$$(y^2 + 10y - 11) + 11 = 0 + 11$$
$$y^2 + 10y = 11.$$

Consider the binomial $y^2 + 10y$. If it is to be a perfect square, its third terms should be $(\frac{10}{2})^2$, or 25. Now we add 25 to each side of the last equation.

$$(y^2 + 10y) + 25 = 11 + 25$$
$$y^2 + 10y + 25 = 36$$
$$(y + 5)^2 = 36.$$

This last equation was solved previously.

The process which led us from

$$y^2 + 10y - 11 = 0$$

to the equivalent expression

$$(y + 5)^2 = 36$$

is called *completing the square*. The following examples illustrate this process.

Example 1 Solve: $x^2 + 6x + 5 = 0$.

An equivalent equation is

$$x^2 + 6x = -5.$$

Now we seek the third term which we will add to $x^2 + 6x$ so that it becomes a perfect square trinomial. The term we are seeking is $(\frac{6}{2})^2$, or 9. We add 9 to each side of the equation.

$$x^2 + 6x + 9 = -5 + 9$$
$$(x + 3)^2 = 4.$$

Now since

$$2^2 = 4 \quad \text{and} \quad (-2)^2 = 4,$$

the last equation is equivalent to

$$x + 3 = 2 \quad \text{or} \quad x + 3 = -2$$
$$x = -1 \quad \text{or} \quad x = -5.$$

Thus, the solution set is $\{-1, -5\}$.

Check.

$x^2 + 6x + 5$	0
$(-1)^2 + 6(-1) + 5$	0
$1 - 6 + 5$	
$-5 + 5$	
0	

$x^2 + 6x + 5$	0
$(-5)^2 + 6(-5) + 5$	0
$25 - 30 + 5$	
$-5 + 5$	
0	

Example 2 Solve: $3x^2 + 6x + 9 = 0$.

$$3(x^2 + 2x + 3) = 0$$
$$x^2 + 2x + 3 = 0$$
$$x^2 + 2x = -3.$$

Notice that before we "complete the square," we find an equivalent equation such that the coefficient of x^2 is 1. Now, what number shall we add to each side of the last equation so that $x^2 + 2x$ will become a perfect square? The number is $(\frac{2}{2})^2$, or 1. Thus,

14.3 Perfect Squares and Quadratic Equations

$$x^2 + 2x + 1 = -3 + 1$$
$$(x + 1)^2 = -2.$$

But there is no real number which when squared is equal to -2. Thus, the solution set of $3x^2 + 6x + 9 = 0$ is ϕ.

Example 3 Solve: $y^2 + 10y + 8 = 0$
$$y^2 + 10y = -8.$$

The number which we will add to each side of the equation is $(\frac{10}{2})^2$, or 25.

$$y^2 + 10y + 25 = -8 + 25$$
$$(y + 5)^2 = 17$$
$$y + 5 = \sqrt{17} \quad \text{or} \quad y + 5 = -\sqrt{17}$$
$$y = -5 + \sqrt{17} \quad \text{or} \quad y = -5 - \sqrt{17}.$$

Therefore, the solution set of $y^2 + 10y + 8 = 0$ is

$$\{-5 + \sqrt{17}, -5 - \sqrt{17}\}.$$

EXERCISE 14.2

1. Square.

   a. $(x + 1)^2$
   b. $(x + 2)^2$
   c. $(x + 4)^2$
   d. $(x - 1)^2$
   e. $(x - 2)^2$
   f. $(x - 3)^2$
   g. $(x + \frac{1}{2})^2$
   h. $(x + \frac{3}{4})^2$
   i. $(a - \frac{1}{3})^2$
   j. $(m - \frac{2}{5})^2$
   k. $(n + \frac{1}{8})^2$
   l. $(t - \frac{4}{5})^2$

2. Replace b, c, or m with a numeral to obtain a sentence which is true for all replacements of the variables.

   a. $(x - 1)^2 = x^2 - 2x + c$
   b. $(x + 5)^2 = x^2 + 10x + c$
   c. $(r + 7)^2 = r^2 + 14r + c$
   d. $(t - 2)^2 = t^2 - 4t + c$
   e. $(x + 4)^2 = x^2 + 8x + c$
   f. $(x + \frac{1}{2})^2 = x^2 + x + c$
   g. $(x + m)^2 = x^2 + 4x + 4$
   h. $(p + m)^2 = p^2 + 8p + 16$
   i. $(x - m)^2 = x^2 - 2x + 1$
   j. $(x - m)^2 = x^2 - 6x + 9$
   k. $(z - m)^2 = z^2 - z + \frac{1}{4}$
   l. $(x - m)^2 = x^2 - \frac{2}{3}x + \frac{1}{9}$
   m. $(x - 2)^2 = x^2 + bx + 4$
   n. $(x - 7)^2 = x^2 - bx + 49$
   o. $(y + 9)^2 = y^2 + by + 81$
   p. $(x + \frac{1}{4})^2 = x^2 + bx + \frac{1}{16}$
   q. $(y - \frac{2}{5})^2 = y^2 - by + \frac{4}{25}$
   r. $(x + .2)^2 = x^2 + bx + .04$

Quadratic Equations

s. $(t + m)^2 = t^2 + bt + 16$ **t.** $(r - m)^2 = r^2 - 12r + c$
u. $(x + m)^2 = x^2 + bx + 36$ **v.** $(x - \frac{3}{4})^2 = x^2 + bx + c$
w. $(y - m)^2 = y^2 - \frac{6}{5}y + c$ **x.** $(z + m)^2 = z^2 + bz + \frac{36}{49}$
y. $(t - m)^2 = t^2 - 16t + c$ **z.** $(x + m)^2 = x^2 + bx + 81$

3. In each case, replace the c with a numeral to obtain an expression which is a perfect square.

Example $x^2 + c \times x + 9$.
    If c is replaced by 6, we have $x^2 + 6x + 9$, which is equivalent to $(x + 3)^2$.

a. $x^2 + cx + 4$ **b.** $y^2 - 10y + c$
c. $z^2 + cz + \frac{9}{4}$ **d.** $x^2 - 6x + c$
e. $r^2 + 8r + c$ **f.** $x^2 + cx + \frac{1}{4}$
g. $t^2 + 3t + c$ **h.** $s^2 + 5s + c$
i. $x^2 - cx + \frac{4}{9}$ **j.** $t^2 - \frac{1}{3}t + c$
k. $x^2 - cx + \frac{1}{9}$ **l.** $x^2 - .2x + c$

4. Determine the solution sets.

a. $(x - 2)^2 = 4$ **b.** $(r - 6)^2 = 9$
c. $(t + 1)^2 = 16$ **d.** $(s - 3)^2 = 25$
e. $(v - 1)^2 = 9$ **f.** $(x + 4)^2 = 36$
g. $(x + 2)^2 = 4$ **h.** $(s - 5)^2 = 1$
i. $(x + \frac{2}{5})^2 = 3$ **j.** $(y + 2)^2 = 8$

5. Determine the solution sets by completing the square.

a. $p^2 - 8p = 0$ **b.** $x^2 + 4x = 0$
c. $x^2 + 4x - 12 = 0$ **d.** $y^2 - 8y + 15 = 0$
e. $t^2 + 2t - 8 = 0$ **f.** $m^2 + 2m - 3 = 0$
g. $n^2 + 10n + 16 = 0$ **h.** $a^2 + 8a - 20 = 0$
i. $h^2 - 6h - 27 = 0$ **j.** $z^2 - 10z - 24 = 0$
k. $x^2 + 5x = 0$ **l.** $x^2 - 5x = 0$
m. $y^2 + 11y = 0$ **n.** $p^2 + 9p + 8 = 0$
o. $x^2 + 3x - 40 = 0$ **p.** $h^2 + h - \frac{3}{4} = 0$
q. $x^2 - 7x + 6 = 0$ **r.** $2a^2 + 6a + 4 = 0$
s. $3x^2 + 9x + 6 = 0$ **t.** $b^2 + \frac{5}{2}b + \frac{9}{16} = 0$
u. $x^2 + 16x + 20 = 0$ **v.** $y^2 + 8y + 3 = 0$
w. $z^2 + 10z + 15 = 0$ **x.** $m^2 + 4m + 1 = 0$

y. $n^2 + 5n - 1 = 0$ **z.** $s^2 + 2s + 2 = 0$
a'. $t^2 - 7t - 2 = 0$ **b'.** $3x^2 + 4x - 1 = 0$

14.4 The Quadratic Formula

Recall that equations of the form

$$ax^2 + bx + c = 0,$$

where a, b, and c are real numbers and $a \neq 0$, are called quadratic equations. The process of completing the square, which we learned in the preceding section, will enable us to derive a *formula* for solving quadratic equations. Before deriving this formula, let us review in detail one example which will lead to the solution of a given quadratic equation. Study the development below step by step, since it will serve as a model for deriving the general formula.

Example Solve: $2m^2 + 3m - 1 = 0$.

$$2m^2 + 3m = 1$$
$$2(m^2 + \tfrac{3}{2}m) = 1$$
$$m^2 + \tfrac{3}{2}m = \tfrac{1}{2}$$
$$m^2 + \tfrac{3}{2}m + \tfrac{9}{16} = \tfrac{1}{2} + \tfrac{9}{16}$$
$$(m + \tfrac{3}{4})^2 = \tfrac{17}{16}.$$

$$m + \tfrac{3}{4} = \sqrt{\tfrac{17}{16}} \quad \text{or} \quad m + \tfrac{3}{4} = -\sqrt{\tfrac{17}{16}}$$
$$m = -\tfrac{3}{4} + \tfrac{\sqrt{17}}{4} \quad \text{or} \quad m = -\tfrac{3}{4} - \tfrac{\sqrt{17}}{4}$$
$$m = \tfrac{-3 + \sqrt{17}}{4} \quad \text{or} \quad m = \tfrac{-3 - \sqrt{17}}{4}.$$

Thus, the solution set is $\left\{ \dfrac{-3 + \sqrt{17}}{4}, \dfrac{-3 - \sqrt{17}}{4} \right\}$.

Now we repeat the steps using the general form of a quadratic equation.

$$ax^2 + bx + c = 0 \qquad\qquad x^2 + \tfrac{b}{a}x + \tfrac{b^2}{4a^2} = -\tfrac{c}{a} + \tfrac{b^2}{4a^2}$$
$$a\left(x^2 + \tfrac{b}{a}x\right) = -c$$
$$x^2 + \tfrac{b}{a}x = -\tfrac{c}{a} \qquad\qquad \left(x + \tfrac{b}{2a}\right)^2 = \tfrac{-4ac + b^2}{4a^2}.$$

$$x + \frac{b}{2a} = \sqrt{\frac{b^2 - 4ac}{4a^2}} \quad \text{or} \quad x + \frac{b}{2a} = -\sqrt{\frac{b^2 - 4ac}{4a^2}}$$

$$x = -\frac{b}{2a} + \frac{\sqrt{b^2 - 4ac}}{2|a|} \quad \text{or} \quad x = -\frac{b}{2a} - \frac{\sqrt{b^2 - 4ac}}{2|a|}$$

$$x = \frac{-b + \sqrt{b^2 - 4ac}}{2|a|} \quad \text{or} \quad x = \frac{-b - \sqrt{b^2 - 4ac}}{2|a|}.$$

The last line above is the *quadratic formula,* which gives us the solution set of any quadratic equation, $ax^2 + bx + c = 0$. The solution set is

$$\left\{ \frac{-b + \sqrt{b^2 - 4ac}}{2|a|}, \frac{-b - \sqrt{b^2 - 4ac}}{2|a|} \right\}.$$

Now we can solve any quadratic equation by replacing a, b, and c in the formula by the appropriate numerals. Study the following examples.

Example 1 Solve: $6x^2 - 5x - 4 = 0$.

In this equation, $a = 6$, $b = -5$, and $c = -4$. Thus, we shall make these replacements in the quadratic formula.

$$x = \frac{-b + \sqrt{b^2 - 4ac}}{2|a|} \quad \text{or} \quad x = \frac{-b - \sqrt{b^2 - 4ac}}{2|a|}$$

$$x = \frac{-(-5) + \sqrt{(-5)^2 - 4(6)(-4)}}{2(6)} \quad \text{or} \quad x = \frac{-(-5) - \sqrt{(-5)^2 - 4(6)(-4)}}{2(6)}$$

$$x = \frac{5 + \sqrt{25 + 96}}{12} \quad \text{or} \quad x = \frac{5 - \sqrt{25 + 96}}{12}$$

$$x = \frac{5 + \sqrt{121}}{12} \quad \text{or} \quad x = \frac{5 - \sqrt{121}}{12}$$

$$x = \frac{5 + 11}{12} \quad \text{or} \quad x = \frac{5 - 11}{12}$$

$$x = \frac{16}{12} \quad \text{or} \quad x = -\frac{6}{12}$$

$$x = \frac{4}{3} \quad \text{or} \quad x = -\frac{1}{2}.$$

Thus, $\{\frac{4}{3}, -\frac{1}{2}\}$ is the solution set of $6x^2 - 5x - 4 = 0$.

Example 2 Solve: $x^2 + 6x + 3 = 0$.

In this equation, $a = 1$, $b = 6$, and $c = 3$.

$$x = \frac{-b + \sqrt{b^2 - 4ac}}{2|a|} \quad \text{or} \quad x = \frac{-b - \sqrt{b^2 - 4ac}}{2|a|}$$

$$x = \frac{-6 + \sqrt{6^2 - 4(1)(3)}}{2(1)} \quad \text{or} \quad x = \frac{-6 - \sqrt{6^2 - 4(1)(3)}}{2(1)}$$

$$x = \frac{-6 + \sqrt{36 - 12}}{2} \quad \text{or} \quad x = \frac{-6 - \sqrt{36 - 12}}{2}$$

$$x = \frac{-6 + \sqrt{24}}{2} \quad \text{or} \quad x = \frac{-6 - \sqrt{24}}{2}$$

$$x = \frac{-6 + 2\sqrt{6}}{2} \quad \text{or} \quad x = \frac{-6 - 2\sqrt{6}}{2}$$

$$x = \frac{2(-3 + \sqrt{6})}{2} \quad \text{or} \quad x = \frac{2(-3 - \sqrt{6})}{2}$$

$$x = -3 + \sqrt{6} \quad \text{or} \quad x = -3 - \sqrt{6}.$$

Thus, $\{-3 + \sqrt{6}, -3 - \sqrt{6}\}$ is the solution set of $x^2 + 6x + 3 = 0$.

It is possible to have quadratic equations whose solution set in the universe of real numbers is the empty set. Consider, for example, this very simple quadratic equation,

$$x^2 = -2.$$

This equation may be turned into the question

What real number squared results in -2?

We know that real numbers consist of positive numbers, negative numbers, and zero. We also know that every positive number squared is a positive number, every negative number squared is a positive number, and zero squared is zero. So, there is no real number whose square is a negative number. Thus, the equation $x^2 = -2$ has no root in the universe of real numbers.

14.5 The Discriminant

For the quadratic equation $ax^2 + bx + c = 0$, the expressions

$$\frac{-b + \sqrt{b^2 - 4ac}}{2|a|} \quad \text{and} \quad \frac{-b - \sqrt{b^2 - 4ac}}{2|a|}$$

give us a way of determining the solutions, or *roots* of the equation.

We have said that many quadratic equations do not have roots among the real numbers. With a little thought we will see that it is not necessary to solve a quadratic equation to discover that it has no roots in the real-number system. All we need to do is examine $\sqrt{b^2 - 4ac}$ in the expression which gives the roots.

If $b^2 - 4ac$ is a positive number, then

$$\frac{-b + \sqrt{b^2 - 4ac}}{2|a|} \quad \text{and} \quad \frac{-b - \sqrt{b^2 - 4ac}}{2|a|}$$

name two different real numbers, and thus, the equation has two real roots.

If $b^2 - 4ac$ is 0, then each of

$$\frac{-b + \sqrt{b^2 - 4ac}}{2|a|} \quad \text{and} \quad \frac{-b - \sqrt{b^2 - 4ac}}{2|a|}$$

yields $\frac{-b}{2|a|}$, and thus, the equation has one real root.

If $b^2 - 4ac$ is a negative number, then $\sqrt{b^2 - 4ac}$ does not exist among real numbers and the equation has no roots in the real-number system.

The expression $b^2 - 4ac$ is called the *discriminant* and is abbreviated by the Greek letter Δ [read: delta].

We summarize the results of the preceding discussion as follows:

◄ $\Delta > 0$; two real roots.

◄ $\Delta = 0$; one real root.

◄ $\Delta < 0$; no real roots.

Let us now describe the nature of the roots of some quadratic equations by examining the discriminant.

Example 1 $3x^2 + 5x + 1 = 0$.

$$\Delta = b^2 - 4ac = 5^2 - 4 \times 3 \times 1 = 25 - 12 = 13.$$

$\Delta > 0$; therefore, the equation has two real roots.

We solve this equation.

$$\frac{-b + \sqrt{\Delta}}{2|a|} = \frac{-5 + \sqrt{13}}{6}$$

$$\frac{-b - \sqrt{\Delta}}{2|a|} = \frac{-5 - \sqrt{13}}{6}$$

Thus, the solution set is $\left\{\frac{-5 + \sqrt{13}}{6}, \frac{-5 - \sqrt{13}}{6}\right\}$ and we see that indeed the equation $3x^2 + 5x + 1 = 0$ has two real roots.

Example 2 $x^2 + 2x + 1 = 0$.

$$\Delta = b^2 - 4ac = 2^2 - 4 \times 1 \times 1 = 4 - 4 = 0.$$

$\Delta = 0$; therefore, the equation has one real root.

This root is $\frac{-2}{2}$ or -1.

14.5 The Discriminant

Example 3 $x^2 + 5x + 20 = 0.$

$$\Delta = b^2 - 4ac = 5^2 - 4 \times 1 \times 20 = 25 - 80 = -55.$$

$\Delta < 0$; therefore, the equation has no real roots.

EXERCISE 14.3

1. Solve each quadratic equation using the quadratic formula.
   a. $x^2 + 2x - 3 = 0$
   b. $x^2 + 4x - 21 = 0$
   c. $x^2 - 11x + 28 = 0$
   d. $x^2 - 14x + 33 = 0$
   e. $x^2 = 3x + 10$
   f. $x^2 - 3x - 5 = 0$
   g. $x^2 + 6x - 3 = 0$
   h. $x^2 + 8x + 5 = 0$
   i. $x^2 - 6x - 1 = 0$
   j. $3x^2 - 2x - 7 = 0$
   k. $3x^2 + 9x - 5 = 0$
   l. $7x^2 + 2x - 2 = 0$
   m. $3x^2 - 5x - 8 = 0$
   n. $-2x^2 + 8x + 5 = 0$
   o. $3x^2 + 5x = 0$
   p. $-6x^2 + 11 = 0$
   q. $2x^2 = 4 - 2x$
   r. $7x = -3 - 2x^2$
   s. $3x^2 + 17x = 6$
   t. $4x^2 = 3x + 1$
   u. $6x^2 - 5x - 4 = 0$
   v. $x^2 + 7x = -4$
   w. $2x^2 - 6 = -5x$
   x. $-8x - 3 = 5x^2$
   y. $3(x - 1) = 2x(x + 5)$
   z. $x(3x + 1) = 5(3 - x)$
   a'. $6(x + 4) = x(x - 3)$
   b'. $5(4 - x) = 7x(x + 2)$
   c'. $-4(x + 3) = 5x(4 - x)$
   d'. $3(2 + x) = -2x(3 - x)$

2. Solve for x in terms of the coefficients.
   a. $mx^2 + nx + p = 0$
   b. $rx^2 - sx - t = 0$

3. Without solving each equation, tell whether it has two real roots, one real root, or no real roots.
   a. $y^2 + y - 1 = 0$
   b. $2m^2 - m + 1 = 0$
   c. $17x^2 + 5x - 10 = 0$
   d. $n^2 + n + 1 = 0$
   e. $3p^2 - \frac{1}{2}p - \frac{1}{4} = 0$
   f. $s^2 - 2s - 3 = 0$
   g. $t^2 = t - 7$
   h. $z = 2z^2 + 5$
   i. $5u - \frac{3}{4}u^2 = \frac{1}{2}$
   j. $2w = -3w^2$
   k. $3 = -r + \frac{1}{2}r^2$
   l. $1.5x^2 - 2.5x + 3.5 = 0$
   m. $\dfrac{y^2 - y}{6} = 7$
   n. $\dfrac{n(n - 1)}{n} = 15$
   o. $(t - 1)(t - 2) = 4$

14.6 Using Equation Properties

In Chapter 6 we learned the Multiplication Property for Equations:

If $x = y$, then $xz = yz$.

An example of the use of this property is

$$\tfrac{1}{2}x + 5 = 4$$
$$(\tfrac{1}{2}x + 5) \times 2 = 4 \times 2$$
$$x + 10 = 8.$$

The equations $x + 10 = 8$ and $\tfrac{1}{2}x + 5 = 4$ are *equivalent* equations, because they have the same solution set, namely $\{-2\}$.

In the illustration above we used the number 2 as a multiplier. As long as we use a nonzero real number as a multiplier, we will get an equivalent equation.

When using as a multiplier a polynomial involving a variable, the situation is quite different. Let us see what may happen.

$$3x = 10.$$

Using x as a multiplier,

$$(3x)x = 10x$$
$$3x^2 = 10x$$
$$3x^2 - 10x = 0$$
$$x(3x - 10) = 0.$$

The solution set of the original equation $3x = 10$ is $\{\tfrac{10}{3}\}$. The solution set of the equation $x(3x - 10) = 0$ is $\{0, \tfrac{10}{3}\}$. Thus, the two equations are *not* equivalent, because they have different solution sets.

We must also exercise caution when squaring. For example, the following statement is true for all x and y.

If $x = y$, then $x^2 = y^2$.

But the statement

if $x^2 = y^2$, then $x = y$

is *not* true for all x and y. Here is an example to prove that it is false.

$$(-3)^2 = 3^2 \text{ is true}$$

but $\qquad\qquad -3 = 3$ is false.

In this case, x is -3 and y is 3.

If we start out with an equation, say $x - 1 = 5$, and obtain a new

14.6 Using Equation Properties

equation, $(x - 1)^2 = 25$, the new equation is not equivalent to the first. The equation $x - 1 = 5$ has for its solution set $\{6\}$. The solution set of the equation $(x - 1)^2 = 25$ is $\{-4, 6\}$.

Whenever we perform operations like those just described, we must compare the solution set of the new equation with that of the original equation before deciding whether the derived equation is equivalent to the original equation. Operations of this kind are performed in the case of fractional equations with a variable in the denominator, and in the case of radical equations. Study the examples below.

Example 1 Solve: $\dfrac{2}{2x - 1} = 4.$

$$\dfrac{2}{2x - 1} \times (2x - 1) = 4(2x - 1)$$
$$2 = 8x - 4$$
$$6 = 8x$$
$$\tfrac{3}{4} = x.$$

Now we will check to verify whether $\{\tfrac{3}{4}\}$ is the solution set.

Check.

$$\begin{array}{c|c} \dfrac{2}{2x - 1} & 4 \\ \hline \dfrac{2}{2(\tfrac{3}{4}) - 1} & 4 \\ \dfrac{2}{\tfrac{3}{2} - \tfrac{2}{2}} & \\ \dfrac{2}{\tfrac{1}{2}} & \\ 4 & \end{array}$$

Thus, $\{\tfrac{3}{4}\}$ is the solution set of $\dfrac{2}{2x - 1} = 4.$

Example 2 Solve: $\dfrac{x^2 - 9}{x - 3} = -2.$

$$\dfrac{(x - 3)(x + 3)}{x - 3} = -2$$
$$x + 3 = -2$$
$$x = -5.$$

Check.

$\frac{x^2 - 9}{x - 3}$	-2
$\frac{(-5)^2 - 9}{-5 - 3}$	-2
$\frac{25 - 9}{-8}$	
$\frac{16}{-8}$	
-2	

Thus, -5 is a solution of the equation.

Now consider the following method of solving the same equation.

$$\frac{x^2 - 9}{x - 3}(x - 3) = -2(x - 3)$$
$$x^2 - 9 = 2x + 6$$
$$x^2 + 2x - 15 = 0$$
$$(x - 3)(x + 5) = 0.$$

$$x - 3 = 0 \quad \text{or} \quad x + 5 = 0$$
$$x = 3 \quad \text{or} \quad x = -5.$$

We know that -5 is a solution. Now let us check 3.

Check.

$\frac{x^2 - 9}{x - 3}$	-2
$\frac{3^2 - 9}{3 - 3}$	-2
$\frac{0}{0}$	

Undefined

Thus, 3 is not a solution of $\frac{x^2 - 9}{x - 3} = -2$. The solution set is $\{-5\}$. Notice that the first method of solving required less work.

Example 3 Solve: $\sqrt{3x + 5} = \sqrt{7x - 1}$.

$$(\sqrt{3x + 5})^2 = (\sqrt{7x - 1})^2$$
$$3x + 5 = 7x - 1$$
$$4x = 6$$
$$x = \tfrac{3}{2}.$$

14.6 Using Equation Properties

Check.

$\sqrt{3x+5}$	$\sqrt{7x-1}$
$\sqrt{3 \times \frac{3}{2} + 5}$	$\sqrt{7 \times \frac{3}{2} - 1}$
$\sqrt{\frac{9}{2} + \frac{10}{2}}$	$\sqrt{\frac{21}{2} - \frac{2}{2}}$
$\sqrt{\frac{19}{2}}$	$\sqrt{\frac{19}{2}}$

Thus, $\{\frac{3}{2}\}$ is the solution set of $\sqrt{3x+5} = \sqrt{7x-1}$.

Example 4 Solve: $2\sqrt{x} + 5 = 1$.

$$2\sqrt{x} = -4$$
$$\sqrt{x} = -2$$
$$(\sqrt{x})^2 = (-2)^2$$
$$x = 4.$$

4 seems to be a solution. Let us check.

Check.

$2\sqrt{x} + 5$	1
$2\sqrt{4} + 5$	1
$2 \times 2 + 5$	
9	

But $9 \neq 1$. Thus, 4 is not a solution, and the solution set of $2\sqrt{x} + 5 = 1$ is ϕ.

There is another way to see that the equation in Example 4 does not have a solution. Note that $\sqrt{x}$ is a nonnegative number. Hence, $\sqrt{x}$ cannot be equal to -2, as is required in the equation $\sqrt{x} = -2$, which is equivalent to the original equation.

EXERCISE 14.4

1. For each pair of equations:
    - i. tell how the second equation in the pair is obtained from the first,
    - ii. find the solution set of each equation,
    - iii. state whether the equations are equivalent.

    a. $x = -5$; $x^2 = -5x$

    b. $\dfrac{1}{x+1} = 3$; $3x + 3 = 1$

    c. $2n = -3$; $4n^2 = 9$

d. $\dfrac{y^2 - 1}{y + 1} = 3; \quad y - 1 = 3$

e. $\dfrac{1}{s} = 1; \quad \dfrac{1}{s^2} = 1$

f. $2t + 5 = 11; \quad 4t^2 + 20t + 25 = 121$

g. $\sqrt{u} = 7; \quad u = 49$

h. $\sqrt{m - 1} = 11; \quad m - 1 = 121$

i. $\sqrt{z^2 + 4} = -12; \quad z^2 + 4 = 144$

j. $\sqrt{\dfrac{v}{4}} = -\dfrac{1}{2}; \quad \dfrac{v}{4} = \dfrac{1}{4}$

2. Find the solution set of each equation.

a. $\dfrac{3}{3x + 5} = 1$

b. $\dfrac{-2}{4x - 1} = -3$

c. $\dfrac{5}{x + 1} = 2$

d. $\sqrt{2x - 1} = 2$

e. $\sqrt{x + 7} = 0$

f. $\sqrt{x + 1} = \sqrt{2x - 1}$

g. $\sqrt{8x + 2} = \sqrt{2x - 4}$

h. $\sqrt{9x - 1} = \sqrt{2 - 3x}$

i. $\dfrac{2}{x + 3} = \dfrac{3}{2x + 1}$

j. $\dfrac{3}{2(x + 6)} = \dfrac{-2}{3(2x - 1)}$

k. $\dfrac{5}{x + 11} = \dfrac{1}{x + 3}$

l. $\dfrac{4}{x^2 - 16} = \dfrac{2}{x - 4}$

m. $\sqrt{x^2 + 2} = \sqrt{2x^2 - x + 1}$

n. $\sqrt{3x^2 + x - 5} = \sqrt{x + 7}$

o. $\dfrac{\sqrt{t + 3}}{8} = 0$

p. $\dfrac{\sqrt{y + 4}}{2} = \dfrac{\sqrt{7y + 1}}{4}$

14.7 Quadratic Equations in Solving Problems

There are many problems which lead to quadratic equations. The roots of the equations may provide answers to these problems. However, some problems may be so simple that we can "guess" the answers. For example, consider this problem:

> I thought of a number. The square of the number is 36. What number did I think of?

You no doubt quickly answer: 6 or −6.

It would be wasteful to go through the following:

$$x^2 = 36.$$

14.7 Quadratic Equations in Solving Problems

$$x = \sqrt{36} \quad \text{or} \quad x = -\sqrt{36}$$
$$x = 6 \quad \text{or} \quad x = -6.$$

Therefore, I thought of 6 or −6.

But there are problems in which it is extremely difficult to "guess" the answer. In such cases, the ability to solve quadratic equations is very helpful.

Example 1 The sum of the squares of two consecutive positive integers is 41. Find the integers.

Let x be the smaller integer. Then $x + 1$ is the other integer. We have the following equation:

$$x^2 + (x + 1)^2 = 41.$$

Solve:

$$x^2 + (x^2 + 2x + 1) = 41$$
$$2x^2 + 2x + 1 = 41$$
$$2x^2 + 2x - 40 = 0$$
$$2(x^2 + x - 20) = 0$$
$$x^2 + x - 20 = 0$$
$$(x - 4)(x + 5) = 0.$$
$$x - 4 = 0 \quad \text{or} \quad x + 5 = 0$$
$$x = 4 \quad \text{or} \quad x = -5.$$

Thus, the solution set of the equation is $\{4, -5\}$.

Since the original problem calls for positive integers, we must reject −5.

Let us check 4. If 4 is the smaller integer, then 5 is the other integer. And $4^2 + 5^2 = 16 + 25 = 41$.

Therefore, the integers are 4 and 5.

Example 2 The product of one less than a certain number and two less than three times this number is 14. Determine all such numbers.

If we let x stand for this number, we have

$$(x - 1)(3x - 2) = 14.$$

Solve:

$$3x^2 - 2x - 3x + 2 = 14$$
$$3x^2 - 5x + 2 = 14$$
$$3x^2 - 5x - 12 = 0$$
$$(x - 3)(3x + 4) = 0.$$

$$x - 3 = 0 \quad \text{or} \quad 3x + 4 = 0$$
$$x = 3 \quad \text{or} \quad x = -\tfrac{4}{3}.$$

Thus, the solution set of the equation is $\{3, -\tfrac{4}{3}\}$.

Let us check 3 in the original problem.
One less than a number: $3 - 1 = 2$.
Two less than three times this number: $3 \times 3 - 2 = 7$.
The product: $2 \times 7 = 14$.
Thus, 3 is indeed one solution.

What about $-\tfrac{4}{3}$?

One less than a number: $-\tfrac{4}{3} - 1 = -2\tfrac{1}{3}$
Two less than three times this number: $-\tfrac{4}{3} \times 3 - 2 = -6$
The product: $-2\tfrac{1}{3} \times (-6) = 14$
Thus, $-\tfrac{4}{3}$ is also a solution.

14.8 Solving Problems about Area

Quadratic equations are useful in geometry for solving area problems. Let us recall some of the basic formulas.

If we know the width w and the length l of the rectangle, we can determine the area using the formula

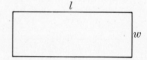

$$A_{\text{rectangle}} = lw.$$

It is easy to see that the area of a square is found by using the formula for the area of a rectangle. For the square, the length and the width are the same. Thus,

$$A_{\text{square}} = s^2,$$

where s stands for the measure of one of its sides.

The area of a triangle is given by the formula

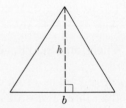

$$A_{\text{triangle}} = \tfrac{1}{2}bh,$$

where b is the measure of a base and h is the measure of the corresponding altitude. Note that in a right triangle, the two sides which are perpendicular represent a base and an altitude.

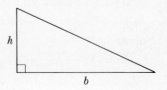

14.8 Solving Problems about Area

Example 1 An altitude of a triangle is 6 in. longer than the corresponding base. The area of the triangle is $13\frac{1}{2}$ sq in. Determine the length of this altitude and the base.

Let x be the measure of the base in inches. Then the measure of the altitude is $(x + 6)$ in. The area of the triangle is $\frac{1}{2} \times x \times (x + 6)$ sq in. We have the equation

$$\tfrac{1}{2}x(x + 6) = 13\tfrac{1}{2}.$$

Solve:

$$\tfrac{1}{2}x^2 + 3x = 13\tfrac{1}{2}$$
$$x^2 + 6x = 27$$
$$x^2 + 6x - 27 = 0$$
$$(x - 3)(x + 9) = 0.$$
$$x - 3 = 0 \quad \text{or} \quad x + 9 = 0$$
$$x = 3 \quad \text{or} \quad x = -9.$$

Since we are concerned here with the measure of line segments, we must reject the negative number. Thus, the length of the base is 3 in. and that of the altitude is 3 + 6, or 9 in.

The area then would be

$$\tfrac{1}{2}(3 \times 9) \text{ sq in.} \quad \text{or} \quad 13\tfrac{1}{2} \text{ sq in.}$$

which agrees with the statement of the problem.

Example 2 Mr. Jones is pouring two different cement slabs. The area of slab A is 693 sq ft greater than the area of slab B. Slab A is four times as long as it is wide. Slab B is 9 ft longer than slab A. The width of slab B is 12 ft less than the width of slab A. Determine the length and the width of each cement slab.

Let x be the width of slab A. Then $4x$ is the length of slab A.

Length of slab B: $4x + 9$
Width of slab B: $x - 12$.

Area of slab A: $4x \times x$
Area of slab B: $(4x + 9)(x - 12)$.

Now we can write an equation for the problem.

$$4x^2 = (4x + 9)(x - 12) + 693.$$

Solve:

$$4x^2 = 4x^2 - 48x + 9x - 108 + 693$$
$$0 = -39x + 585$$
$$39x = 585$$
$$x = 15.$$

Thus, the width of slab A is 15 ft. Then the length of slab A is 60 ft, the width of slab B is 3 ft, and the length of slab B is 69 ft.

Area of slab A: 15×60 or 900 sq ft
Area of slab B: 3×69 or 207 sq ft.

Thus, the answers meet the conditions of the problem.

EXERCISE 14.5

Solve each problem.

1. Each of three consecutive integers is squared. The three results are added and the sum 29 is obtained. What are the integers?

2. The sum of the squares of two consecutive positive integers is equal to 113. What are the two integers?

3. The product of one greater than a given integer and nine less than two times the integer is equal to -12. Determine the given integer.

4. The sum of the squares of two consecutive positive integers is 313. What are the two integers?

5. The sum of the square of a number and six times the number is 16. Determine all such numbers.

6. Five hundred pine seedlings were planted in rows, so that the number of seedlings in each row was five greater than the number of rows. How many rows were there?

7. The sum of a number and its reciprocal is equal to $2\sqrt{2}$. What is the number?

8. Determine the area of a square whose perimeter is 44 in.

9. One rectangle has a width of 5 in. and length of 7 in. Another rectangle has a width of 5 in. and a length of 14 in. Determine the difference between the areas of the two rectangles.

10. How does the area of a rectangle change if its length is doubled while the width remains the same? its width is doubled while the length remains the same? Work out your answers using the general formula.

11. How does the area of a rectangle change if its length is tripled while

14.8 Solving Problems about Area

the width remains the same? its width is tripled while the length remains the same?

12. How does the area of a rectangle change if its length and the width are both doubled? both tripled?

13. If the measure of one side of a rectangle is multiplied by n, what change takes place in the area?

14. If the measure of both the width and the length of a rectangle is multiplied by n, what changes take place in the area of the rectangle?

15. How does the area of a square change if we double the length of each of its sides? How does the area change if we triple the length of each of its sides?

16. Determine the area of triangle ACD pictured in Figure 14.1.

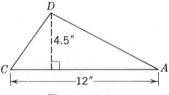

Figure 14.1

17. Rectangle $ABGC$ is pictured in Figure 14.2. If a triangular piece of the corner (EGD) is removed, what is the area of the remaining pentagon $ABDEC$?

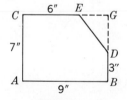

Figure 14.2

18. In the picture in Figure 14.3 $ARST$ is a rectangle. Determine the area of triangle TSR.

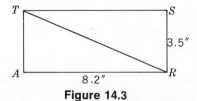

Figure 14.3

19. In the picture in Figure 14.4 the area of the square $VTSR$ is 84.6 sq in. Determine the measure of $\overline{RS}$.

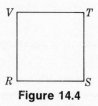

Figure 14.4

20. Rectangles $ABCD$ and $FGHK$ are pictured in Figure 14.5. If the measure of $\overline{HG}$ is 4 in. and the measure of $\overline{FG}$ is 7 in., determine the area of the shaded region.

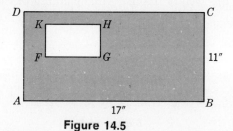

Figure 14.5

21. In the picture in Figure 14.6 the area of triangle AGH is 72 sq in. and the measure of $\overline{AG}$ is 12 in. What is the measure of altitude $\overline{HK}$?

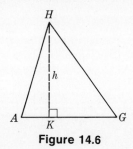

Figure 14.6

22. If a rectangle is 13 in. long and 8 in. wide, what is the measure of the sides of a square of equal area?

23. If the areas of the square and the rectangle in Figure 14.7 are equal, determine l.

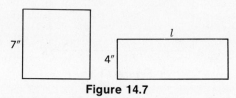

Figure 14.7

14.8 Solving Problems about Area

24. The measures of the altitude $\overline{CK}$ and of side $\overline{AB}$ of triangle ABC in Figure 14.8 are equal, and the area of triangle ABC is 24.5 sq in. Determine the measure of $\overline{AB}$.

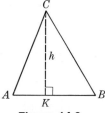

Figure 14.8

25. If the perimeter of a rectangle is 72 in. and its area is 155 sq in., determine the length and the width of the rectangle.

26. If the length of a rectangle exceeds its width by 12 in. and the area is 28 sq in., determine the length and the width of the rectangle.

27. The length of a rectangle is 3 in. longer than its width. A square whose side is as long as the length of the rectangle has an area 48 sq in. greater than the area of the rectangle. Find the dimensions of the rectangle.

28. Mr. Woody cut out two pieces of wood in the shape of rectangles. In piece A, the length was five times the width. In piece B, the width was 4 in. less than the length of piece A, and the length was 2 in. more than the width of piece A. The areas of both rectangles are the same. Find the dimensions of each rectangle.

29. The side of one square is 5 ft longer than the side of another square. The area of the square with the longer sides is 153 sq ft greater than the area of the other square. Find the length of the sides of each square.

30. In the diagram in Figure 14.9, if the area enclosed by the large rectangle is twice the area of the small rectangle, compute the width of the shaded strip.

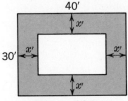

Figure 14.9

31. Subtract 11 from the number of square inches in the area of a square, and you get twice the number of inches in the perimeter. Determine the measure of the side of the square.

32. The length of the rectangular solid in Figure 14.10 is one unit more than its width. Its height is equal to its width. Its total area is $\frac{13}{4}$ square units. Determine the measures of the edges.

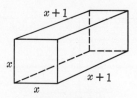

Figure 14.10

33. The equation $d = st - 16t^2$ gives the vertical distance d in feet that an object is from the starting point t seconds after being projected vertically upward with the beginning speed of s feet per second. How long will it take an object to reach a height of 616 ft above the starting point, if it is projected vertically upward at the speed of 200 ft per sec?

34. Solve the equation $d = st - 16t^2$ for t in terms of d and s. From this solution state the relation between d and s for which t has *one* value.

GLOSSARY

Delta (Δ): For the equation $ax^2 + bx + c = 0$, $\Delta = b^2 - 4ac$.

Discriminant: The same as *delta*.

Quadratic formula: For the equation $ax^2 + bx + c = 0$, $x = \frac{-b \pm \sqrt{b^2 - 4ac}}{2|a|}$.

CHAPTER REVIEW PROBLEMS

1. Solve each quadratic equation by factoring.
- **a.** $x^2 - 6x + 5 = 0$
- **b.** $x^2 - 4x - 21 = 0$
- **c.** $x^2 + 7x + 10 = 0$
- **d.** $2x^2 - 7x - 4 = 0$
- **e.** $x^2 - 4 = 0$
- **f.** $4x^2 - 9 = 0$
- **g.** $-6x^2 - 19x + 7 = 0$
- **h.** $-8x^2 - 18x + 35 = 0$

2. Determine the solution sets by completing the square.
- **a.** $n^2 + 6n = 0$
- **b.** $b^2 + \frac{3}{2}b - \frac{7}{16} = 0$
- **c.** $3x^2 - 9x - 30 = 0$
- **d.** $y^2 + 7y + 4 = 0$

Chapter Review Problems

3. Determine the solution sets by the use of the quadratic formula.
   a. $x^2 - 8x + 15 = 0$
   b. $2x^2 + 5x - 3 = 0$
   c. $6x^2 - x - 1 = 0$
   d. $12x^2 + 11x + 2 = 0$
   e. $15x^2 + 7x - 2 = 0$
   f. $14x^2 + 5x - 1 = 0$

4. Compute the discriminant and tell whether each equation has two real roots, one real root, or no real roots.
   a. $x^2 + 3x + 4 = 0$
   b. $x^2 + x + 1 = 0$
   c. $3x^2 - x - 2 = 0$
   d. $2x^2 + x + 1 = 0$
   e. $-x^2 + 3x - 4 = 0$
   f. $x^2 + 2x + 1 = 0$

5. Without solving, tell whether the following equations have two real roots, one real root, or no real roots.
   a. $x = -x^2 - 1$
   b. $\dfrac{y}{y-1} = y + 2$
   c. $2z = 3z^2 + 8$
   d. $\dfrac{1}{2} + m + 2m^2 = 3m$
   e. $t(t+1) = 3t(2-4t)$
   f. $\dfrac{u+2}{2u+3} = \dfrac{3u+4}{4u+5}$

6. Solve each quadratic equation using any method you wish.
   a. $x^2 + x - 2 = 0$
   b. $x^2 = x + 12$
   c. $2 = 5x^2 + x$
   d. $\dfrac{x+4}{x+9} = \dfrac{x}{3x-1}$
   e. $\dfrac{x+1}{x+3} + 2 = \dfrac{x+2}{x+1} + \dfrac{1}{3}$
   f. $\dfrac{2}{2x+1} = 2$
   g. $\dfrac{-10}{1-3x} = 5$
   h. $\sqrt{26x+3} = 4$
   i. $\sqrt{3x-1} = \sqrt{5x+2}$

7. Solve each problem.
   a. An airplane is flying at a certain number of miles per hour. If this number is doubled and added to the square of the speed, 436,920 is obtained. What is the speed of the plane?
   b. The difference between the square of a positive integer and the square of one-half of that integer is 192. What is the integer?
   c. The sum of a number and its reciprocal is $\tfrac{13}{6}$. What is the number?
   d. The product of a number decreased by 3 and the same number increased by 1 is 2. What is the number? How many such numbers are there?
   e. A number squared is equal to that number increased by 12. What is the number? How many such numbers are there?

f. A number added to the square of the number results in 3.75. Determine all such numbers.

g. A triangle and a square have equal areas. Each side of the square is 12 in. long. If a base of the triangle is 18 in. long, what is the length of the altitude?

h. The length of a rectangle is 3 more than twice its width. Its area is 65. Determine the length and the width of the rectangle.

i. One leg of a right triangle is twice as long as the other leg. The area of the triangle is 49. What are the lengths of the two legs?

j. The altitude of a triangle is 6 in. longer than its base and the area of the triangle is 13.5 sq in. Determine the measure of the base and the altitude.

k. A triangle has a base twice as long as the altitude. The area is 30.25 sq in. Compute the length of the base and the altitude. How many different triangles with these dimensions are there?

TABLE OF ROOTS AND POWERS

No.	Sq.	Sq. Root	Cube	Cu. Root	No.	Sq.	Sq. Root	Cube	Cu. Root
1	1	1.000	1	1.000	51	2,601	7.141	132,651	3.708
2	4	1.414	8	1.260	52	2,704	7.211	140,608	3.733
3	9	1.732	27	1.442	53	2,809	7.280	148,877	3.756
4	16	2.000	64	1.587	54	2,916	7.348	157,564	3.780
5	25	2.236	125	1.710	55	3,025	7.416	166,375	3.803
6	36	2.449	216	1.817	56	3,136	7.483	175,616	3.826
7	49	2.646	343	1.913	57	3,249	7.550	185,193	3.849
8	64	2.828	512	2.000	58	3,364	7.616	195,112	3.871
9	81	3.000	729	2.080	59	3,481	7.681	205,379	3.893
10	100	3.162	1,000	2.154	60	3,600	7.746	216,000	3.915
11	121	3.317	1,331	2.224	61	3,721	7.810	226,981	3.936
12	144	3.464	1,728	2.289	62	3,844	7.874	238,328	3.958
13	169	3.606	2,197	2.351	63	3,969	7.937	250,047	3.979
14	196	3.742	2,744	2.410	64	4,096	8.000	262,144	4.000
15	225	3.875	3,375	2.466	65	4,225	8.062	274,625	4.021
16	256	4.000	4,096	2.520	66	4,356	8.124	287,496	4.041
17	289	4.123	4,913	2.571	67	4,489	8.185	300,763	4.062
18	324	4.243	5,832	2.621	68	4,624	8.246	314,432	4.082
19	361	4.359	6,859	2.668	69	4,761	8.307	328,509	4.102
20	400	4.472	8,000	2.714	70	4,900	8.357	343,000	4.121
21	441	4.583	9,261	2.759	71	5,041	8.426	357,911	4.141
22	484	4.690	10,648	2.802	72	5,184	8.485	373,248	4.160
23	529	4.796	12,167	2.844	73	5,329	8.544	389,017	4.179
24	576	4.899	13,824	2.884	74	5,476	8.602	405,224	4.198
25	625	5.000	15,625	2.924	75	5,625	8.660	421,875	4.217
26	676	5.099	17,576	2.962	76	5,776	8.718	438,976	4.236
27	729	5.196	19,683	3.000	77	5,929	8.775	456,533	4.254
28	784	5.292	21,952	3.037	78	6,084	8.832	474,552	4.273
29	841	5.385	24,389	3.072	79	6,241	8.888	493,039	4.291
30	900	5.477	27,000	3.107	80	6,400	8.944	512,000	4.309
31	961	5.568	29,791	3.141	81	6,561	9.000	531,441	4.327
32	1,024	5.657	32,768	3.175	82	6,724	9.055	551,368	4.344
33	1,089	5.745	35,937	3.208	83	6,889	9.110	571,787	4.362
34	1,156	5.831	39,304	3.240	84	7,056	9.165	592,704	4.380
35	1,225	5.916	42,875	3.271	85	7,225	9.220	614,125	4.397
36	1,296	6.000	46,656	3.302	86	7,396	9.274	636,056	4.414
37	1,369	6.083	50,653	3.332	87	7,569	9.327	658,503	4.431
38	1,444	6.164	54,872	3.362	88	7,744	9.381	681,472	4.448
39	1,521	6.245	59,319	3.391	89	7,921	9.434	704,969	4.465
40	1,600	6.325	64,000	3.420	90	8,100	9.487	729,000	4.481
41	1,681	6.403	68,921	3.448	91	8,281	9.539	753,571	4.498
42	1,764	6.481	74,088	3.476	92	8,464	9.592	778,688	4.514
43	1,849	6.557	79,507	3.503	93	8,649	9.644	804,357	4.531
44	1,936	6.633	85,184	3.530	94	8,836	9.695	830,584	4.547
45	2,025	6.708	91,125	3.557	95	9,025	9.747	857,375	4.563
46	2,116	6.782	97,336	3.583	96	9,216	9.798	884,736	4.579
47	2,209	6.856	103,823	3.609	97	9,409	9.849	912,673	4.595
48	2,304	6.928	110,592	3.634	98	9,604	9.899	941,192	4.610
49	2,401	7.000	117,649	3.659	99	9,801	9.950	970,299	4.626
50	2,500	7.071	125,000	3.684	100	10,000	10.000	1,000,000	4.642

Index

Index

Abscissa, 255-256
Absolute value, 36-38
  equations involving, 152-155
  inequalities, 241-244
  square root and, 381
Addition, associative property of, 72, 88, 128
  commutative property of, 71, 88
  equation properties, 155-158
  inequalities, 235-237
  of integers, 38-41
  method, 345-348
  open rational expressions, 203-208
  property of zero, 78, 88
Additive identity, 88
Additive inverse, 87, 88, 89, 100
  of a difference, 171
  of a sum, 170-171
Alphation, 86
Angle measure, 349-353
Angles, base, 350
  complementary, 349

degree-measure of, 350
right, 349
supplementary, 349
vertex, 350
Area problems, 444-450
Associative properties, 70-73
  of addition, 72, 88, 128
  of multiplication, 72-73, 88

Base, 98, 350
Base angles, 350
Betation, 86
Binary operations, 14, 16, 128
Binomials, 120, 170-171, 176-180
  multiplication of, 387-394
  patterns in squaring, 427-428
  rationalizing denominators, 412-416
  special products, 391-394
  subtraction of, 176-180
Boole, George, 1
Boolean Algebra, 1

Cantor, Georg, 1
Cartesian plane, 256
Cartesian sets, 254
Closure, 73–76
Coefficients, 340
Commutative properties, 70–73
 of addition, 71, 88
 of multiplication, 71–72, 88, 127
Comparison method, 339–342
Complement of a set, 20–23
Complementary angles, 349
Consecutive integers, 329–330
Coordinate, plane, 254–259
 x-, 256
 y-, 256
Cube root, 363

Decimal numeral, repeating, 43–47
 terminating, 43
Degree-measure of angles, 350
Denominators, rationalizing, 370–376, 412–416
Dependent system, 308–314
Descartes, René, 256
Directed numbers, 32
Discriminant, 435–437
Disjoint sets, 11–14
Distance problems, 332–334
Distributive properties, 76–77, 88, 127, 136
Division, definition of, 84
 open rational expressions, 196–203
 polynomials, 417–421
 property for equations, 159–160
 property of one, 78
 of real numbers, 82–86

Empty set, 15
Equality, 9–11
 symmetric property of, 76
Equations, 142–167, 218–234, 301–314
 addition properties for, 155–158
 applications of, 146–152
 dependent systems, 308–314
 division property for, 159–160

equivalent, 438
formulas as, 227–228
graphing, 267–269, 304–307
inconsistent systems, 308–314
independent systems, 308–314
involving absolute value, 152–155
linear, 267, 306
 standard form of, 303
mathematics use in science, 222–227
mutliplication properties for, 158–162
open rational expressions, 213–215
problems leading to systems of, 348–349
quadratic, 383, 424–452
 the discriminant, 435–437
 patterns in squaring binomials, 427–428
 perfect squares and, 428–433
 the quadratic formula, 433–435
 solving, 424–427
 in solving problems, 442–450
 using properties of, 438–442
radical, 381–384
with rational expressions, 213–215
to solve problems, 229–234
solving, 218–222
subtraction properties for, 156–157
in two variables, 301–304
using all properties, 163–164
from word sentences to, 143–146
writing equivalent expressions, 164–167
Equiangular triangles, 350
Equilateral triangles, 350
Equivalent equations, 438
Equivalent expressions, 122
 proving, 126
 writing, 164–167
Equivalent sets, 1–6
Exponents, 98–119
 of the form $\frac{1}{n}$, 376–377
 mathematical phrases, 116–117
 negative integers as, 108–110
 power of a power, 102
 power of a product, 102–104

Index

power of a quotient, 104-105
product of powers, 98-102
quotient of powers, 105-108
rational-number, 113-116, 377-380
scientific notation, 113-116
word phrases, 116-117
zero as, 108-110
Expressions, equivalent, 122
  proving, 126
  writing, 164-167
  open, 120-141
    distributivity, 127-134
    multiplication, 134-138
    proving equivalence of, 126
    rearrangement, 127-134
    subtraction, 134-138
  open rational, 185-217
    addition, 203-208
    complex, 209-213
    defined, 185
    division, 196-203
    equations with, 213-215
    multiplication, 185-191
    Multiplicative Identity Theorem, 192-196
    Multiplicative Inverses Theorem, 88, 191-192
    subtraction, 203-208
  rational, 213-215, 410
  simplification of, 171-176, 370-376

Factoring, complete, 402-404
  over a set of integers, 398
  prime, 402
  to simplify rational expressions, 410
  techniques, 404-410
  trinomials, 394-402
Field, 87
Finite sets, 1-6, 17-20
Formulas, as equations, 227-228
  quadratic, 433-435
Functions, 252-300
  defined, 263
  inverse of, 284
  linear, 281-283, 288
    defined, 283

naming, 292-293
quadratic, 293-297
  defined, 293

Graphing, equations, 267-269, 304-307
  inequalities, 314-322
    intersections of sets, 319-322
    unions, 319-322
  weakness of the method, 338-339

Horizontal lines, 277-281

Inconsistent systems, 308-314
Independent systems, 308-314
Inequalities, 167-180, 234-248, 314-334
  absolute value in, 241, 244
  addition, 235-237
  binomials, 170-171, 176-180
  distance problems, 332-334
  graphing, 314-322
    intersections of sets, 319-322
    unions, 319-322
  linear programming, 322-329
  multiplication, 237-241
  number-theory problems, 329-332
  rate problems, 332-334
  simple, 167-170
  simplification of expressions, 171-176
  subtraction, 235-237
Infinite sets, 1-6, 17-20
Integers, 32-36
  addition of, 38-41
  consecutive, 329-330
  factoring over a set of, 398
  negative, 108-110
  zero, 108-110
Intercepts, 274-276
  method, 275
  -slope form, 276-277
  x-, 275, 276
  y-, 275, 276
Inverses, 283-292
  additive, 87, 88, 89, 100

Inverses (*continued*)
   of a difference, 171
   of a sum, 170–171
   defined, 284
   of the function, 284
   multiplicative, 88, 191–192
   operations, 58, 83
   reflection of, 284
Irrational numbers, 48–50
Isomorphism, 66
Isosceles triangles, 349–350

Legs, 350
Linear equations, 267, 306
   standard form of, 303
Linear functions, 281–283, 288
   defined, 283
Linear programming, 322–329
Lines, horizontal, 277–281
   vertical, 277–281

Mathematical phrases, 91–93, 116–117
Mathematics, as a study of patterns, 66–67
   uses in science, 222–227
Membership, 1–6
Mixture problems, 355–358
Monomials, 120
Multiplication, associative property of, 72–73, 88
   of binomials, 387–394
   commutative property of, 71–72, 88, 127
   distributive properties of, 76–77, 88, 127, 136
   inequalities, 237–241
   open expressions, 134–138
   open rational expressions, 185–191
   properties for equations, 158–162
   property of one for, 78, 88
   property of zero for, 78
   of real numbers, 65–70
Multiplicative Identity Theorem, 192–196
Multiplicative Inverses Theorem, 88, 191–192

Negative integers, 108–110
Negative numbers, 68–70, 362
Number line, the, 31–32, 50–56
Number-theory problems, 329–332
Numbers, 6–9
   directed, 32
   irrational, 48–50
   negative, 68–70, 362
   positive, 66–68, 362
   rational, 41–43
      exponents, 113–116, 377–380
   See also Real numbers
Numerals, 6–9
   decimal repeating, 43–47
      terminating, 43
   simplifications in writing, 86–87

One, 78–82, 88
   division property of, 78
   multiplication property of, 78, 88
Open expressions (*see* Expressions, open)
Open rational expressions (*see* Expressions, open rational)
Order of operations, 29–31
Ordered numbered pairs, 254–259
Ordered pairs, 253–254
Ordinates, 256
Origin, 255

Perfect squares, 428–433
Perimeters, 353–355
Plane, Cartesian, 256
   coordinate, 254–259
   points in, 254–255
Polygon, perimeter of, 353–355
Polygonal regions, 323
Polynomials, 416–417
   dividing, 417–421
   form, 121
   over the real numbers, 120
   prime, 403
Positive numbers, 66–68, 362
Power, of a power, 102
   product of, 98–102
   of a product, 102–104

Index

quotient of, 105-108, 110
  of a quotient, 104-105
Prime polynomials, 403
Product, power of, 102-104
  of powers, 98-102
Programming, linear, 322-329
Proof, 121, 126
Properties, addition for equations, 155-158
  of Additive Inverses, 89
  associative, 70-73
    of addition, 72, 88, 128
    of multiplication, 72-73, 88
  commutative, 70-73
    of addition, 71, 88
    of multiplication, 71-72, 88, 127
  distributive, 76-77
    of multiplication, 76-77, 88, 127, 136
  division for equations, 159-160
  multiplication for equations, 158-162
  of Multiplicative Inverses, 88, 191-192
  of one, 78-82, 88
    division, 78
    multiplication, 78, 88
  of opposites, 88
  power of a power, 102
  power of a quotient, 105
  product of powers, 99-100
  quadratic equations, 438-442
  quotient of powers, 105-108, 110
  of radicals, 365, 373
  rearrangement, 129-130
    factor, 130-131
  of reciprocals, 84, 88
  subtraction for equations, 156-157
  symmetric, 76
  of zero, 78-82, 88
    addition, 78, 88
    multiplication, 78
Pythagorean Theorem, 51

Quadratic equations, 383, 424-452
  the discriminant, 435-437
  patterns in squaring binomials, 427-428
  perfect squares and, 428-433
  the quadratic formula, 433-435
  solving, 424-427
  in solving problems, 442-450
  using properties of, 438-442
Quadratic formulas, 433-435
Quadratic functions, 293-297
  defined, 293
Quotient, power of, 104-105
  of powers, 105-108

Radicals, 364
  equations, 381-384
  properties of, 365, 373
  signs, 364
  simplifying, 367
Radicands, 364
Rate problems, 332-334
Rational approximations, 50
Rational expressions, equations with, 213-215
  factoring to simplify, 410
Rational numbers, 41-43
  exponents of, 113-116, 377-380
Rationalizing denominators, 370-376, 412-416
Real numbers, 29-97
  absolute value, 36-38
  associative properties, 70-73
  closure, 73-76
  commutative properties, 70-73
  distributive properties, 76-77
  division of, 82-86
  integers, 32-36
    addition of, 38-41
  irrational numbers, 48-50
  mathematical phrases, 91-93
  monomials over, 120
  multiplication of, 65-70
  nonnegative, 364
  the number line, 31-32, 50-56
  order of operations, 29-31
  pairs, 254-255
  polynomials over, 120
  rational numbers, 41-43

Real numbers (continued)
  repeating decimal numerals, 43–47
  set of, as a system, 87–91
  simplifications in writing, 86–87
  subtraction of, 56–61
  word phrases, 91–93
Rearrangement properties, 129–130
  factor, 130–131
Reciprocals, 84, 88
Rectangular coordinate system, 256
Relations, 252–300
  defined, 262–263
  domain of, 263
  range of, 263
Replacement sets, 252
Right angles, 349
Right triangles, 349
Roots, 148, 435
  cube, 363
  products of, 366–370
  square, 362–366
    absolute value and, 381
    positive, 362
    principal, 362
    product of, 366
  third, 363

Scientific notation, 113–116
Sentences, false, 142
  open, 143, 167
    solution sets of, 252–253
    with two variables, 259–262
  true, 142
  from word sentences to equations, 143–146
Sets, 1–28
  Cartesian, 254
  complement of, 20–23
  disjoint, 11–14
  empty, 15
  equality, 9–11
  equivalent, 1–6
  finite, 1–6, 17–20
  graphing intersections of, 319–322
  infinite, 1–6, 17–20
  isomorphic, 66

  membership, 1–6
  numbers, 6–9
  numerals, 6–9
  operations with, 14–17
  of real numbers as a system, 78–91
  replacement, 252
  subsets, 9–11, 17–20
  union of, 14
  universal, 9, 252
  Venn diagrams, 11–14
  See also Solution sets
Slope, 269–274
  defined, 270
  -intercept form, 276–277
Solution, 146
Solution sets, 142–184
  equations, 142–167
    addition properties for, 155–158
    applications of, 146–152
    division property for, 159–160
    involving absolute value, 152–155
    multiplication properties for, 158–162
    subtraction property for, 156–157
    using all properties, 163–164
    from word sentences to, 143–146
    writing equivalent expressions, 164–167
  inequalities, 167–180
    binomials, 170–171, 176–180
    simple, 167–170
    simplification of expressions, 171–176
  open, 168
  of open sentences, 252–253
  semiclosed, 168
  semiopen, 168
Square root, 362–366
  absolute value and, 381
  positive, 362
  principal, 362
  product of, 366
Squares, perfect, 428–433
Subsets, 9–11, 17–20
Substitution method, 342–345
Subtraction, binomials, 176–180

Index

inequalities, 235–237
open expressions, 134–138
open rational expressions, 203–208
property for equations, 156–157
real numbers, 56–61
Supplementary angles, 349
Symmetric property of equality, 76

Terms, 120
Theorem, defined, 127
Third root, 363
Triangles, equiangular, 350
equilateral, 350
isosceles, 349–350
right, 349
Trinomials, factoring, 394–402

Union of sets, 14
Unions, graphing, 319–322
Universal sets, 9, 252

Universe, the, 9

Variables, equations in two, 301–304
open sentences with two, 259–262
Venn diagrams, 11–14
Vertex angle, 350
Vertical lines, 277–281

Word phrases, 91–93, 116–117

x-coordinate, 256
x-intercept, 275, 276

y-coordinate, 256
y-intercept, 275, 276

Zero, 78–82, 88
addition property, 78, 88
integers, 108–110
multiplication property, 78